Cooperative Control of Multi-Agent Systems

A CONSENSUS REGION APPROACH

Zhongkui Li • Zhisheng Duan

CRC Press

Taylor & Francis Group

Boca Raton London New York

CRC Press is an imprint of the
Taylor & Francis Group, an **informa** business

AUTOMATION AND CONTROL ENGINEERING
A Series of Reference Books and Textbooks

Series Editors

FRANK L. LEWIS, Ph.D.,
Fellow IEEE, Fellow IFAC
Professor
The Univeristy of Texas Research Institute
The University of Texas at Arlington

SHUZHI SAM GE, Ph.D.,
Fellow IEEE
Professor
Interactive Digital Media Institute
The National University of Singapore

PUBLISHED TITLES

Cooperative Control of Multi-agent Systems: A Consensus Region Approach,
Zhongkui Li; Zhisheng Duan

Nonlinear Control of Dynamic Networks,
Tengfei Liu; Zhong-Ping Jiang; David J. Hill

Modeling and Control for Micro/Nano Devices and Systems,
Ning Xi; Mingjun Zhang; Guangyong Li

Linear Control System Analysis and Design with MATLAB®, Sixth Edition,
Constantine H. Houpis; Stuart N. Sheldon

Real-Time Rendering: Computer Graphics with Control Engineering,
Gabriyel Wong; Jianliang Wang

Anti-Disturbance Control for Systems with Multiple Disturbances,
Lei Guo; Songyin Cao

Tensor Product Model Transformation in Polytopic Model-Based Control,
Péter Baranyi; Yeung Yam; Péter Várlaki

Fundamentals in Modeling and Control of Mobile Manipulators, *Zhijun Li;*
Shuzhi Sam Ge

Optimal and Robust Scheduling for Networked Control Systems, *Stefano Longo;*
Tingli Su; Guido Herrmann; Phil Barber

Advances in Missile Guidance, Control, and Estimation, *S.N. Balakrishna;*
Antonios Tsourdos; B.A. White

End to End Adaptive Congestion Control in TCP/IP Networks,
Christos N. Houmkozlis; George A Rovithakis

Robot Manipulator Control: Theory and Practice, *Frank L. Lewis;*
Darren M Dawson; Chaouki T. Abdallah

Quantitative Process Control Theory, *Weidong Zhang*

Classical Feedback Control: With MATLAB® and Simulink®, Second Edition,
Boris Lurie; Paul Enright

Intelligent Diagnosis and Prognosis of Industrial Networked Systems,
Chee Khiang Pang; Frank L. Lewis; Tong Heng Lee; Zhao Yang Dong

Synchronization and Control of Multiagent Systems, *Dong Sun*

Subspace Learning of Neural Networks, *Jian Cheng; Zhang Yi; Jiliu Zhou*

Reliable Control and Filtering of Linear Systems with Adaptive Mechanisms,
Guang-Hong Yang; Dan Ye

Reinforcement Learning and Dynamic Programming Using Function Approximators, *Lucian Busoniu; Robert Babuska; Bart De Schutter; Damien Ernst*

Modeling and Control of Vibration in Mechanical Systems, *Chunling Du; Lihua Xie*

Analysis and Synthesis of Fuzzy Control Systems: A Model-Based Approach, *Gang Feng*

Lyapunov-Based Control of Robotic Systems, *Aman Behal; Warren Dixon; Darren M. Dawson; Bin Xian*

System Modeling and Control with Resource-Oriented Petri Nets, *MengChu Zhou; Naiqi Wu*

Sliding Mode Control in Electro-Mechanical Systems, Second Edition, *Vadim Utkin; Juergen Guldner; Jingxin Shi*

Autonomous Mobile Robots: Sensing, Control, Decision Making and Applications, *Shuzhi Sam Ge; Frank L. Lewis*

Linear Control Theory: Structure, Robustness, and Optimization, *Shankar P. Bhattacharyya; Aniruddha Datta; Lee H.Keel*

Optimal Control: Weakly Coupled Systems and Applications, *Zoran Gajic*

Deterministic Learning Theory for Identification, Recognition, and Control, *Cong Wang; David J. Hill*

Intelligent Systems: Modeling, Optimization, and Control, *Yung C. Shin; Myo-Taeg Lim; Dobrila Skataric; Wu-Chung Su; Vojislav Kecman*

FORTHCOMING TITLES

Modeling and Control Dynamic Sensor Network, *Silvia Ferrari; Rafael Fierro; Thomas A. Wettergren*

Optimal Networked Control Systems, *Jagannathan Sarangapani; Hao Xu*

CRC Press
Taylor & Francis Group
6000 Broken Sound Parkway NW, Suite 300
Boca Raton, FL 33487-2742

First issued in paperback 2017

© 2015 by Taylor & Francis Group, LLC
CRC Press is an imprint of Taylor & Francis Group, an Informa business

No claim to original U.S. Government works

Version Date: 20140916

ISBN 13: 978-1-4665-6994-2 (hbk)
ISBN 13: 978-1-138-07362-3 (pbk)

Library of Congress Cataloging-in-Publication Data

Li, Zhongkui.
    Cooperative control of multi-agent systems : a consensus region approach / authors, Zhongkui Li, Zhisheng Duan.
        pages cm. -- (Automation and control engineering)
    Includes bibliographical references and index.
    ISBN 978-1-4665-6994-2 (hardback)
    1. Cooperating objects (Computer systems)--Automatic control. 2. Machine-to-machine communications. 3. Multiagent systems. I. Duan, Zhisheng. II. Title.

TK5105.67.L52 2014
004.6--dc23   2014034791

Visit the Taylor & Francis Web site at
http://www.taylorandfrancis.com

and the CRC Press Web site at
http://www.crcpress.com

Contents

Preface   **vii**

1 Introduction and Mathematical Background   **1**
  1.1 Introduction to Cooperative Control of Multi-Agent Systems   1
    1.1.1 Consensus   4
    1.1.2 Formation Control   6
    1.1.3 Flocking   7
  1.2 Overview of This Monograph   7
  1.3 Mathematical Preliminaries   9
    1.3.1 Notations and Definitions   9
    1.3.2 Basic Algebraic Graph Theory   11
    1.3.3 Stability Theory and Technical Tools   16
  1.4 Notes   18

2 Consensus Control of Linear Multi-Agent Systems: Continuous-Time Case   **19**
  2.1 Problem Statement   20
  2.2 State Feedback Consensus Protocols   21
    2.2.1 Consensus Condition and Consensus Value   22
    2.2.2 Consensus Region   24
    2.2.3 Consensus Protocol Design   29
      2.2.3.1 The Special Case with Neutrally Stable Agents   30
      2.2.3.2 The General Case   31
      2.2.3.3 Consensus with a Prescribed Convergence Rate   33
  2.3 Observer-Type Consensus Protocols   35
    2.3.1 Full-Order Observer-Type Protocol I   35
    2.3.2 Full-Order Observer-Type Protocol II   40
    2.3.3 Reduced-Order Observer-Based Protocol   41
  2.4 Extensions to Switching Communication Graphs   43
  2.5 Extension to Formation Control   45
  2.6 Notes   50

3 Consensus Control of Linear Multi-Agent Systems: Discrete-Time Case   **53**
  3.1 Problem Statement   54

3.2   State Feedback Consensus Protocols   54
      3.2.1   Consensus Condition   55
      3.2.2   Discrete-Time Consensus Region   57
      3.2.3   Consensus Protocol Design   59
              3.2.3.1   The Special Case with Neutrally Stable Agents   60
              3.2.3.2   The General Case   62
3.3   Observer-Type Consensus Protocols   65
      3.3.1   Full-Order Observer-Type Protocol I   65
      3.3.2   Full-Order Observer-Type Protocol II   67
      3.3.3   Reduced-Order Observer-Based Protocol   68
3.4   Application to Formation Control   69
3.5   Discussions   71
3.6   Notes   72

4   H_∞ and H_2 Consensus Control of Linear Multi-Agent Systems   73
4.1 H_∞ Consensus on Undirected Graphs   74
      4.1.1   Problem Formulation and Consensus Condition   74
      4.1.2 H_∞ Consensus Region   77
      4.1.3 H_∞ Performance Limit and Protocol Synthesis   80
4.2 H_2 Consensus on Undirected Graphs   83
4.3 H_∞ Consensus on Directed Graphs   85
      4.3.1   Leader-Follower Graphs   85
      4.3.2   Strongly Connected Directed Graphs   87
4.4   Notes   92

5   Consensus Control of Linear Multi-Agent Systems Using Distributed Adaptive Protocols   93
5.1   Distributed Relative-State Adaptive Consensus Protocols . .   95
      5.1.1   Consensus Using Edge-Based Adaptive Protocols . . .   96
      5.1.2   Consensus Using Node-Based Adaptive Protocols . . . 100
      5.1.3   Extensions to Switching Communication Graphs . . . 101
5.2   Distributed Relative-Output Adaptive Consensus Protocols . 103
      5.2.1   Consensus Using Edge-Based Adaptive Protocols . . . 104
      5.2.2   Consensus Using Node-Based Adaptive Protocols . . . 107
      5.2.3   Simulation Examples 109
5.3   Extensions to Leader-Follower Graphs 111
5.4   Robust Redesign of Distributed Adaptive Protocols 114
      5.4.1   Robust Edge-Based Adaptive Protocols 115
      5.4.2   Robust Node-Based Adaptive Protocols 119
      5.4.3   Simulation Examples 121
5.5   Distributed Adaptive Protocols for Graphs Containing Directed Spanning Trees 123
      5.5.1   Distributed Adaptive Consensus Protocols 123

  5.5.2 Robust Redesign in the Presence of External Distur-
        bances 129
5.6 Notes 135

**6 Distributed Tracking of Linear Multi-Agent Systems with a
Leader of Possibly Nonzero Input   137**
6.1 Problem Statement 138
6.2 Distributed Discontinuous Tracking Controllers 139
    6.2.1 Discontinuous Static Controllers 139
    6.2.2 Discontinuous Adaptive Controllers 142
6.3 Distributed Continuous Tracking Controllers 144
    6.3.1 Continuous Static Controllers 144
    6.3.2 Adaptive Continuous Controllers 147
6.4 Distributed Output-Feedback Controllers 152
6.5 Simulation Examples 156
6.6 Notes 158

**7 Containment Control of Linear Multi-Agent Systems with
Multiple Leaders   159**
7.1 Containment of Continuous-Time Multi-Agent Systems with
    Leaders of Zero Inputs 160
    7.1.1 Dynamic Containment Controllers 161
    7.1.2 Static Containment Controllers 164
7.2 Containment Control of Discrete-Time Multi-Agent Systems
    with Leaders of Zero Inputs 164
    7.2.1 Dynamic Containment Controllers 165
    7.2.2 Static Containment Controllers 168
    7.2.3 Simulation Examples 168
7.3 Containment of Continuous-Time Multi-Agent Systems with
    Leaders of Nonzero Inputs 170
    7.3.1 Distributed Continuous Static Controllers 171
    7.3.2 Adaptive Continuous Containment Controllers 176
    7.3.3 Simulation Examples 180
7.4 Notes 182

**8 Distributed Robust Cooperative Control for Multi-Agent
Systems with Heterogeneous Matching Uncertainties   183**
8.1 Distributed Robust Leaderless Consensus 184
    8.1.1 Distributed Static Consensus Protocols 185
    8.1.2 Distributed Adaptive Consensus Protocols 190
8.2 Distributed Robust Consensus with a Leader of Nonzero Con-
    trol Input 197
8.3 Robustness with Respect to Bounded Non-Matching Distur-
    bances 202

8.4 Distributed Robust Containment Control with Multiple Leaders 205

8.5 Notes 205

9 Global Consensus of Multi-Agent Systems with Lipschitz Nonlinear Dynamics   **207**

9.1 Global Consensus of Nominal Lipschitz Nonlinear Multi-Agent Systems 208

    9.1.1 Global Consensus without Disturbances 208

    9.1.2 Global H_∞ Consensus Subject to External Disturbances 211

    9.1.3 Extensions to Leader-Follower Graphs 214

    9.1.4 Simulation Example 216

9.2 Robust Consensus of Lipschitz Nonlinear Multi-Agent Systems with Matching Uncertainties 219

    9.2.1 Distributed Static Consensus Protocols 219

    9.2.2 Distributed Adaptive Consensus Protocols 224

    9.2.3 Adaptive Protocols for the Case without Uncertainties 230

    9.2.4 Simulation Examples 231

9.3 Notes 232

Bibliography   **235**

Index   **251**

Preface

In the past two decades, rapid advances in miniaturizing of computing, communication, sensing, and actuation have made it feasible to deploy a large number of autonomous agents to work cooperatively to accomplish civilian and military missions. This, compared to a single complex agent, has the capability to significantly improve the operational effectiveness, reduce the costs, and provide additional degrees of redundancy. Having multiple autonomous agents to work together efficiently to achieve collective group behaviors is usually referred to as cooperative control of multi-agent or multi-vehicle systems. Due to its potential applications in various areas such as satellite formation flying, distributed computing, robotics, surveillance and reconnaissance systems, electric power systems, cooperative attack of multiple missiles, and intelligent transportation systems, cooperative control of multi-agent systems has received compelling attention from various scientific communities, especially the systems and control community.

For a cooperative control problem, the main task is to design appropriate controllers to achieve the desired group objective. Due to the large number of agents, the spatial distribution of actuators, limited sensing capability of sensors, and short wireless communication ranges, it is considered too expensive or even infeasible in practice to implement centralized controllers. Thus, distributed control, depending only on local information of the agents and their neighbors, appears to be a promising tool for handling multi-agent systems. Designing appropriate distributed controllers is generally a challenging task, especially for multi-agent systems with complex dynamics, due to the interconnected effect of the agent dynamics, the interaction graph among agents, and the cooperative control laws.

In this book, we address typical cooperative control problems, including consensus, distributed tracking, and containment control, for multi-agent systems with general linear agent dynamics, linear agent dynamics with uncertainties, and Lipschitz nonlinear agent dynamics. Different from the existing works, we present a systematic consensus region approach to designing distributed cooperative control laws. One benefit of the proposed consensus region is that it decouples the design of the feedback gain matrices of the consensus protocols from the communication graph and can serve in a certain sense as a measure for the robustness of the consensus protocols to variations of the communication graph. We give algorithms built on consensus regions to design the consensus protocols. In particular, by exploiting the decoupling feature of the consensus region approach, we present several distributed adaptive consensus protocols, which adaptively tunes the weights on the communication graph and can be designed by each agent in a fully distributed fashion. This is one unique feature of the book.

This book consists of nine chapters. Chapter 1 introduces the background of cooperative control and reviews graph theory and some technical tools. Chapters 2 and 3 are concerned with the consensus control problem for continuous-time and discrete-time linear multi-agent systems, respectively, and present several fixed-gain consensus protocols. Chapter 4 studies the H_∞ and H_2 consensus problems for linear multi-agent systems subject to external disturbances. Chapter 5 continues Chapter 2 to investigate the consensus protocol design problem for continuous-time linear multi-agent systems and presents several fully distributed adaptive consensus protocols. Chapter 6 considers the distributed tracking control problem for linear multi-agent systems having a leader with nonzero control input. Chapter 7 studies the distributed containment control problem for the case with multiple leaders. Chapter 8 is concerned with the robust cooperative control problem for multi-agent systems with linear nominal agent dynamics subject to heterogeneous matching uncertainties. The global consensus problem for Lipschitz nonlinear multi-agent systems is finally discussed in Chapter 9.

The results in this book would not have been possible without the efforts and help of our colleagues, collaborators, and students. In particular, we are indebted to Professor Lin Huang, a member of the Chinese Academy of Sciences, the supervisor and also a mentor of both of us, for leading us to the research field of control and for his enduring support. We are also indebted to Professors Guanrong Chen, Wei Ren, Lihua Xie, Gang Feng, and Frank L. Lewis for their constant support and inspirations. We as well acknowledge Guanghui Wen, Yu Zhao, Jingyao Wang, and Miao Diao, former and current students of the second author, for their assistance in reviewing parts of the manuscript. We would like to extend our thanks to Professor Frank L. Lewis for inviting us to write this book and to Nora Konopka, Karen Simon, and Joselyn Banks at Taylor & Francis for their professionalism. We wish to thank our families for their support, patience, and endless love.

Finally, we gratefully acknowledge the financial support from the National Natural Science Foundation of China under grants 61104153, 11332001, 61225013, and a Foundation for the Author of National Excellent Doctoral Dissertation of PR China. This work is also partially supported by the State Key Laboratory for Turbulence and Complex Systems, College of Engineering, Peking University.

All the numerical simulations are done using MATLAB. MATLAB® is a registered trademark of The MathWorks, Inc. For product information please contact

The MathWorks, Inc.
3 Apple Hill Drive, Natick, MA, 01760-2098 USA
Tel: 508-647-7000; Fax: 508-647-7001
E-mail: info@mathworks.com; Web: www.mathworks.com

1

Introduction and Mathematical Background

CONTENTS

1.1	Introduction to Cooperative Control of Multi-Agent Systems		1
	1.1.1	Consensus	4
	1.1.2	Formation Control	6
	1.1.3	Flocking	6
1.2	Overview of This Monograph		7
1.3	Mathematical Preliminaries		9
	1.3.1	Notations and Definitions	9
	1.3.2	Basic Algebraic Graph Theory	11
	1.3.3	Stability Theory and Technical Tools	16
1.4	Notes		18

1.1 Introduction to Cooperative Control of Multi-Agent Systems

The subject of this monograph is cooperative control of multi-agent systems, where an agent denotes a dynamical system which can be a ground/underwater vehicle, an aircraft, a satellite, a smart sensor with microprocessors, and so forth. In some literature, multi-agent systems are also called multi-vehicle systems, with efforts to avoid causing confusion with the multi-agent systems in computer science. The multi-agent systems concerned in this monograph and in the systems and control community are quite different from those in the computer science community, regarding the meanings, the objectives, and the commonly used tools, even though they share the same name.

The objective of cooperative control is to have multiple autonomous agents work together efficiently to achieve collective group behavior via local interaction. Recent advances in miniaturizing of computing, communication, sensing, and actuation have made it feasible to deploy a large number of autonomous agents to work cooperatively to accomplish civilian and military missions. This, compared to a single complex agent, has the capability to significantly improve the operational effectiveness, reduce the costs, and provide additional degrees of redundancy. Cooperative control of multi-agent systems consists of many different and yet related research topics and problems. Typical coopera-

1

tive control problems include consensus, formation control, flocking, coverage control, distributed estimation, task and role assignment, among others.

The study of cooperative control of multi-agent systems is stimulated by at least the following two reasons.

- Observations and descriptions of collective behavior in nature: For instance, flocking is the collective motion regularly seen in a flock of birds, a school of fish, or a swarm of insects [118]. A group of birds or fishes moving in flocks are believed to have the advantage of seeking foods, migrating, or avoiding predators and obstacles. Flocking behavior was first simulated on a computer by Reynolds [140] in the 1980s by proposing three simple rules. The mathematical study of the flocking behavior can be well carried out in the framework of cooperative control of multi-agent systems. Several algorithms consisting of consensus components and artificial potential functions have been presented in the literature, e.g., in [29, 119, 162, 165], which have been theoretically shown to produce the flocking behavior observed by Reynolds. These algorithms only require individuals to coordinate with their nearby neighbors and do not involve any central coordination. Recently, ten autonomous drones flying in formation and coordinating flight patterns without central control have been reported as the first outdoor demonstration of how biologically inspired control rules can be used to create resilient yet dynamic flocks [184].

- Motivation and stimulus of many industrial applications and social systems: We take satellite formation flying as a typical example. Satellite formation flying is the concept that multiple satellites work together in a group to accomplish the objective of one larger, usually more expensive, satellite. Coordinating smaller satellites has several advantages, including unprecedented image resolution, reduced cost, better reconfigurability, and higher redundancy [9, 167]. Some potential applications for satellite formation flying include space interferometers and military surveillance instruments. Satellite formation flying missions, e.g., the space interferometry for the Terrestrial Planet Finder mission [6], generally require autonomous precise attitude and position maintenance of multiple satellites. These problems can be formulated as formation control and attitude synchronization, which are typical cooperative control problems in multi-agent systems. Another example is wireless sensor networks, for which two important problems are clock synchronization [153] and deployment in an unknown environment [59]. One efficient protocol for synchronizing the clocks of a wireless sensor network is based on a cascade of two consensus algorithms, which is shown to be adaptive to time-varying clock drifts and changes of the communication topology, and robust to packet drop and sensor node failure [145]. Deploying a mobile sensor network to cover an unknown environment is known as coverage control [27], which is an extensively studied cooperative control problem [61, 76]. Apart from the aforementioned two examples, coordination of multi-agent systems has many other potential applications, such as distributed computing [170], robotics [113], surveillance and reconnaissance systems [49], electric power systems [12], cooperative attack of multiple missiles [66], and intelligent transportation systems [73].

Given a group objective, the cooperative control problem of multi-agent systems is mainly composed of three components, namely, the agent dynamics, the interactions among the agents, and the cooperative control laws required to achieve the group objective. The configuration of these three components is depicted in FIGURE 1.1. The selection of the cooperative control laws depends on the agent dynamics and the interaction topology. The interplay of these three components generally renders the design of the control cooperative laws troublesome, especially for the case with complex agent dynamics.

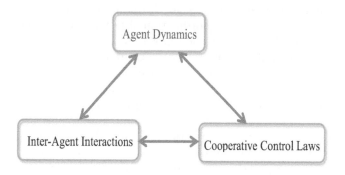

FIGURE 1.1: Configuration of multi-agent systems.

For different scenarios, the dynamics of the employed agents may also be different. The simplest agent dynamics are those modeled by first-order integrators, which can be adopted if the velocities of the agents such as mobile sensors and robots can be directly manipulated. When the accelerations rather than the velocities of the agents can be directly manipulated, it is common to use second-order integrators to describe the agent dynamics. Most existing works on cooperative control of multi-agent systems focus on first-order and second-order integrator agent dynamics. More complicated agent dynamics are described by high-order linear systems, which contain integrator-type agents as special cases and can also be regarded as the linearized models of nonlinear agents dynamics. If we are concerned with cooperative control of multiple robotic manipulators or attitude synchronization of satellite formation flying, the agents are generally described by second-order nonlinear Euler-Lagrangian equations. In some cases, the agents may be perturbed by parameter uncertainties, unknown parameters, or external disturbances, which should also be taken into account.

For multi-agent systems concerned by the systems and control community,

the agents are generally dynamically decoupled from each other, which implies the necessity of cooperation in terms of information exchange between the
agents to achieve collective behavior. Specifically, each agent needs to receive
information from other agents via a direct sensing or communication network.
The interaction (information exchange) topology among the agents is usually
represented as a graph, where each agent is a node and the information exchanges between the agents are represented as edges. Detailed knowledge of
the graph theory is summarized in Section 1.3. An edge in a graph implies
that there exists information flow between the nodes on this edge. The information flows among the agents can be directed or undirected. The interaction
graph among the agents can be static or dynamic. For static graphs, the edges
are not time varying, i.e., the graph remains unchanged throughout the whole
process. In contrast, the edges in dynamic graphs are time-varying, due to
disturbances and/or subject to communication range limitations.

For the cooperative control problem, the main task is to design appropriate
controllers to achieve a desired coordination objective. For cooperative control
design, two approaches, namely, the centralized approach and the distributed
approach, are commonly adopted. The centralized approach assumes that at
least a central hub (which can be one of the agents) is available and has the
ability to collect information from and send control signals to all the other
agents. The distributed approach is based on local interactions only, i.e., each
agent exchanges information with its neighbors. Due to the large number of
agents, the spatial distribution of actuators, limited sensing capability of sensors, and short wireless communication ranges, it is considered too expensive
or even infeasible in practice to implement centralized controllers. Thus, distributed control, depending only on local information of the agents and their
neighbors, appears to be a promising resolution for multi-agent systems. In
some cases, only the relative states or output information between neighboring
agents, rather than the absolute states or outputs of the agents, is accessible
for controllers design. One typical case is that the agents are equipped with
only ultrasonic range sensors. Another instance is deep-space formation flying,
where the measurement of the absolute position of the spacecraft is accurate
only in the order of kilometers, which is thus useless for control purposes [155].

In the past decade, cooperative control of multi-agent systems has been
extensively studied and quite a large amount of research works has been
reported; see the recent surveys [4, 20, 115, 120, 134], the monographs
[8, 74, 112, 127, 138, 139], and the references therein. In the following subsections, we will briefly introduce some typical cooperative control problems:
consensus, formation control, and flocking.

1.1.1   Consensus

In the area of cooperative control of multi-agent systems, consensus is an
important and fundamental problem, which is closely related to formation
control, flocking, distributed estimation, and so on. Consensus means that a

team of agents reaches an agreement on a common value by interacting with each other via a sensing or communication network.

Consensus problems have a long history in computer science and form the foundation of distributed computing. Pioneering works on the consensus problems for distributed decision making and parallel computing include [14, 170]. The emerging interest in consensus control in the last decade is mainly stimulated by [65] and [121]. In particular, a theoretical explanation was provided in [65] for the alignment behavior observed in the Vicsek model [174] and a general framework of the consensus problem for networks of integrators was proposed in [121]. Since then, a large body of works has been reported on the consensus problems; refer to the surveys [4, 20, 120, 134].

In the following, we briefly introduce some recent works on consensus. In [133], a sufficient condition was derived to achieve consensus for first-order integrator multi-agent systems with jointly connected communication graphs. Consensus of networks of second- and high-order integrators was studied in [68, 97, 130, 132, 135, 202]. The consensus problem of multi-agent systems with general linear dynamics was addressed in [83, 86, 93, 106, 146, 173, 185, 195]. Consensus of multiple rigid bodies or spacecraft described as Euler-Lagrange systems was investigated in [7, 24, 25, 79, 109, 142, 176]. In [90, 156, 189], conditions were derived for achieving consensus for multi-agent systems with Lipschitz-type nonlinearity. Consensus algorithms were designed in [22, 75] for multi-agent systems with quantized communication links. The effect of various communication delays and input delays on consensus was considered in [13, 97, 114, 122, 166]. Sampled-data control protocols were proposed to achieve consensus for fixed and switching agent networks in [17, 47]. Distributed consensus protocols with event-triggered communications were designed in [34, 149]. Distributed H_∞ consensus and control problems were investigated in [85, 99] for networks of agents subject to external disturbances.

Existing consensus algorithms can be roughly categorized into two classes, namely, consensus without a leader (i.e., leaderless consensus) and consensus with a leader. The latter is also called leader-follower consensus or distributed tracking. The authors in [56, 57] designed a distributed neighbor-based estimator to track an active leader. Distributed tracking algorithms were proposed in [131] for a network of agents with first-order dynamics. In [21], discontinuous controllers were studied in the absence of velocity or acceleration measurements. The authors in [109] addressed the distributed coordinated tracking problem for multiple Euler-Lagrange systems with a dynamic leader. The distributed tracking of general linear multi-agent systems with a leader of nonzero control input was addressed in [92, 95].

The distributed tracking problem deals with only one leader. However, in some practical applications, there might exist more than one leader in agent networks. In the presence of multiple leaders, the containment control problem arises, where the followers are to be driven into a given geometric space spanned by the leaders [67]. The study of containment control has been motivated by many potential applications. For instance, a group of autonomous vehicles (designated as leaders) equipped with necessary sensors to detect the hazardous obstacles can be used to safely maneuver another group of vehicles (designated as followers) from one target to another, by ensuring that the followers are contained within the moving safety area formed by the leaders [19, 67]. A hybrid containment control law was proposed in [67] to drive the followers into the convex hull spanned by the leaders. Distributed containment control problems were studied in [16, 18, 19] for a group of first-order or second-order integrator agents under fixed or switching directed communication topologies. The containment control was considered in [104] for second-order multi-agent systems with random switching topologies. A hybrid model predictive control scheme was proposed in [46] to solve the containment and distributed sensing problems in leader-follower multi-agent systems. The authors in [36, 110, 111] studied the containment control problem for a collection of Euler-Lagrange systems. The containment control for multi-agent systems with general linear dynamics was considered in [88, 94].

1.1.2   Formation Control

Compared with the consensus problem where the final states of all agents typically become a singleton, the final states of all agents can be more diversified under the formation control scenario. The objective of formation control is to stabilize the relative distances/positions among the agents to prescribed values. Formation control finds natural applications in satellite formation flying, cooperative transportation, sensor networks, and so forth. Roughly speaking, formation control can be categorized into formation producing and formation tracking. Formation producing refers to the algorithm design for a group of agents to reach some pre-desired geometric pattern in the absence of a group reference while formation tracking refers to the same task following a pre-designated group reference [20]. Due to the existence of the group reference, formation tracking is usually much more challenging than formation producing.

The consensus algorithms lay a firm basis for the formation control problem. By adding appropriate offset variables, the consensus protocols can be modified to produce desired formation structures [44, 86, 136]. Also, as an extension of the consensus algorithms, some coupling matrices were introduced [100, 123, 137] for systems with single-integrator kinematics and double-integrator dynamics to offset the corresponding control inputs by some angles. The collective motions for nonholonomic mobile robots were studied in [147, 148] for both all-to-all and limited communications. The inverse agreement problem was studied in [35] and was used to force a team of agents to disperse in space. Motivated by the graph rigidity, research has been conducted in [45, 54, 188] to drive a group of agents to a desired inter-agent configuration by ensuring that a certain number of edge distances are identical to the desired ones. Compared with other formation producing algorithms which require edge vector information, the formation control algorithms based on rigidity only require edge distance information. As a tradeoff, some unstable equilibria, rather than the desired inter-agent configuration, might exist. Formation control in the presence of a group reference was studied in [38, 40] for a group of nonholonomic mobile robots and in [141] for multiple rigid bodies. More works on formation control were reported in [20].

1.1.3 Flocking

Flocking is a typical collective behavior in autonomous multi-agent systems, which can be widely observed in nature in the scenes of grouping of birds, schooling of fish, swarming of bacteria, and so on. In practice, understanding the mechanisms responsible for the emergence of flocking in animal groups can help develop many artificial autonomous systems such as formation control of unmanned air vehicles, motion planning of mobile robots, and scheduling of automated highway systems.

The classical flocking model proposed by Reynolds [140] in the 1980s consists of three heuristic rules: 1) collision avoidance: attempt to avoid collisions with nearby flockmates; 2) velocity matching: attempt to match velocity with nearby flockmates; and 3) flock centering: attempt to stay close to nearby flockmates. These three rules are usually known as separation, alignment, and cohesion, respectively. In 1995, Vicsek et al. proposed a simple model for phase transition of a group of self-driven particles and numerically observed the collective behavior of synchronization of multiple agents' headings [174]. In a sense, Vicsek's model can be viewed as a special case of Reynolds' flocking model, because only velocity matching is considered in the former. To systematically investigate the flocking control problem in multi-agent systems, distributed algorithms consisting of a local artificial potential function and a velocity consensus component were proposed in [119, 165]. The local artificial potential function guarantees separation and cohesion whereas the velocity consensus component ensures alignment in such multi-agent systems. This kind of three-rule combined flocking algorithms were then extended to deal with obstacles avoidance in [28]. The flocking control problem of multiple nonholonomic mobile robots was investigated in [39] by using tools from nonlinear control theory. Distributed protocols to achieve flocking in the presence of a single or multiple virtual leaders were designed in [160, 161, 162]. The flocking control problem of multi-agent systems with nonlinear velocity couplings was investigated in [52, 179]. Distributed hybrid controllers were proposed in [194], which can ensure that the flocking behavior emerges and meanwhile retain the network connectivity.

1.2   Overview of This Monograph

The objective of this monograph is to address typical cooperative control problems of multi-agent systems, including consensus, distributed tracking, and containment control. Although there exist several monographs [8, 74, 112, 127, 138, 139] on recent advances in cooperative control of multi-agent systems, it is worth noting that the agent dynamics considered in those monographs are assumed to be first-order, second-order integrators, Euler-Lagrange systems, or systems with passivity properties. These assumptions might be too restrictive under many circumstances. This motivates us to write this monograph, which focuses on cooperative control of multi-agent systems with more general and complex agent dynamics.

In this monograph, we address the consensus, distributed tracking and containment problems for multi-agent systems with general linear agent dynamics, linear agent dynamics with uncertainties, or Lipschitz nonlinear agent dynamics. Differing from the existing works, in this monograph we present a systematic consensus region approach to designing distributed cooperative control laws, which is a quite challenging task due to the interplay of the agent dynamics, the interaction graph among agents, and the cooperative control laws. The proposed consensus region decouples the design of the feedback gain matrices of the consensus protocols from the communication graph and can serve in a certain sense as a measure for the robustness of the consensus protocols to variations of the communication graph. By exploiting the decoupling feature of the proposed consensus region approach, distributed adaptive consensus protocols are proposed, which adaptively tune the weights on the communication graph and can be designed and implemented by each agent in a fully distributed fashion. This is one main contribution of this monograph. The results in this monograph also extend many existing results in the area of cooperative control.

This monograph consists of nine chapters. A brief introduction of these chapters is given as follows.

- Chapter 1 introduces the background of cooperative control and reviews the basic graph theory and some technical tools which will be used later.

- Chapter 2 is concerned with the consensus control problem of multi-agent systems with general continuous-time linear agent dynamics. The notion of continuous-time consensus region is introduced. Algorithms based on the consensus region will be presented to design several consensus protocols based on relative states or relative outputs of neighboring agents. Extensions of consensus algorithms to formation control are also discussed.

- Chapter 3 presents the discrete-time counterpart of Chapter 2. The notion of discrete-time consensus region is introduced and algorithms are presented to design consensus protocols based on relative states or relative outputs of neighboring agents. The main differences between continuous-time and discrete-time consensus problems are highlighted.

- Chapter 4 studies the H_∞ and H_2 consensus problems for linear multi-agent systems subject to external disturbances. The consensus regions with H_∞ and H_2 performance indices are introduced and algorithms to design consensus protocols to achieve consensus with desired H_∞ and H_2 performances are presented.

- Chapter 5 continues Chapter 2 to investigate the consensus protocol design problem for continuous-time linear multi-agent systems. By exploiting the decoupling feature of the proposed consensus region approach, distributed adaptive consensus protocols are proposed, which can be designed by each agent in a fully distributed fashion. The robustness issue of the adaptive protocols is also discussed.

- Chapter 6 considers the distributed tracking control problem for linear multi-agent systems with a leader of nonzero control input. Several distributed discontinuous and continuous tracking controllers are designed, which contains additional terms to deal with the effect of the leader's nonzero control input.

- Chapter 7 studies the distributed containment control problem for linear multi-agent systems with multiple leaders. The containment control problem reduces to distributed tracking for the case with only one leader. Several distributed controllers are designed by using local information of neighboring agents.

- Chapter 8 is concerned with the robust cooperative control problem for multi-agent systems with linear nominal agent dynamics but subject to heterogeneous matching uncertainties. Under some further assumptions on the uncertainties, distributed controllers are designed for both the cases with and without a leader having nonzero control input.

- Chapter 9 considers the global consensus problem for Lipschitz nonlinear multi-agent systems. Both the cases with and without matching uncertainties are investigated, for which distributed protocols are designed.

1.3   Mathematical Preliminaries

1.3.1   Notations and Definitions

The notations used in this book are standard: $\mathbf{R}^{n \times n}$ and $\mathbf{C}^{n \times n}$ denote the set of $n \times n$ real matrices and complex matrices, respectively. $\mathbf{R}^+$ denotes the set of positive real numbers. The superscript T means transpose for real matrices and H means conjugate transpose for complex matrices. I_N represents the identity matrix of dimension N. Matrices, if not explicitly stated, are assumed to have compatible dimensions. Denote by $\mathbf{1}$ a column vector with all entries equal to one. $\|x\|$ denotes its 2-norm of a vector x and $\|A\|$ denotes the induced 2-norm of a real matrix A. For $\zeta \in \mathbf{C}$, denote by $\mathrm{Re}(\zeta)$ its real part and by $\mathrm{Im}(\zeta)$ its imaginary part. $\mathrm{rank}(A)$ denotes the rank of a matrix A. $\mathrm{diag}(A_1, \cdots, A_n)$ represents a block-diagonal matrix with matrices A_i, $i = 1, \cdots, n$, on its diagonal. $\det(B)$ denotes the determinant of a matrix B. For real symmetric matrices X and Y, $X > (\geq)Y$ means that $X - Y$ is positive (semi-)definite. For a symmetric matrix A, $\lambda_{\min}(A)$ and $\lambda_{\max}(A)$ denote, respectively, the minimum and maximum eigenvalues of A. $\mathrm{range}(B)$ denotes the column space of a matrix B, i.e, the span of its column vectors. ι denotes the imaginary unit.

Definition 1 *A matrix $A \in \mathbf{C}^{n \times n}$ is neutrally stable in the continuous-time sense if it has no eigenvalue with positive real part and the Jordan block corresponding to any eigenvalue on the imaginary axis is of size one, while is Hurwitz if all of its eigenvalues have strictly negative real parts.*

Definition 2 *A matrix $A \in \mathbf{C}^{n \times n}$ is neutrally stable in the discrete-time sense if it has no eigenvalue with magnitude larger than 1 and the Jordan block corresponding to any eigenvalue with unit magnitude is of size one, while is Schur stable if all of its eigenvalues have magnitude less than 1.*

Definition 3 *A matrix $A \in \mathbf{R}^{n \times n}$ is a Hermite matrix if $A = A^H$. Real Hermite matrices are called symmetric matrix. A is a unitary matrix if $A^H A = A A^H = I$. Real unitary matrices are called orthogonal matrices.*

Definition 4 *Matrix $P \in \mathbf{R}^{n \times n}$ is an orthogonal projection onto a subspace $\mathcal{S}$, if $\mathrm{range}(P) = \mathcal{S}$, $P^2 = P$, and $P^T = P$. The orthogonal projection onto a subspace is unique. For an orthogonal projection P onto a subspace $\mathcal{S}$, if the columns of $C \in \mathbf{R}^{n \times m}$ are an orthonormal basis for $\mathcal{S}$, then $P = C^T C$ and $\mathcal{S} = \mathrm{range}(P)$ [168].*

Definition 5 *The Kronecker product of matrices $A \in \mathbf{R}^{m \times n}$ and $B \in \mathbf{R}^{p \times q}$ is defined as*

$$A \otimes B = \begin{bmatrix} a_{11}B & \cdots & a_{1n}B \\ \vdots & \ddots & \vdots \\ a_{m1}B & \cdots & a_{mn}B \end{bmatrix},$$

which satisfies the following properties:

$$(A \otimes B)(C \otimes D) = (AC) \otimes (BD),$$
$$(A \otimes B)^T = A^T \otimes B^T,$$
$$A \otimes (B + C) = A \otimes B + A \otimes C,$$

where the matrices are assumed to be compatible for multiplication.

Definition 6 *A square matrix $A \in \mathbf{R}^{n \times n}$ is called a singular (nonsingular) M-matrix, if all its off-diagonal elements are non-positive and all eigenvalues of A have nonnegative (positive) real parts.*

Definition 7 *Let C be a set in a real vector space $V \subseteq \mathbf{R}^p$. The set C is called convex if, for any x and y in C, the point $(1 - z)x + zy$ is in C for any $z \in [0,1]$. The convex hull for a set of points $X = \{x_1, \cdots, x_n\}$ in V is the minimal convex set containing all points in X, given by $\{\sum_{i=1}^{N} \alpha_i x_i \mid x_i \in X, \alpha_i \in \mathbf{R} \geq 0, \sum_{i=1}^{N} \alpha_i = 1\}$.*

1.3.2 Basic Algebraic Graph Theory

A team of agents interacts with each other via communication or sensing networks to achieve collective objectives. It is convenient to model the information exchanges among agents by directed or undirected graphs. A directed graph $\mathcal{G}$ is a pair $(\mathcal{V}, \mathcal{E})$, where $\mathcal{V} = \{v_1, \cdots, v_N\}$ is a nonempty finite node set and $\mathcal{E} \subseteq \mathcal{V} \times \mathcal{V}$ is an edge set of ordered pairs of nodes, called edges. A weighted graph associates a weight with every edge in the graph. A subgraph $\mathcal{G}_s = (\mathcal{V}_s, \mathcal{E}_s)$ of $\mathcal{G}$ is a graph such that $\mathcal{V}_s \subseteq \mathcal{V}$ and $\mathcal{E}_s \subseteq \mathcal{E}$. Self loops in the form of (v_i, v_i) are excluded unless otherwise indicated. The edge (v_i, v_j) in the edge set $\mathcal{E}$ denotes that agent v_j can obtain information from agent v_i, but not necessarily vice versa. For an edge (v_i, v_j), node v_i is called the parent node, v_j is the child node, and v_i is a neighbor of v_j. The set of neighbors of node v_i is denoted as $\mathcal{N}_i$, whose cardinality is called the in-degree of node v_i.

A graph is defined as being balanced when it has the same number of ingoing and outgoing edges for all the nodes (Edge (v_i, v_j) is said to be outgoing with respect to node v_i and incoming with respect to v_j). A graph with the property that $(v_i, v_j) \in \mathcal{E}$ implies $(v_j, v_i) \in \mathcal{E}$ for any $v_i, v_j \in \mathcal{V}$ is said to be undirected, where the edge (v_i, v_j) denotes that agents v_i and v_j can obtain information from each other. Clearly, an undirected graph is a special balanced graph.

A directed path from node v_{i_1} to node v_{i_l} is a sequence of ordered edges of the form $(v_{i_k}, v_{i_{k+1}})$, $k = 1, \cdots, l - 1$. An undirected path in an undirected graph is defined analogously. A cycle is a directed path that starts and ends at the same node. A directed graph is strongly connected if there is a directed path from every node to every other node. Note that for an undirected graph, strong connectedness is simply termed connectedness. A directed graph is complete if there is an edge from every node to every other node. A undirected tree is an undirected graph where all the nodes can be connected by the way of a single undirected path. A (rooted) directed tree is a directed graph in which every node has exactly one parent except for one node, called the root, which has no parent and has directed paths to all other nodes. A directed tree is defined as spanning when it connects all the nodes in the graph. It can be demonstrated that this implies that there is at least one root node connected

with a simple path to all the other nodes. A graph is said to have or contain
a directed spanning tree if a subset of the edges forms a directed spanning
tree. This is equivalent to saying that the graph has at least one node with
directed paths to all other nodes. For undirected graphs, the existence of a
directed spanning tree is equivalent to being connected. However, in directed
graphs, the existence of a directed spanning tree is a weaker condition than
being strongly connected. A strongly connected graph contains at least one
directed spanning tree.

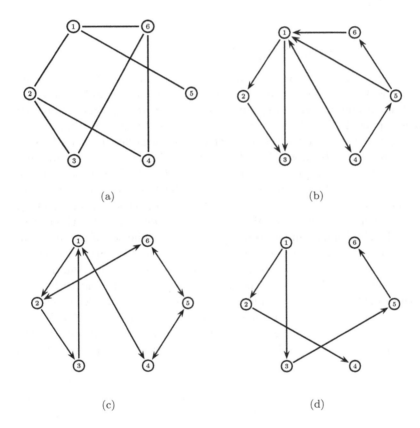

FIGURE 1.2: Different types of graphs of six nodes: (a) an undirected connected graph, (b) a strongly connected graph, (c) a balanced and strongly
connected graph, (d) a directed spanning tree.

In FIGURE 1.2 are several examples of different types of graphs with six
nodes. In this book, a directed edge in a graph is denoted by a line with
a directional arrow while an undirected edge is denoted by a line with a
bidirectional arrow or no arrows.

The adjacency matrix $\mathcal{A} = [a_{ij}] \in \mathbf{R}^{N \times N}$ associated with the directed graph $\mathcal{G}$ is defined such that $a_{ii} = 0$, a_{ij} is a positive value if $(v_j, v_i) \in \mathcal{E}$ and $a_{ij} = 0$ otherwise. The adjacency matrix of an undirected graph is defined analogously except that $a_{ij} = a_{ji}$ for all $i \neq j$. Note that a_{ij} denotes the weight for the edge $(v_j, v_i) \in \mathcal{E}$. If the weights are not relevant, then a_{ij} is set equal to 1 if $(v_j, v_i) \in \mathcal{E}$. The Laplacian matrix $\mathcal{L} = [\mathcal{L}_{ij}] \in \mathbf{R}^{N \times N}$ of the graph $\mathcal{G}$ is defined as $\mathcal{L}_{ii} = \sum_{j \neq i} a_{ij}$ and $\mathcal{L}_{ij} = -a_{ij}$, $i \neq j$. The Laplacian matrix can be written into a compact form as $\mathcal{L} = \mathcal{D} - \mathcal{A}$, where $\mathcal{D} = \text{diag}(d_1, \cdots, d_N)$ is the degree matrix with d_i as the in-degree of the i-th node.

Example 1 *For the graphs depicted in FIGURE 1.2, we compute the corresponding Laplacian matrices as*

$$
\mathcal{L}_a = \begin{bmatrix} 3 & -1 & 0 & 0 & -1 & -1 \\ -1 & 3 & -1 & -1 & 0 & 0 \\ 0 & -1 & 2 & 0 & 0 & -1 \\ 0 & -1 & 0 & 2 & 0 & -1 \\ -1 & 0 & 0 & 0 & 1 & 0 \\ -1 & 0 & -1 & -1 & 0 & 3 \end{bmatrix}, \mathcal{L}_b = \begin{bmatrix} 3 & 0 & 0 & -1 & -1 & -1 \\ -1 & 1 & 0 & 0 & 0 & 0 \\ -1 & -1 & 2 & 0 & 0 & 0 \\ -1 & 0 & 0 & 1 & 0 & 0 \\ 0 & 0 & 0 & -1 & 1 & 0 \\ 0 & 0 & 0 & 0 & -1 & 1 \end{bmatrix},
$$

$$
\mathcal{L}_c = \begin{bmatrix} 2 & 0 & -1 & -1 & 0 & 0 \\ -1 & 2 & 0 & 0 & 0 & -1 \\ 0 & -1 & 1 & 0 & 0 & 0 \\ -1 & 0 & 0 & 2 & -1 & 0 \\ 0 & 0 & 0 & -1 & 2 & -1 \\ 0 & -1 & 0 & 0 & -1 & 2 \end{bmatrix}, \mathcal{L}_d = \begin{bmatrix} 0 & 0 & 0 & 0 & 0 & 0 \\ -1 & 1 & 0 & 0 & 0 & 0 \\ -1 & 0 & 1 & 0 & 0 & 0 \\ 0 & -1 & 0 & 1 & 0 & 0 \\ 0 & 0 & -1 & 0 & 1 & 0 \\ 0 & 0 & 0 & 0 & -1 & 1 \end{bmatrix}.
$$

From the definition of the Laplacian matrix and also the above example, it is easy to see that $\mathcal{L}$ is diagonally dominant and has nonnegative diagonal entries. Since $\mathcal{L}$ has zero row sums, 0 is an eigenvalue of $\mathcal{L}$ with an associated eigenvector **1**. According to Gershgorin's disc theorem [58], all nonzero eigenvalues of $\mathcal{L}$ are located within a disk in the complex plane centered at $d_{\max}$ and having radius of $d_{\max}$, where $d_{\max}$ denotes the maximum in-degree of all nodes. According to the definition of M-matrix in the last subsection, we know that the Laplacian matrix $\mathcal{L}$ is a singular M-matrix.

Lemma 1 ([2, 133]) *The Laplacian matrix $\mathcal{L}$ of a directed graph $\mathcal{G}$ has at least one zero eigenvalue with **1** as a corresponding right eigenvector and all nonzero eigenvalues have positive real parts. Furthermore, zero is a simple eigenvalue of $\mathcal{L}$ if and only if $\mathcal{G}$ has a directed spanning tree. In addition, there exists a nonnegative left eigenvector r of $\mathcal{L}$ associated with the zero eigenvalue, satisfying $r^T \mathcal{L} = 0$ and $r^T \mathbf{1} = 1$. Moreover, r is unique if $\mathcal{G}$ has a directed spanning tree.*

Lemma 2 ([121]) *For an undirected graph, zero is a simple eigenvalue of $\mathcal{L}$ if and only if the graph is connected. The smallest nonzero eigenvalue λ_2 of $\mathcal{L}$ satisfies $\lambda_2 = \min_{x \neq 0, \mathbf{1}^T x = 0} \frac{x^T \mathcal{L} x}{x^T x}$.*

For an undirected graph, it is known that $\mathcal{L} \geq 0$, i.e., $y^T \mathcal{L} y \geq 0, \forall y \in \mathbf{R}^N$. In this case, it is not difficult to verify that $\mathcal{L}$ satisfies the following sum-of-squares (SOS) property:

$$y^T \mathcal{L} y = \frac{1}{2} \sum_{i=1}^{N} \sum_{j=1}^{N} a_{ij} (y_i - y_j)^2,$$

where y_i denotes the i-th item of y. For strongly connected and balanced directed graphs, $y^T \mathcal{L} y \geq 0, \forall y \in \mathbf{R}^N$, still holds, due to the property that $\frac{1}{2}(\mathcal{L} + \mathcal{L}^T)$ represents the Laplacian matrix of the undirected graph obtained from the original directed graph by replacing the directed edges with undirected ones [121]. However, for general directed graphs which are not balanced, the Laplacian matrix $\mathcal{L}$ is not symmetric and $y^T \mathcal{L} y$ can be sign-indefinite. Fortunately, by exploiting the property of $\mathcal{L}$ being a singular M-matrix, we have the following result.

Lemma 3 ([10, 127, 189]) *Suppose that $\mathcal{G}$ is strongly connected. Then, there is a positive left eigenvector $r = [r_1, \cdots, r_N]^T$ of $\mathcal{L}$ associated with the zero eigenvalue and $R\mathcal{L} + \mathcal{L}^T R \geq 0$, where $R = \mathrm{diag}(r_1, \cdots, r_N)$.*

The smallest nonzero eigenvalue λ_2 of the Laplican matrix $\mathcal{L}$ for a connected undirected graph is usually called the Fiedler eigenvalue, also known as the graph algebraic connectivity. For a strongly connected directed graph, a generalized algebraic connectivity can be defined as follows.

Lemma 4 ([189]) *For a strongly connected graph $\mathcal{G}$ with Laplacian matrix $\mathcal{L}$, define its generalized algebraic connectivity as $a(\mathcal{L}) = \min\limits_{r^T x = 0, x \neq 0} \frac{x^T (R\mathcal{L} + \mathcal{L}^T R) x}{2 x^T R x}$, where r and R are defined as in Lemma 3. Then, $a(\mathcal{L}) > 0$. For balanced graphs, $a(\mathcal{L}) = \lambda_2(\frac{\mathcal{L} + \mathcal{L}^T}{2})$, where $\lambda_2(\frac{\mathcal{L} + \mathcal{L}^T}{2})$ denotes the smallest nonzero eigenvalue of $\frac{\mathcal{L} + \mathcal{L}^T}{2}$.*

The Laplacian matrix is very useful in the study of consensus of continuous-time multi-agent systems.

Lemma 5 ([133, 138]) *Suppose that $z = [z_1, \cdots, z_N]^T$ with $z_i \in \mathbf{R}$. Let $\mathcal{A} \in \mathbf{R}^{N \times N}$ and $\mathcal{L} \in \mathbf{R}^{N \times N}$ be, respectively, the adjacency matrix and the Laplacian matrix associated with the directed graph $\mathcal{G}$. Then, consensus is reached in the sense of $\lim_{t \to \infty} \|z_i(t) - z_j(t)\| = 0, \forall i, j = 1, \cdots, N$, for the closed-loop system $\dot{z} = -\mathcal{L} z$ or equivalently $\dot{z}_i = -\sum_{j=1}^{N} a_{ij}(x_i - x_j)$, where a_{ij} denotes the (i, j)-th entry of $\mathcal{A}$, if and only if $\mathcal{G}$ has a directed spanning tree. Furthermore, the final consensus value is given by $r^T z(0)$, where r is the normalized left eigenvector of $\mathcal{L}$ associated with the zero eigenvalue.*

Apart from the Laplacian matrix $\mathcal{L}$ defined as above, another form of

Laplacian matrix is $\widehat{\mathcal{L}} = I - \widehat{\mathcal{A}} = I - \mathcal{D}^{-1}A$. In the event that $\mathcal{D}$ is singular due to a vertex v_i with zero in-degree, we set $d_i^{-1} = 0$ to complete the definition. Clearly, $\widehat{\mathcal{L}} = \mathcal{D}^{-1}\mathcal{L}$. In order to avoid confusion, we refer to $\widehat{\mathcal{L}}$ as the normalized Laplacian matrix and $\widehat{\mathcal{A}}$ as the normalized adjacency matrix. From the definitions of $\mathcal{D}$ and $\mathcal{A}$, it follows that the (i, j)-th entry $\hat{a}_{ij}$ of $\widehat{\mathcal{A}}$ is equal to $\frac{a_{ij}}{\sum_{j=1}^{N} a_{ij}}$. Note that even for undirected graphs, $\widehat{\mathcal{L}}$ is not necessarily symmetric. Regarding the spectrum of $\widehat{\mathcal{L}}$, we have the following lemma.

Lemma 6 ([44]) *Zero is an eigenvalue of $\widehat{\mathcal{L}}$ with $\mathbf{1}$ as a corresponding right eigenvector and all nonzero eigenvalues lie in a disk of radius 1 centered at the point $1 + \iota 0$ in the complex plane. Furthermore, zero is a simple eigenvalue of $\mathcal{L}$ if and only if $\mathcal{G}$ has a directed spanning tree.*

The stochastic matrix is more convenient to deal with the consensus problem for discrete-time multi-agent systems, as detailed in Chapter 3.

For a directed graph $\mathcal{G}$ with N nodes, the row-stochastic matrix $\mathcal{D} \in \mathbf{R}^{N \times N}$ is defined as $d_{ii} > 0$, $d_{ij} > 0$ if $(j, i) \in \mathcal{E}$ but 0 otherwise, and $\sum_{j=1}^{m} d_{ij} = 1$. The stochastic matrix for undirected graphs can be defined analogously except that $d_{ij} = d_{ji}$ for all $i \neq j$. For undirected graphs, the stochastic matrix $\mathcal{D}$ is doubly stochastic, since it is symmetric and all its row sums and column sums equal to 1. Because the stochastic matrix $\mathcal{D}$ has unit row sums, 1 is clearly an eigenvalue of $\mathcal{D}$ with an associated eigenvector $\mathbf{1}$. According to Gershgorin's disc theorem [58], all of the eigenvalues of $\mathcal{D}$ are located in the unit disk centered at the origin.

Lemma 7 ([133]) *All of the eigenvalues of the stochastic matrix $\mathcal{D}$ are either in the open unit disk centered at the origin or equal to 1. Furthermore, 1 is a simple eigenvalue of $\mathcal{D}$ if and only if the graph $\mathcal{G}$ contains a directed spanning tree.*

Lemma 8 ([120, 133]) *Suppose that $z = [z_1, \cdots, z_N]^T$ with $z_i \in \mathbf{R}$. Then, consensus is reached in the sense of $\lim_{k \to \infty} \|z_i(k) - z_j(k)\| = 0$, $\forall i, j = 1, \cdots, N$, for the closed-loop system $z(k + 1) = (I - \epsilon\mathcal{L})z$ or equivalently $z_i(k+1) = z_i(k) - \sum_{j=1}^{N} a_{ij}(x_i(k) - x_j(k))$, where a_{ij} denotes the (i, j)-th entry of the adjacency matrix $\mathcal{A}$, $\epsilon \in (0, d_{\max}]$, and $d_{\max}$ denotes the maximum in-degree of all nodes, if and only if $\mathcal{G}$ has a directed spanning tree. Furthermore, the final consensus value is given by $\hat{r}^T z(0)$, where $\hat{r}$ is the normalized left eigenvector of $I - \epsilon\mathcal{L}$ associated with the one eigenvalue.*

Note that in the above lemma $I - \epsilon\mathcal{L}$ with $\epsilon \in (0, d_{\max}]$ is actually a row-stochastic matrix, since $I - \epsilon\mathcal{L}$ has nonnegative entries and has row sums equal to 1 which follows from the fact that $(I - \epsilon\mathcal{L})\mathbf{1} = \mathbf{1}$.

1.3.3   Stability Theory and Technical Tools

Lemma 9 (Theorem 4.2 in [71]) *Consider the autonomous system*

$$\dot{x} = f(x), \tag{1.1}$$

where $f : D \to \mathbf{R}^n$ is a locally Lipschitz map from a domain $D \in \mathbf{R}^n$ into $\mathbf{R}^n$. Suppose that $x = 0$ is an equilibrium point of (1.1). Let $V : \mathbf{R}^n \to \mathbf{R}$ be a continuously differentiable function such that $V(0) = 0$ and $V(x) > 0$, $\forall x \neq 0$, $V(x)$ is radially unbounded (i.e., $V(x) \to \infty$ as $\|x\| \to \infty$), and $\dot{V} < 0$, $\forall x \neq 0$. Then, $x = 0$ is globally asymptotically stable.

Lemma 10 (LaSalle's Invariance Principle, Theorem 4.4 in [71]) *Let $\Omega \subset D$ be a compact set that is positively invariant with respect to (1.1). Let $V : D \to \mathbf{R}$ be a continuously differentiable function such that $\dot{V} \leq 0$ in Ω. Let E be the set of all points in Ω where $\dot{V} = 0$ and M be the largest invariant set in E. Then, every solution starting in Ω approaches M as $t \to \infty$.*

Note that in many applications the construction of $V(x)$ will guarantee the existence of a set Ω. In particular, if $\Omega_c = \{x \in \mathbf{R}^n : V(x) \leq c\}$ is bounded and $\dot{V} \leq 0$, then we can choose $\Omega = \Omega_c$.

The following lemma is a corollary of Lemma 10.

Lemma 11 (Corollary 4.2 in [71]) *Let $x = 0$ be an equilibrium point for (1.1). Let $V : \mathbf{R}^n \to \mathbf{R}$ be a continuously differentiable, radially unbounded, positive definite function such that $\dot{V} \leq 0$ for all $x \in \mathbf{R}^n$. Let $S = \{x \in \mathbf{R}^n : \dot{V}(x) = 0\}$ and suppose that no solution can stay identically in S, other than the trivial solution $x(t) \equiv 0$. Then, the origin is globally asymptotically stable.*

Lemma 12 (Theorems 4.8 and 4.9 in [71]) *Consider the nonautonomous system*

$$\dot{x} = f(x, t), \tag{1.2}$$

where $f : [0, \infty) \times D \to \mathbf{R}^n$ is a piecewise continuous in t and locally Lipschitz in x on $[0, \infty) \times D$, and $D \in \mathbf{R}^n$ is a domain that contains the origin $x = 0$. Suppose that $x = 0$ is an equilibrium point of (1.2). Let $V : [0, \infty) \times D \to \mathbf{R}$ be a continuously differentiable function such that

$$W_1(x) \leq V(t, x) \leq W_2(x),$$
$$\dot{V}(t, x) = \frac{\partial V}{\partial t} + \frac{\partial V}{\partial x} f(t, x) \leq 0 \tag{1.3}$$

for any $t \geq 0$ and $x \in D$, where $W_1(x)$ and $W_2(x)$ are continuous positive definite functions on D. Then, $x = 0$ is uniformly stable. Suppose that $V(t, x)$ satisfies (1.3) and

$$\dot{V}(t, x) \leq -W_3(x)$$

for any $t \geq 0$ and $x \in D$, where $W_3(x)$ is a continuous positive definite function on D. Then, $x = 0$ is uniformly asymptotically stable. Moreover, if $D = \mathbf{R}^n$ and $W_1(x)$ is radially unbounded, then $x = 0$ is globally uniformly asymptotically stable.

Lemma 13 (Barbalat's Lemma, Lemma 4.2 in [154]) *If the differentiable function $f(t)$ has a finite limit as $t \to \infty$, and if $\dot{f}$ is uniformly continuous, then $\dot{f} \to 0$ as $t \to \infty$.*

A simple sufficient condition for a differentiable function to be uniformly continuous is that its derivative is bounded.

Lemma 14 (LaSalle–Yoshizawa Theorem, Theorem A.8 in [72]) *Suppose that $x = 0$ is an equilibrium point of (1.2) and f is locally Lipschitz in x uniformly in t. Let $V : [0, \infty) \times \mathbf{R}^n \to \mathbf{R}$ be a continuously differentiable function such that*

$$W_1(x) \le V(t, x) \le W_2(x)$$
$$\dot{V}(t, x) \le -W_3(x) \le 0$$

for any $t \ge 0$ and $x \in \mathbf{R}^n$, where $W_1(x)$ and $W_2(x)$ are continuous positive definite functions and $W_1(x)$ is radically unbounded. Then, all the solutions of (1.2) are globally uniformly bounded and satisfy

$$\lim_{t \to \infty} W_3(x(t)) = 0.$$

Lemma 15 ([26]) *For a system $\dot{x} = f(x, t)$, where $f(x, t)$ is locally Lipschitz in x and piecewise continuous in t, assume that there exists a continuously differentiable function $V(x, t)$ such that along any trajectory of the system,*

$$\begin{aligned} \alpha_1(\|x\|) &\le V(x, t) \le \alpha_2(\|x\|), \\ \dot{V}(x, t) &\le -\alpha_3(\|x\|) + \epsilon, \end{aligned} \tag{1.4}$$

where $\epsilon > 0 \in \mathbf{R}$, α_1 and α_2 are class $\mathcal{K}_\infty$ functions, and α_3 is a class $\mathcal{K}$ function [1]. Then, the solution $x(t)$ of $\dot{x} = f(x, t)$ is uniformly ultimately bounded.

For the special case of autonomous system $\dot{x} = f(x)$, the condition (1.4) can be modified to that there exists a continuous positive definite and radically unbounded function $V(x)$ such that $\dot{V}(x) \le -W(x) + \epsilon$, where $W(x)$ is also positive definite and radically unbounded.

Lemma 16 (Comparison Lemma, Lemma 3.4 in [71]) *Consider the scalar differential equation*

$$\dot{u} = f(t, u), \ u(t_0) = u_0,$$

where $f(t, u)$ is continuous in t and locally Lipschitz in u, for all $t \ge 0$ and all $u \in J \subset \mathbf{R}$. Let $[t_0, T)$ (T could be infinity) be the maximal interval of

---

[1]A continuous function $\alpha : [0, a) \to [0, \infty)$ is said to belong to class $\mathcal{K}$ if it is strictly increasing and $\alpha(0) = 0$. It is said to belong to class $\mathcal{K}_\infty$ if $a = \infty$ and $a(r) \to \infty$ as $\to \infty$ [72].

existence of the solution $u(t)$, and suppose $u(t) \in J$ for all $t \in [t_0, T)$. Let $v(t)$ be a continuous function whose upper right-hand derivative $D^+ v(t)$ [2] satisfies the differential inequality

$$D^+ v(t) = f(t, v(t)), \quad v(t_0) = u_0,$$

where $v(t) \in J$ for all $t \in [t_0, T)$. Then, $v(t) \leq u(t)$ for all $t \in [t_0, T)$.

Lemma 17 (Young's Inequality, [11]) *If a and b are nonnegative real numbers and p and q are positive real numbers such that $1/p + 1/q = 1$, then $ab \leq \frac{a^p}{p} + \frac{b^q}{q}$.*

Lemma 18 ([58]) *For any given vectors $x, y \in \mathbf{R}^n$ and positive-definite matrix $P \in \mathbf{R}^{n \times n}$, the following inequality holds:*

$$2x^T y \leq x^T P x + y^T P^{-1} y.$$

Lemma 19 (Schur Complement Lemma, [15]) *For any constant symmetric matrix $S = \begin{bmatrix} S_{11} & S_{12} \\ S_{12}^T & S_{22} \end{bmatrix}$, the following statements are equivalent:*

(1) $S < 0$,

(2) $S_{11} < 0$, $S_{22} - S_{12}^T S_{11}^{-1} S_{12} < 0$,

(3) $S_{22} < 0$, $S_{11} - S_{12} S_{22}^{-1} S_{12}^T < 0$.

Lemma 20 (Finsler's Lemma, [15, 64]) *For $P = P^T \in \mathbf{R}^{n \times n}$ and $H \in \mathbf{R}^{m \times n}$, the following two statements are equivalent:*

(1) $P - \sigma H^T H < 0$ holds for some scalar $\sigma \in \mathbf{R}$,

(2) There exists a $X \in \mathbf{R}^{n \times m}$ such that $P + XH + H^T X^T < 0$.

1.4 Notes

The materials of Section 1.1 are mainly based on [4, 8, 20, 74, 112, 120, 134, 138, 139]. Section 1.3.2 is mainly based on [48, 134, 138]. Section 1.3.3 is mainly based on [15, 71, 72, 154].

---

[2]$D^+ v(t)$ is defined by $D^+ v(t) = \lim\limits_{h \to 0^+} \sup \frac{v(t+h) - v(t)}{h}$.

2

Consensus Control of Linear Multi-Agent Systems: Continuous-Time Case

CONTENTS

2.1   Problem Statement 20
2.2   State Feedback Consensus Protocols 21
        2.2.1   Consensus Condition and Consensus Value 22
        2.2.2   Consensus Region 24
        2.2.3   Consensus Protocol Design 29
                 2.2.3.1   The Special Case with Neutrally Stable Agents .. 30
                 2.2.3.2   The General Case 31
                 2.2.3.3   Consensus with a Prescribed Convergence Rate . 33
2.3   Observer-Type Consensus Protocols 35
        2.3.1   Full-Order Observer-Type Protocol I 35
        2.3.2   Full-Order Observer-Type Protocol II 40
        2.3.3   Reduced-Order Observer-Based Protocol 41
2.4   Extensions to Switching Communication Graphs 43
2.5   Extension to Formation Control 45
2.6   Notes 50

In the area of cooperative control of multi-agent systems, consensus is an important and fundamental problem, which is closely related to formation control, flocking, and other problems. Consensus means that a team of agents reaches an agreement on a common value by interacting with each other via a sensing or communication network. For the consensus control problem, the main task is to design appropriate distributed controllers, usually called consensus protocols, based on local information of neighboring agents to achieve consensus.

Two pioneer papers on consensus control are [65] and [121]. A theoretical explanation was provided in [65] for the alignment behavior observed in the Vicsek model [174] and a general framework of the consensus problem for networks of integrators was proposed in [121]. Since then, the consensus problem has been extensively studied by numerous researchers from various perspectives. Please refer to Section 1.1.1 of the previous chapter and the surveys [4, 20, 120, 134] for recent works on consensus. In most existing works on consensus, the agent dynamics are restricted to be first-order, second-order, or high-order integrators, which might be restrictive in many circumstances.

In this chapter, we consider the consensus problem for multi-agent systems with each node being a general continuous-time linear system. The importance of studying consensus for general linear multi-agent systems lies in at least two aspects. First, the general linear agents are more representative, which contain the first-order, second-order, and high-order integrators as special cases. Second, consensus of general linear multi-agent systems can be regarded as local consensus of nonlinear multi-agent systems, and it paves a way for studying more complex multi-agents systems, e.g., those with linear nominal dynamics but subject to different types of uncertainties.

In this chapter, we present a consensus region approach to systematically designing distributed consensus protocols for multi-agent systems with general continuous-time linear agents. In Section 2.2, a distributed static consensus protocol based on the relative states of neighboring agents is proposed, under which the condition for reaching consensus depends on three correlated factors: the agent dynamics, the design parameters of the consensus protocol, and the nonzero eigenvalues of the Laplacian matrix of the communication graph. To facilitate the consensus protocol design, the notion of consensus region is introduced and analyzed. In light of the consensus region methodology, the design of the consensus protocol can be decoupled into two steps: 1) Determine the feedback gain matrix in order to yield a desirable consensus region; 2) adjust the coupling gain to deal with the effect of the communication graph on consensus. It is also demonstrated that the consensus region should be large enough in order to be robust with respect to variations of the communication graph. Necessary and sufficient conditions are derived to ensure a desirable unbounded consensus region and multi-step algorithms built on the consensus region framework are presented to design the static consensus protocol. The consensus problem with a prescribed convergence rate is further studied. It is shown that for an arbitrarily large convergence rate of reaching consensus, a consensus protocol can be designed if each agent is controllable.

When the relative outputs, rather than the relative states, of neighboring agents are accessible, distributed full-order and reduced-order observer-type consensus protocols are proposed in Section 2.3. The observer-type consensus protocols can be viewed as an extension of the traditional observer-based controller for a single system to one for multi-agent systems. The separation principle of the traditional observer-based controllers still holds in the multi-agent setting. Multi-step algorithms to design the observer-type protocols are also given. It is worth mentioning that for the observer-type protocols the exchange of the protocol's internal states or virtual outputs between neighboring agents via the communication graph is required in order to maintain the separation principle. Extensions to the case with switching communication graphs, which are connected at every time instant, are discussed in Section 2.4. The consensus protocols are modified to solve the formation control problem in Section 2.5.

2.1 Problem Statement

Consider a network of N identical agents with general continuous-time linear dynamics, which may also be regarded as the linearized model of some nonlinear systems. The dynamics of the i-th agent are described by

$$\dot{x}_i = Ax_i + Bu_i,$$
$$y_i = Cx_i, \quad i = 1, \cdots, N, \tag{2.1}$$

where $x_i \in \mathbf{R}^n$, $u_i \in \mathbf{R}^p$, and $y_i \in \mathbf{R}^q$ are, respectively, the state, the control input, and the output of the i-th agent, and A, B, C, are constant matrices with compatible dimensions.

The communication topology among the agents is represented by a directed graph $\mathcal{G}$ (knowledge of basic graph theory can be found in Chapter 1). An edge (i, j) in $\mathcal{G}$ means that the j-th agent can obtain information from the i-th agent, but not conversely. The information exchange between neighboring agents can be done via sensing or/and communication networks. Note that each agent can have access to only local information, i.e., the state or output information from itself and its neighbors.

The objective of this chapter is to solve the consensus problem for the N agents described by (2.1), i.e., to design distributed control laws using only local information to ensure that the N agents described by (2.1) achieve consensus in the sense of $\lim_{t\to\infty} \|x_i(t) - x_j(t)\| = 0$, $\forall i, j = 1, \cdots, N$. The distributed control laws are usually called consensus protocols. We refer to a consensus protocol dynamic if it depends on the protocol's internal state, otherwise we call it static.

Intuitively, if the communication graph $\mathcal{G}$ can be decomposed into some disconnected components, consensus is impossible to reach. Thus, $\mathcal{G}$ should be connected in certain sense in order to achieve consensus. As shown in Lemma 5, a necessary and sufficient condition for a network of single integrators to achieve consensus is that $\mathcal{G}$ contains a directed spanning tree. Throughout this chapter, the following assumption is needed.

Assumption 2.1 *The communication graph $\mathcal{G}$ contains a directed spanning tree.*

In the following sections, we shall design several static and dynamic consensus protocols to solve the consensus problem for the agents in (2.1).

2.2 State Feedback Consensus Protocols

In this section, we consider the case where each agent can have access to its own state and the states from its neighbors. A distributed static consensus protocol in this case is proposed as

$$u_i = cK \sum_{j=1}^{N} a_{ij}(x_i - x_j), \quad i = 1, \cdots, N, \tag{2.2}$$

where $c > 0 \in \mathbf{R}$ is called the (constant) coupling gain or weight, $K \in \mathbf{R}^{p \times n}$ is the feedback gain matrix, and a_{ij} is the (i,j)-th entry of the adjacency matrix associated with $\mathcal{G}$. Both c and K in (2.2) are to be determined. A distinct feature of the consensus protocol (2.2) is that it includes a positive scalar c, which can be regarded as a scaling factor or a uniform weight on the communication graph $\mathcal{G}$. The usefulness of introducing c here (without merging it into K) will be highlighted later.

Let $x = [x_1^T, \cdots, x_N^T]^T$. With (2.1) and (2.2), the closed-loop network dynamics can be obtained as

$$\dot{x} = (I_N \otimes A + c\mathcal{L} \otimes BK)x, \tag{2.3}$$

where $\mathcal{L}$ is the Laplacian matrix of $\mathcal{G}$. The network (2.3) can be also regarded as a large-scale interconnected system [151].

Let $r = [r_1, \cdots, r_N]^T \in \mathbf{R}^N$ be the left eigenvector of $\mathcal{L}$ associated with the eigenvalue 0 and satisfy $r^T \mathbf{1} = 1$. Introduce a new variable $\delta \in \mathbf{R}^{Nn \times Nn}$ as follows:

$$\delta = [(I_N - \mathbf{1}r^T) \otimes I_n]x. \tag{2.4}$$

By the definition of r, it is easy to verify that δ satisfies

$$(r^T \otimes I_n)\delta = 0. \tag{2.5}$$

Also, by the definition of r, it is not difficult to see that 0 is a simple eigenvalue of $I_N - \mathbf{1}r^T$ with $\mathbf{1}$ as the corresponding right eigenvector and 1 is another eigenvalue with multiplicity $N - 1$. Then, it follows from (2.4) that $\delta = 0$ if and only if $x_1 = \cdots = x_N$. That is, the consensus problem is solved if and only if $\lim_{t \to \infty} \delta(t) = 0$. Hereafter, we refer to δ as the consensus error. By using the facts that $\mathcal{L}\mathbf{1} = 0$ and $r^T\mathcal{L} = 0$, it follows from (2.3) and (2.4) that δ evolves according to the following dynamics:

$$\dot{\delta} = (I_N \otimes A + c\mathcal{L} \otimes BK)\delta. \tag{2.6}$$

2.2.1   Consensus Condition and Consensus Value

The following presents a necessary and sufficient condition to achieve consensus.

Theorem 1 *Supposing Assumption 2.1 holds, the distributed consensus protocol (2.2) solves the consensus problem for the agents in (2.1) if and only if all the matrices $A + c\lambda_i BK$, $i = 2, \cdots, N$, are Hurwitz, where $\lambda_i, i = 2, \cdots, N$, are the nonzero eigenvalues of the Laplacian matrix $\mathcal{L}$.*

Proof 1 *Since $\mathcal{G}$ contains a directed spanning, it follows from Lemma 1 that 0 is a simple eigenvalue of $\mathcal{L}$ and the other eigenvalues have positive real part. It is well known that there exists a nonsingular matrix T such that $T^{-1}\mathcal{L}T$ is in the Jordan canonical form [58]. Since the left and right eigenvectors of $\mathcal{L}$ corresponding to the zero eigenvalue are r and $\mathbf{1}$, respectively, we can choose*

$$T = \begin{bmatrix} \mathbf{1} & Y_1 \end{bmatrix} \text{ and } T^{-1} = \begin{bmatrix} r^T \\ Y_2 \end{bmatrix} \tag{2.7}$$

with $Y_1 \in \mathbf{C}^{N \times (N-1)}$ and $Y_2 \in \mathbf{C}^{(N-1) \times N}$ such that

$$T^{-1}\mathcal{L}T = J = \begin{bmatrix} 0 & 0 \\ 0 & \Delta \end{bmatrix}, \tag{2.8}$$

where $\Delta \in \mathbf{C}^{(N-1) \times (N-1)}$ is a upper-triangular matrix whose the diagonal entries are the nonzero eigenvalues of $\mathcal{L}$. Let $\tilde{\delta} = (T^{-1} \otimes I_n)\delta$ with $\tilde{\delta} = [\tilde{\delta}_1^T, \cdots, \tilde{\delta}_N^T]^T$. Then, (2.6) can be represented in terms of $\tilde{\delta}$ as

$$\dot{\tilde{\delta}} = (I_N \otimes A + cJ \otimes BK)\tilde{\delta}. \tag{2.9}$$

From (2.5) and (2.6), it is easy to see that

$$\tilde{\delta}_1 = (r^T \otimes I_n)\delta \equiv 0. \tag{2.10}$$

Note that the elements of the state matrix of (2.9) are either block diagonal or block upper-triangular. Hence, $\tilde{\delta}_i$, $i = 2, \cdots, N$, converge asymptotically to zero if and only if the $N-1$ subsystems along the diagonal, i.e.,

$$\dot{\tilde{\delta}}_i = (A + c\lambda_i BK)\tilde{\delta}_i, \quad i = 2, \cdots, N, \tag{2.11}$$

are asymptotically stable. Therefore, the stability of the matrices $A + c\lambda_i BK$, $i = 2, \cdots, N$, is equivalent to that the state δ of (2.6) converges asymptotically to zero, i.e., the consensus problem is solved.

Remark 1 *The importance of this theorem lies in that it converts the consensus problem of a large-scale multi-agent network under the protocol (2.2) to the simultaneous stabilization problem of a set of matrices with the same dimension as a single agent, thereby significantly reducing the computational complexity. The effects of the communication topology on consensus are characterized by the nonzero eigenvalues of the corresponding Laplacian matrix $\mathcal{L}$, which may be complex, rendering the matrices in Theorem 1 complex-valued.*

The following theorem presents the explicit form for the final consensus value.

Theorem 2 *Suppose Assumption 2.1 holds. If the consensus protocol (2.2) satisfies Theorem 1, then*

$$x_i(t) \to \varpi(t) \triangleq (r^T \otimes e^{At}) \begin{bmatrix} x_1(0) \\ \vdots \\ x_N(0) \end{bmatrix}, \quad \text{as } t \to \infty, \tag{2.12}$$

where $r \in \mathbf{R}^N$ is such that $r^T \mathcal{L} = 0$ and $r^T \mathbf{1} = 1$.

Proof 2 *The solution of (2.3) can be obtained as*

$$
\begin{aligned}
x(t) &= e^{(I_N \otimes A + c\mathcal{L} \otimes BK)t} x(0) \\
&= (T \otimes I_n) e^{(I_N \otimes A + cJ \otimes BK)t} (T^{-1} \otimes I_n) x(0) \\
&= (T \otimes I_n) \begin{bmatrix} e^{At} & 0 \\ 0 & e^{(I_{N-1} \otimes A + c\Delta \otimes BK)t} \end{bmatrix} (T^{-1} \otimes I_n) x(0),
\end{aligned} \tag{2.13}
$$

where matrices T, J and Δ are defined in (2.7) and (2.8). By Theorem 1, $I_{N-1} \otimes A + c\Delta \otimes BK$ is Hurwitz. Thus,

$$
\begin{aligned}
e^{(I_N \otimes A + c\mathcal{L} \otimes BK)t} &\to (\mathbf{1} \otimes I_n) e^{At} (r^T \otimes I_n) \\
&= (\mathbf{1} r^T) \otimes e^{At}, \ \ as \ t \to \infty.
\end{aligned}
$$

It then follows from (2.13) that

$$
x(t) \to (\mathbf{1} r^T) \otimes e^{At} x(0), \ \ as \ t \to \infty,
$$

which directly leads to the assertion.

Remark 2 *From (2.12), we can see that the final consensus value $\varpi(t)$ depends on the state matrix A, the communication topology $\mathcal{L}$, and the initial states of the agents. Here we briefly discuss how the agent dynamics affect $\varpi(t)$. If A is Hurwitz, then $\varpi(t) \to 0$ as $t \to \infty$, i.e., the consensus problem in this case is trivial; If A has eigenvalues along the imaginary axis but no eigenvalues with positive real parts, then $\varpi(t)$ is bounded; If A is unstable, i.e., it has eigenvalues with positive real parts, then $\varpi(t)$ will tend to infinity exponentially. The usefulness of studying the last case is that the linear agents in (2.1) sometimes denote the linearized dynamics of nonlinear agents (e.g., chaotic systems) whose solution, even though A is unstable, can still be bounded, implying the final consensus value may also be bounded.*

2.2.2 Consensus Region

From Theorem 1, it can be noticed that consensus of the agents (2.1) under the protocol (2.2) depends on the feedback gain matrix K, the coupling gain c, and the nonzero eigenvalues λ_i of $\mathcal{L}$ associated with the communication graph $\mathcal{G}$. These three factors are actually coupled with each other. Hence, it is useful to analyze the correlated effects of c, K, and $\mathcal{L}$ on consensus. To this end, the notion of consensus region is introduced.

Definition 8 *The region $\mathcal{S}$ of the parameter σ belonging to the open right half plane, such that the matrix $A + \sigma BK$ is Hurwitz, is called the (continuous-time) consensus region of the network (2.3).*

By definition, the shape of the consensus region relies on only the agent dynamics and the feedback gain matrix K of (2.2). For directed communication graphs, the consensus regions can be roughly classified into five types, namely, bounded consensus region, consensus region with unbounded real part but bounded imaginary part, consensus region with unbounded imaginary part but bounded real part, consensus region with both unbounded real and imaginary parts, and disconnected consensus regions. For undirected communication graphs, there are only three different types of consensus regions, namely, bounded consensus region, unbounded consensus region, and disconnected consensus region.

It should be mentioned that the notion of consensus region is similar to some extent to the synchronization region issue of complex dynamical networks, studied in [41, 42, 101, 105, 124]. The types of bounded, unbounded, disconnected synchronization regions are discussed in [41, 101, 105] for undirected graphs and in [42] for directed graphs. For the case where the inner linking matrix, equivalently, the matrix BK in (2.3), is an identity matrix, then an unbounded synchronization region naturally arises [101].

The key differences between the consensus region in this chapter and the synchronization region in [41, 42, 101, 105, 124] are

- For the synchronization region in [41, 42, 101, 105, 124], the inner linking matrix can be arbitrarily chosen. It is not the case here, since B is a priori given, implying that the consensus region in this chapter is more complicated to analyze.

- We are more interested in how the consensus region is related to the consensus protocol and the communication topology, and more importantly, how to design consensus protocols to yield desirable consensus regions.

The following result is a direct consequence of Theorem 1.

Corollary 1 *The agents described by (2.1) reach consensus under the protocol (2.2) if and only if $c\lambda_i \in \mathcal{S}$, $i = 2, \cdots, N$, where λ_i, $i = 2, \cdots, N$, are the nonzero eigenvalues of $\mathcal{L}$* [1].

In light of the consensus region notion, the design of the consensus protocol (2.2) can be divided into two steps:

1) Determine the feedback gain matrix K to yield a desirable consensus region;

2) Then adjust the coupling gain c such that $c\lambda_i$, $i = 2, \cdots, N$, belong to the consensus region.

The consensus protocol design based on the consensus region as above has a desirable decoupling feature. Specifically, in step 1) the design of the

---

[1] Actually, $c\lambda_i \in \mathcal{S}$, $i = 2, \cdots, N$, are the nonzero eigenvalues of the weighted Laplacian matrix $c\mathcal{L}$.

feedback gain matrix K of the consensus protocol relies on only the agent dynamics, independent of the communication topology, while the effect of the communication topology on consensus is handled in step 2) by manipulating the coupling gain c.

It is worth noting that the consensus region $\mathcal{S}$ can be seen as the stability region of the matrix pencil $A + \sigma BK$ with respect to the complex parameter σ. Thus, tools from the stability of matrix pencils will be utilized to analyze the consensus region. Before moving on, the lemma below is needed.

Lemma 21 ([53]) *Given a complex-coefficient polynomial*

$$p(s) = s^2 + (a + \iota b)s + c + \iota d,$$

where $a, b, c, d \in \mathbf{R}$, and $\iota = \sqrt{-1}$ denotes the imaginary unit, $p(s)$ is stable if and only if $a > 0$ and $abd + a^2 c - d^2 > 0$.

In the above lemma, only second-order polynomials are considered. Similar results for high-order complex-coefficient polynomials can also be given [53]. However, in the latter case, the analysis will be more complicated.

In what follows, we give three examples to depict different types of consensus regions.

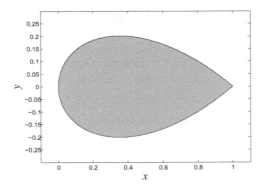

FIGURE 2.1: Bounded consensus region.

Example 2 *(Bounded consensus region)*
 The agent dynamics and the consensus protocol are given by (2.1) and (2.2), respectively, with

$$A = \begin{bmatrix} -2 & -1 \\ 2 & 1 \end{bmatrix}, \ B = \begin{bmatrix} 1 \\ 0 \end{bmatrix}, \ K = \begin{bmatrix} 1 & -1 \end{bmatrix}, \ \sigma = x + \iota y.$$

The characteristic polynomial of $A + \sigma BK$ is

$$\det(sI - (A + \sigma BK)) = s^2 + (1 - x - \iota y)s + 3(x + \iota y).$$

By Lemma 21, $A + \sigma BK$ is Hurwitz if and only if

$$1 - x > 0, \quad -3y^2(1 - x) + 3x(1 - x)^2 - 9y^2 > 0.$$

The consensus region $\mathcal{S}$ is depicted in FIGURE 2.1, which is bounded.

Example 3 *(Consensus region with unbounded real part but bounded imaginary part)*

The agent dynamics and the consensus protocol are given by (2.1) and (2.2), respectively, with

$$A = \begin{bmatrix} -2 & -1 \\ 2 & 1 \end{bmatrix}, \quad B = \begin{bmatrix} 1 & 0 \\ 0 & 1 \end{bmatrix}, \quad K = \begin{bmatrix} 0 & -1 \\ 1 & 0 \end{bmatrix}, \quad \sigma = x + \iota y.$$

The characteristic polynomial of $A + \sigma BK$ in this case is

$$\det(sI - (A + \sigma BK)) = s^2 + s - 2 + (x + 1)(x + 2) - y^2 + \iota(2y + 3)y.$$

By Lemma 21, $A + \sigma BK$ is Hurwitz if and only if $x(x+3) - y^2 - (2x+3)^2 y^2 > 0$. The consensus region $\mathcal{S}$ in this case is depicted in FIGURE 2.2, which is unbounded in the real coordinate but bounded in the imaginary coordinate.

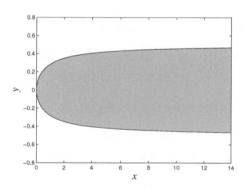

FIGURE 2.2: Consensus region with unbounded real part.

Example 4 *(Disconnected consensus region)*

The agent dynamics and the consensus protocol are given by (2.1) and (2.2), respectively, with

$$A = \begin{bmatrix} -1.4305 & 12.5142 & 3.3759 \\ 1 & -1 & 1 \\ -0.3911 & -5.5845 & -2.3369 \end{bmatrix},$$

$$B = \begin{bmatrix} 0.8 & -0.3 \\ 9.6 & -5 \\ 2.6 & -1 \end{bmatrix}, \quad K = \begin{bmatrix} -0.8 & 0 & 0.3 \\ 9.6 & 0 & 5 \end{bmatrix}.$$

For simplicity, we consider here the case of undirected communication graphs. In this case, $\sigma \in \mathbf{R}^+$. In light of the well-known Hurwitz Criterion [23], we can get that the consensus region $\mathcal{S} = (0, 1.4298) \cup (2.2139, 7.386)$, which clearly consists of two disjoint subregions.

Actually, it is theoretically proved in [41] that for undirected graphs and any natural number n, there exists a network in the form of (2.3) with n disconnected synchronized regions.

In the following, we will give a simple example to show that the consensus region can serve in a certain sense as a measure for the robustness of the consensus protocol to variations of the communication topology.

Example 5 *The agent dynamics and the consensus protocol are given by (2.1) and (2.2), respectively, with*

$$A = \begin{bmatrix} -2.4188 & 1 & 0.0078 \\ -0.3129 & -1 & 1.1325 \\ -0.0576 & 1 & -0.9778 \end{bmatrix}, \quad B = \begin{bmatrix} 1 & 0 \\ 0 & 1 \\ 3.7867 & -0.0644 \end{bmatrix},$$

$$K = \begin{bmatrix} -0.85 & 0 & 0.35 \\ -9.6 & 0 & 6.01 \end{bmatrix}.$$

For simplicity, we here consider the case where the communication graph $\mathcal{G}$ is undirected. The characteristic polynomial of $A + \sigma BK$ can be obtained as

$$\det(sI - (A + \sigma LC)) = s^3 + (0.15\sigma + 4.3966)s^2 + (0.06\sigma^2 - 1.8928\sigma$$
$$+ 4.9426)s + 4.33751\sigma^2 + 4.4978\sigma - 0.0001.$$

By Hurwitz criterion [23], $A + \sigma BK$ is Hurwitz if and only if

$$\sigma > 0, \quad 0.06\sigma^2 - 1.8928\sigma + 4.9426 > 0, \quad 4.33751\sigma^2 + 4.4978\sigma - 0.0001 > 0,$$
$$(0.15\sigma + 4.3966)(0.06\sigma^2 - 1.8928\sigma + 4.9426) > 4.33751\sigma^2 + 4.4978\sigma - 0.0001.$$

From the above inequalities, we can obtain that the consensus region $\mathcal{S} = (1.037, 3.989)$, which is clearly bounded. Assume that the communication graph $\mathcal{G}$ is given by FIGURE 2.3. The Laplacian matrix of $\mathcal{G}$ is equal to

$$\mathcal{L} = \begin{bmatrix} 4 & -1 & -1 & -1 & -1 & 0 \\ -1 & 3 & -1 & 0 & 0 & -1 \\ -1 & -1 & 2 & 0 & 0 & 0 \\ -1 & 0 & 0 & 2 & -1 & 0 \\ -1 & 0 & 0 & -1 & 3 & -1 \\ -1 & 0 & 0 & 0 & -1 & 2 \end{bmatrix},$$

whose nonzero eigenvalues are $1.382, 1.6972, 3.618, 4, 5.3028$. It follows from Corollary 1 that consensus is achieved if and only if the coupling gain c satisfies that $0.7504 < c < 0.7522$.

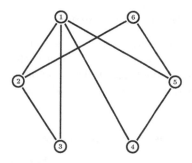

FIGURE 2.3: An undirected communication graph.

In many practical situations, due to the existence of obstacles, communication constraints, or communication link failures, the communication graph may change during the process. Thus, the consensus protocol should maintain certain degree of robustness with respect to the variations of the communication graph. Let us see how modifications of the communication graph affect consensus by considering the following simple cases.

- *An edge between node 3 and node 6 is added, i.e., more information exchanges exist inside the network. The minimal and maximal eigenvalues of the resulting Laplacian matrix are 1.4384 and 5.5616, respectively. By Corollary 1, it can be verified that the protocol (2.2) fails to solve the consensus problem for any $c \in \mathbf{R}^+$.*

- *The edge between node 4 and node 5 is removed, i.e., there are less information exchanges in the network. The minimal and maximal eigenvalues of the resulting Laplacian matrix become 0.8817 and 5.2688, respectively. The protocol (2.2) in this case also fails to solve the consensus problem.*

- *An edge between node 2 and node 6 is added and meanwhile the edge between node 4 and node 5 is removed. In this case, the number of information exchanges remains the same. The minimal and maximal eigenvalues of the resulting Laplacian matrix are 0.7639 and 5.2361, respectively. Once again, the protocol (2.2) fails to solve the consensus problem.*

For a bounded consensus region, it is required that both $c\lambda_2$ and $c\lambda_N$, where λ_2 and λ_N denotes, respectively, the minimal and maximal nonzero eigenvalues of $\mathcal{L}$, belong to the region. The reason for which the aforementioned three cases fail to solve the consensus problem lies in that $\frac{\lambda_N}{\lambda_2}$ does not grow monotonically with the number of added or removed edges. The above sample cases imply, for bounded consensus regions, that the consensus protocol (2.2), if not well designed, can be quite fragile to variations of the communication topology. In other words, it is desirable for the consensus region to be large enough in order to ensure that the consensus protocol maintains a desired robustness margin with respect to the communication topology.

2.2.3   Consensus Protocol Design

It was shown in the last subsection that the consensus protocol should have a large enough consensus region so as to be robust with respect to the communication topology. One convenient and desirable choice is to design the protocol yielding a consensus region which is unbounded in both real and imaginary coordinates.

2.2.3.1   The Special Case with Neutrally Stable Agents

In this subsection, we consider a special case where the matrix A is neutrally stable. It is shown that an unbounded consensus region in the form of the open right-half plane can be achieved. A constructive design algorithm for the protocol (2.2) is then presented.

First, the following lemmas are needed.

Lemma 22 ([117]) *A complex matrix $A \in \mathbf{C}^{n \times n}$ is Hurwitz if and only if there exist a positive-definite matrix $Q = Q^H$ and a matrix $B \in \mathbf{C}^{n \times m}$ such that (A, B) is controllable and $A^H Q + Q A = -B B^H$.*

Lemma 23 *For matrices $S \in \mathbf{R}^{n \times n}$ and $H \in \mathbf{R}^{n \times m}$, where S is skew-symmetric (i.e., $S + S^T = 0$) and (S, H) is controllable, the matrix $S - (x + \iota y) H H^T$ is Hurwitz for any $x > 0$ and $y \in \mathbf{R}$.*

Proof 3 *Let $\tilde{S} = S - (x + \iota y) H H^T$. Then,*

$$
\begin{aligned}
\tilde{S} + \tilde{S}^H &= S - (x + \iota y) H H^T + S^T - (x - \iota y) H H^T \\
&= -2x H H^T \le 0, \ \forall x > 0.
\end{aligned}
\tag{2.14}
$$

Obviously, since (S, H) is controllable, then $(\tilde{S}, H)$ is controllable. By Lemma 22, (2.14) directly leads to the assertion.

Next, an algorithm for constructing the protocol (2.2) is presented, which will be used later.

Algorithm 1 *Given that A is neutrally stable and that (A, B) is stabilizable, the consensus protocol (2.2) can be constructed as follows:*

1) Choose $U \in \mathbf{R}^{n_1 \times n}$ and $W \in \mathbf{R}^{(n-n_1) \times n}$ such that [2]

$$
\begin{bmatrix} U \\ W \end{bmatrix} A \begin{bmatrix} U \\ W \end{bmatrix}^{-1} = \begin{bmatrix} S & 0 \\ 0 & X \end{bmatrix},
\tag{2.15}
$$

where $S \in \mathbf{R}^{n_1 \times n_1}$ is skew-symmetric and $X \in \mathbf{R}^{(n-n_1) \times (n-n_1)}$ is Hurwitz.

2) Let $K = -B^T U^T U$.

---

[2]The matrices U and W can be derived by rendering matrix A into the real Jordan canonical form [58].

3) Select the coupling gain $c > 0$.

The following is the main result of this subsection.

Theorem 3 *Given that A is neutrally stable and that Assumption 2.1 holds, there exists a distributed protocol in the form of (2.2) that solves the consensus problem and meanwhile yields an unbounded consensus region $(0, \infty) \times (-\infty, \infty)$ if and only if (A, B) is stabilizable. One such consensus protocol is given by Algorithm 1.*

Proof 4 *(Necessity) It is trivial by Theorem 1.*

(Sufficiency) Let the related variables be defined as in Algorithm 1. Construct the protocol (2.2) by Algorithm 1 and let $H = UB$. Since (A, B) is stabilizable, then it is easy to see that (S, H) is controllable. Let $U^\dagger \in \mathbf{R}^{n \times n_1}$ and $W^\dagger \in \mathbf{R}^{n \times (n - n_1)}$ be such that $[U^\dagger \ W^\dagger] = [\begin{smallmatrix} U \\ W \end{smallmatrix}]^{-1}$, where $UU^\dagger = I$, $WW^\dagger = I$, $WU^\dagger = 0$, and $UW^\dagger = 0$. It can be verified by some algebraic manipulations that

$$
\begin{bmatrix} U \\ W \end{bmatrix} (A + (x + \iota y)BK) \begin{bmatrix} U \\ W \end{bmatrix}^{-1}
$$

$$
= \begin{bmatrix} S + (x + \iota y)UBKU^\dagger & (x + \iota y)UBKW^\dagger \\ (x + \iota y)WBKU^\dagger & X + (x + \iota y)WBKW^\dagger \end{bmatrix} \tag{2.16}
$$

$$
= \begin{bmatrix} S - (x + \iota y)HH^T & 0 \\ -(x + \iota y)WBH^T & X \end{bmatrix},
$$

which implies that the matrix $A + (x + iy)BK$ is Hurwitz for all $x > 0$ and $y \in \mathbf{R}$, because, by Lemma 22, $S - (x + iy)HH^T$ is Hurwitz for any $x > 0$ and $y \in \mathbf{R}$. Hence, by Theorem 1, the protocol given by Algorithm 1 solves the consensus problem with an unbounded consensus region $(0, \infty) \times (-\infty, \infty)$.

Remark 3 *The consensus region $(0, \infty) \times (-\infty, \infty)$ achieved by the protocol constructed by Algorithm 1 means that such a protocol can reach consensus for any communication topology containing a directed spanning tree and for any positive coupling gain. However, the communication topology and the coupling gain do affect the performances of consensus, e.g., the convergence rate, which will be discussed later.*

Remark 4 *It is worth mentioning that similar results of Theorem 3 are also reported in [173]. Nevertheless, the method leading to Theorem 3 here is quite different from that in [173] and is comparatively much simpler.*

2.2.3.2   The General Case

In this subsubsection, we consider the general case without assuming that A is neutrally stable. In this case, we will see that the consensus region we can get is generally smaller than the case where A is neutrally stable.

Proposition 1 *For the agents described by (2.1), there exists a matrix K such that $A + (x + \iota y)BK$ is Hurwitz for all $x \in [1, \infty)$ and $y \in (-\infty, \infty)$, if and only if (A, B) is stabilizable.*

Proof 5 *(Necessity) It is trivial by letting $x = 1$, $y = 0$.*

(Sufficiency) Since (A, B) is stabilizable, there exists a matrix K such that $A + BK$ is Hurwitz, i.e., there exists a matrix $P > 0$ such that

$$(A + BK)P + P(A + BK)^T < 0.$$

Let $KP = Y$. Then, the above inequality becomes

$$AP + PA^T + BY + Y^T B^T < 0.$$

By Lemma 20 (Finsler's Lemma), there exists a matrix Y satisfying the above inequality if and only if there exists a scalar $\tau > 0$ such that

$$AP + PA^T - \tau BB^T < 0. \tag{2.17}$$

Since the matrix P is to be determined, let $\tau = 2$ without loss of generality. Then,

$$AP + PA^T - 2BB^T < 0. \tag{2.18}$$

Obviously, when (2.18) holds, for any $\tau \geq 2$, (2.17) holds. Take $Y = -B^T$, i.e., $K = -B^T P^{-1}$. Then in light of (2.18) we can get that

$$\begin{aligned}
P[A + (x + \iota y)BK]^H &+ [A + (x + \iota y)BK]P \\
&= P[A + (x - \iota y)BK]^T + [A + (x + \iota y)BK]P \\
&= AP + PA^T - 2xBB^T < 0,
\end{aligned}$$

for all $x \in [1, \infty)$ and $y \in (-\infty, \infty)$. That is, $A + (x + \iota y)BK$ is Hurwitz for all $x \in [1, \infty)$ and $y \in (-\infty, \infty)$.

Theorem 1 and the above proposition together lead to the following result.

Theorem 4 *Assuming that Assumption 2.1 holds, there exists a protocol (2.2) that solves the consensus problem for the agents described by (2.1) and meanwhile yields an unbounded consensus region $\mathcal{S} \triangleq [1, \infty) \times (-\infty, \infty)$ if and only if (A, B) is stabilizable.*

Remark 5 *Compared to the case when A is neutrally stable, where the consensus region is the open right half plane, for the case where A is not neutrally stable, the consensus region can be achieved are generally smaller. This is consistent with the intuition that unstable behaviors are more difficult to synchronize than stable behaviors.*

Next, a two-step consensus protocol design procedure based on the consensus region notion is presented.

Algorithm 2 *Given that (A, B) is stabilizable and Assumption 2.1 holds, a protocol (2.2) solving the consensus problem can be constructed according to the following steps.*

1) *Solve the LMI (2.18) to get one solution $P > 0$. Then, choose the feedback gain matrix $K = -B^T P^{-1}$.*

2) *Select the coupling gain c not less than the threshold value c_{th}, given by*

$$c_{th} = \frac{1}{\min\limits_{i=2,\cdots,N} \text{Re}(\lambda_i)}, \tag{2.19}$$

where λ_i, $i = 2, \cdots, N$, are the nonzero eigenvalues of the Laplacian matrix $\mathcal{L}$.

Remark 6 *This design algorithm has a favorable decoupling feature, as explained in the paragraph below Corollary 1. In step 2) the coupling gain c generally has to be larger than a threshold value, which is related to the specific communication topology. For the case when the agent number N is large, for which the eigenvalues of the corresponding Laplacian matrix are hard to determine or even troublesome to estimate, we need to choose the coupling gain c to be large enough, which of course involves conservatism.*

Remark 7 *The feedback gain matrix K of (2.2) can also be designed by solving the algebraic Riccati equation: $A^T Q + QA + I - QBB^T Q = 0$, as in [171, 195]. In this case, K can be chosen as $K = -B^T Q$. The solvability of the above Ricatti equation is equal to that of the LMI (2.18).*

2.2.3.3 Consensus with a Prescribed Convergence Rate

In this subsection, the protocol (2.2) is designed to achieve consensus with a prescribed convergence rate.

From the proof of Theorem 1, it is easy to see that the convergence rate of the N agents in (2.1) reaching consensus under the protocol (2.2) is equal to the minimal decay rate of the $N - 1$ matrices $A + c\lambda_i BK$, $i = 2, \cdots, N$. The decay rate of the system $\dot{x} = Ax$ is defined as the maximum of negative real parts of the eigenvalues of A [15]. Thus, the convergence rate of the agents (2.1) reaching consensus can be manipulated by properly assigning the eigenvalues of the matrices $A + c\lambda_i BK$, $i = 2, \cdots, N$.

Lemma 24 ([15]) *The decay rate of the system $\dot{x} = Ax$ is larger than $\alpha > 0$ if and only if there exists a matrix $Q > 0$ such that $A^T Q + QA + 2\alpha Q < 0$.*

Proposition 2 *Given the agents in (2.1), there exists a matrix K such that $A + (x + \iota y)BK$ is Hurwitz with a decay rate larger than α for all $x \in [1, \infty)$ and $y \in (-\infty, \infty)$, if and only if there exists a $Q > 0$ such that*

$$AQ + QA^T - 2BB^T + 2\alpha Q < 0. \tag{2.20}$$

Proof 6 *This proof can be completed by following similar steps as in the proof of Proposition 1, which is omitted for conciseness.*

The above proposition and Theorem 1 lead to the following result.

Theorem 5 *Assume that Assumption 2.1 holds. There exists a protocol (2.2) that solves the consensus problem for the agents in (2.1) with a convergence rate larger than α and yields an unbounded consensus region $[1, \infty) \times (-\infty, \infty)$ if and only if there exists a K such that $A + BK$ is Hurwitz with a decay rate larger than α.*

Then Algorithm 2 can be modified as follows.

Algorithm 3 *For $\mathcal{G}$ satisfying Assumption 2.1, a protocol (2.2) solving the consensus problem with a convergence rate larger than α can be constructed as follows:*

1) *Choose the feedback gain matrix $K = -B^T Q^{-1}$, where $Q > 0$ is one solution to (2.20).*

2) *Select the coupling gain $c \geq c_{th}$, where c_{th} is defined as in (2.19).*

Remark 8 *Note that a sufficient condition satisfying Theorem 5 is that (A, B) is controllable. It is well known that the eigenvalues of $A - BK$ can be arbitrarily assigned provided that complex conjugate eigenvalues are assigned in pairs if (A, B) is controllable [23]. However for the consensus problem with a prescribed convergence rate of linear multi-agent systems in this subsection, the eigenvalues of $A + c\lambda_i BK$, $i = 2, \cdots, N$, cannot be arbitrarily assigned due to the fact that the consensus protocol has to be distributed (depending on local information), i.e., the protocol in this case is structural. This indicates one constraint and difficulty brought by distributed control. Fortunately, a stronger result is obtained that the real parts of the eigenvalues of $A + c\lambda_i BK$, $i = 2, \cdots, N$, which indicates the convergence rate of reaching consensus can be arbitrarily assigned if (A, B) is controllable.*

Remark 9 *Under the condition that (A, B) is controllable, the protocol achieving consensus with a arbitrarily large convergence rate α can be constructed by Algorithm 3. However, larger α implies higher feedback gains in the consensus protocol (2.2). Thus, a tradeoff has to be made between the convergence rate and the cost of the consensus protocol.*

Remark 10 *Similarly as in Remark 7, the feedback gain matrix K of (2.2) solving the consensus problem with a convergence rate larger than α can also be designed by solving the algebraic Riccati equation: $A^T \tilde{Q} + \tilde{Q} A + I - \tilde{Q} B B^T \tilde{Q} + 2\alpha \tilde{Q} = 0$. In this case, K can be chosen as $K = -B^T \tilde{Q}$. A sufficient condition for the solvability of the above Ricatti equation is that (A, B) is controllable.*

2.3   Observer-Type Consensus Protocols

In the previous section, the consensus protocol (2.2) rely on the relative states of neighboring agents. However, in many circumstances the state information might not be available. Instead, each agent can have access to the local output information, i.e., its own output and the outputs from its neighbors. In this section, we will design consensus protocols using local output information.

2.3.1   Full-Order Observer-Type Protocol I

In this subsection, we assume that each agent can get the relative output information of itself with respect to its neighbors. Based on the relative output information, we propose an observer-type consensus protocol as follows:

$$\dot{v}_i = (A + BF)v_i + cL \sum_{j=1}^{N} a_{ij}[C(v_i - v_j) - (y_i - y_j)],$$

$$u_i = Fv_i, \quad i = 1, \cdots, N, \tag{2.21}$$

where $v_i \in \mathbf{R}^n$, $i = 1, \cdots, N$, are the protocol's internal states, a_{ij} is the (i,j)-th entry of the adjacency matrix associated with $\mathcal{G}$, $c > 0$ denotes the coupling gain, $L \in \mathbf{R}^{n \times q}$ and $F \in \mathbf{R}^{p \times n}$ are the feedback gain matrices. The term $\sum_{j=1}^{N} a_{ij} C(v_i - v_j)$ in (2.21) means that the agents need to transmit the virtual outputs of their corresponding controllers to their neighbors via the communication topology $\mathcal{G}$.

Let $z_i = [x_i^T, v_i^T]^T$ and $z = [z_1^T, \cdots, z_N^T]^T$. Then, the closed-loop network dynamics resulting from (2.1) and (2.21) can be obtained as

$$\dot{z} = (I_N \otimes \mathcal{A} + c\mathcal{L} \otimes \mathcal{H})z, \tag{2.22}$$

where $\mathcal{L} \in \mathbf{R}^{N \times N}$ is the Laplacian matrix of $\mathcal{G}$, and

$$\mathcal{A} = \begin{bmatrix} A & BF \\ 0 & A + BF \end{bmatrix}, \quad \mathcal{H} = \begin{bmatrix} 0 & 0 \\ -LC & LC \end{bmatrix}.$$

The following presents a sufficient condition for the consensus problem using the observer-type protocol (2.21).

Theorem 6 *Suppose that Assumption 2.1 holds. The observer-type protocol (2.21) solves the consensus problem for the agents described by (2.1) if all the matrices $A + BF$ and $A + c\lambda_i LC$, $i = 2, \cdots, N$, are Hurwitz, where $\lambda_i, i = 2, \cdots, N$, are the nonzero eigenvalues of the $\mathcal{L}$. Furthermore, the final consensus value is given by*

$$x_i(t) \to \varpi(t), \quad v_i(t) \to 0, \ i = 1, \cdots, N, \quad \text{as } t \to \infty, \tag{2.23}$$

where $\varpi(t)$ is defined as in (2.12).

Proof 7 *Similarly as in the previous section, let $r \in \mathbf{R}^N$ be the left eigenvector of $\mathcal{L}$ associated with the eigenvalue 0, satisfying $r^T \mathbf{1} = 1$. Define*

$$\xi(t) = [(I_N - \mathbf{1}r^T) \otimes I_{2n}]z(t), \tag{2.24}$$

which satisfies $(r^T \otimes I_{2n})\xi = 0$. Following similar steps as in the proof of Theorem 1, we can see that the consensus problem of (2.22) can be reduced to the asymptotical stability problem of ξ, which evolves according to the following dynamics:

$$\dot{\xi} = (I_N \otimes \mathcal{A} + c\mathcal{L} \otimes \mathcal{H})\xi. \tag{2.25}$$

Let T be defined as in (2.7) such that $T^{-1}\mathcal{L}T = J = \left[\begin{smallmatrix} 0 & 0 \\ 0 & \Delta \end{smallmatrix}\right]$, where the diagonal entries of Δ are the nonzero eigenvalues of $\mathcal{L}$. Let $\tilde{\xi} = [\tilde{\xi}_1^T, \cdots, \tilde{\xi}_N^T]^T = (T^{-1} \otimes I_{2n})\xi$. Then, (2.24) can be rewritten as

$$\dot{\tilde{\xi}} = (I_N \otimes \mathcal{A} + cJ \otimes \mathcal{H})\tilde{\xi}. \tag{2.26}$$

Note that $\tilde{\xi}_1 = (r^T \otimes I_{2n})\xi \equiv 0$ and that the elements of the state matrix of (2.26) are either block diagonal or block upper-triangular. Hence, $\tilde{\xi}_i$, $i = 2, \cdots, N$, converge asymptotically to zero if and only if the $N - 1$ subsystems along the diagonal, i.e.,

$$\dot{\tilde{\xi}}_i = (\mathcal{A} + c\lambda_i\mathcal{H})\tilde{\xi}_i, \quad i = 2, \cdots, N, \tag{2.27}$$

are asymptotically stable. It is easy to check that matrices $\mathcal{A} + c\lambda_i\mathcal{H}$ are similar to $\left[\begin{smallmatrix} A + c\lambda_i LC & 0 \\ -c\lambda_i LC & A + BF \end{smallmatrix}\right]$, $i = 2, \cdots, N$. Therefore, the stability of the matrices $A + BF$, $A + c\lambda_i LC$, $i = 2, \cdots, N$, is equivalent to that the state ξ of (2.25) converges asymptotically to zero, i.e., the consensus problem is solved.

By following similar steps as in the proof of Theorem 2, we can obtain the solution of (2.22) as

$$z(t) = e^{(I_N \otimes \mathcal{A} + c\mathcal{L} \otimes \mathcal{H})t}z(0)$$
$$= (T \otimes I_{2n}) \begin{bmatrix} e^{\mathcal{A}t} & 0 \\ 0 & e^{(I_{N-1} \otimes \mathcal{A} + c\Delta \otimes \mathcal{H})t} \end{bmatrix} (T^{-1} \otimes I_{2n})z(0), \tag{2.28}$$

Note that $I_{N-1} \otimes \mathcal{A} + c\Delta \otimes \mathcal{H}$ is Hurwitz. Thus, we can get from (2.28) that

$$z(t) \to (\mathbf{1}r^T) \otimes e^{\mathcal{A}t}z(0), \quad \text{as } t \to \infty,$$

implying that

$$z_i(t) \to (r^T \otimes e^{\mathcal{A}t}) \begin{bmatrix} z_1(0) \\ \vdots \\ z_N(0) \end{bmatrix}, \quad \text{as } t \to \infty, \tag{2.29}$$

for $i = 1, \cdots, N$. Since $A + BK$ is Hurwitz, (2.29) directly leads to (2.23).

Remark 11 *The observer-type consensus protocol (2.21) can be seen as an extension of the traditional observer-based controller for a single system to one for multi-agent systems. The Separation Principle of the traditional observer-based controllers still holds in the multi-agent setting.*

Remark 12 *By comparing Theorems 2 and 6, it is observed that no matter the consensus protocol takes the dynamic form (2.21) or the static form (2.2), the final consensus value reached by the agents will be the same, which relies on only the communication topology, the initial states, and the agent dynamics.*

For the observer-type protocol (2.21), the consensus region can be similarly defined. Assuming that $A + BF$ is Hurwitz, the region of the parameter σ belonging to the open right half plane, such that $A + \sigma LC$ is Hurwitz, is the consensus region of the network (2.22).

In what follows, we will show how to design the observer-type protocol (2.21). First, consider the special case where A is neutrally stable.

Algorithm 4 *Given that $A \in \mathbf{R}^{n \times n}$ is neutrally stable and that (A, B, C) is stabilizable and detectable, the consensus protocol (2.21) can be constructed as follows:*

1) Let F be such that $A + BF$ is Hurwitz.

2) Choose $\hat{U} \in \mathbf{R}^{n \times n_1}$ and $\hat{W} \in \mathbf{R}^{n \times (n - n_1)}$ such that

$$\begin{bmatrix} \hat{U} & \hat{W} \end{bmatrix}^{-1} A \begin{bmatrix} \hat{U} & \hat{W} \end{bmatrix} = \begin{bmatrix} \hat{S} & 0 \\ 0 & \hat{X} \end{bmatrix},$$

where $\hat{S} \in \mathbf{R}^{n_1 \times n_1}$ is skew-symmetric and $\hat{X} \in \mathbf{R}^{(n - n_1) \times (n - n_1)}$ is Hurwitz.

3) Let $L = -\hat{U}\hat{U}^T C^T$.

4) Select the coupling gain $c > 0$.

Theorem 7 *Given that A is neutrally stable and that Assumption 2.1 holds, there exists an observer-type protocol (2.21) that solves the consensus problem and meanwhile yields an unbounded consensus region $(0, \infty) \times (-\infty, \infty)$ if and only if (A, B, C) is stabilizable and detectable. One such protocol can be constructed by Algorithm 4.*

Proof 8 *(Necessity) It is trivial.*
(Sufficiency) let $\hat{H} = C\hat{U}$. Then, it follows from the detectability of (A, C) that $(\hat{S}, \hat{H})$ is observable. Let $\hat{U}^\dagger \in \mathbf{R}^{n_1 \times n}$ and $\hat{W}^\dagger \in \mathbf{R}^{(n - n_1) \times n}$ be such that $\begin{bmatrix} \hat{U}^\dagger \\ \hat{W}^\dagger \end{bmatrix} = [\hat{U} \ \hat{W}]^{-1}$, where $\hat{U}^\dagger \hat{U} = I$, $\hat{W}^\dagger \hat{W} = I$, $\hat{U}^\dagger \hat{W} = 0$, and $\hat{W}^\dagger \hat{U} = 0$. It

can be verified by some algebraic manipulations that

$$
\begin{aligned}
&[\hat{U} \quad \hat{W}]^{-1} (A + (x + \iota y)LC) [\hat{U} \quad \hat{W}] \\
&= \begin{bmatrix} \hat{S} + (x + \iota y)\hat{U}^{\dagger} LC\hat{U} & (x + \iota y)\hat{U}^{\dagger} LC\hat{W} \\ (x + \iota y)\hat{W}^{\dagger} LC\hat{U} & \hat{X} + (x + \iota y)\hat{W}^{\dagger} LCW \end{bmatrix} \\
&= \begin{bmatrix} \hat{S} - (x + \iota y)\hat{H}^T \hat{H} & -(x + \iota y)\hat{H}^T C\hat{W} \\ 0 & \hat{X} \end{bmatrix}.
\end{aligned}
\tag{2.30}
$$

By Lemma 23, we can get that $\hat{S} - (x + \iota y)\hat{H}^T \hat{H}$ is Hurwitz for any $x > 0$ and $y \in \mathbf{R}$ (which actually is dual to Lemma 23). Then, it follows from (2.30) that $A + (x + \iota y)LC$ is Hurwitz for all $x > 0$ and $y \in \mathbf{R}$. Hence, by Theorem 6, the protocol given by Algorithm 4 solves the consensus problem with an unbounded consensus region $(0, \infty) \times (-\infty, \infty)$.

Next, consider the general case without requiring A neutrally stable. The dual of Proposition 1 is presented as follows.

Proposition 3 *Given the agent dynamics (2.1), there exists a matrix L such that $A + (x + \iota y)LC$ is Hurwitz for all $x \in [1, \infty)$ and $y \in (-\infty, \infty)$, if and only if (A, C) is detectable.*

Theorem 6 and the above proposition together lead to the following result.

Theorem 8 *Assume that Assumption 2.1 holds. There exists an observer-type protocol (2.21) that solves the consensus problem and meanwhile yields an unbounded consensus region $S \triangleq [1, \infty) \times (-\infty, \infty)$ if and only if (A, B, C) is stabilizable and detectable.*

The following algorithm is presented to construct an observer-type protocol (2.21) satisfying Theorem 8.

Algorithm 5 *Assuming that (A, B, C) is stabilizable and detectable, the observer-type protocol (2.21) can be constructed according to the following steps.*

1) Choose the feedback gain matrix F such that $A + BF$ is Hurwitz.

2) Solve the following LMI:

$$
AQ + QA^T - 2C^T C < 0
\tag{2.31}
$$

to get one solution $Q > 0$. Then, choose the feedback gain matrix $L = -Q^{-1}C^T$.

3) Select the coupling gain $c > c_{th}$, with c_{th} given in (2.19).

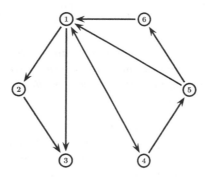

FIGURE 2.4: A directed communication graph.

From the proof of Theorem 6, it is easy to see that the convergence rate of the N agents in (2.1) reaching consensus under the protocol (2.21) is equal to the minimal decay rate of the $N-1$ systems in (2.27). Thus, the convergence rate of the agents (2.1) reaching consensus can be manipulated by properly assigning the eigenvalues of matrices $A+BF$ and $A+c\lambda_i LC$, $i=2,\cdots,N$. The following algorithm present a multi-step procedure to design (2.21) solving the consensus problem with a prescribed convergence rate.

Algorithm 6 *Assuming that (A, B, C) is controllable and observable, the observer-type protocol (2.21) solving the consensus problem with a convergence rate larger than α can be constructed as follows:*

1) Get the feedback gain matrix F, e.g., by using the Ackermann's formula [70], such that the poles of matrix $A+BF$ lie in the left-half plane of $x=-\alpha$.

2) Choose the feedback gain matrix $L=-Q^{-1}C^T$, where $Q>0$ is one solution to

$$A^TQ + QA - 2C^TC + 2\alpha Q < 0. \tag{2.32}$$

3) Select the coupling gain $c > c_{th}$, with c_{th} given in (2.19).

Example 6 *The agent dynamics are given by (2.1), with*

$$A = \begin{bmatrix} 0 & -1 & 0 \\ 1 & 0 & 0 \\ 0 & 1 & 0 \end{bmatrix}, \ B = \begin{bmatrix} 1 \\ 0 \\ 0 \end{bmatrix}, \ C = \begin{bmatrix} 0 & 0 & 1 \end{bmatrix}.$$

Obviously, the matrix A is neutrally stable and (A, B, C) is controllable and observable. A third-order consensus protocol is in the form of (2.2).

By the function `place` *of* MATLAB®*, the feedback gain matrix F of (2.21) is given as $F = \begin{bmatrix} -4.5 & -5.5 & -3 \end{bmatrix}$ such that the poles of $A+BF$ are $-1, -1.5, -2$. The matrix T such that $T^{-1}AT = J$ is of the real real Jordan canonical*

form is $T = \begin{bmatrix} 0 & 0.5774 & 0 \\ 0 & 0 & -0.5774 \\ 1 & -0.5774 & 0 \end{bmatrix}$, *with* $J = \begin{bmatrix} 0 & 0 & 0 \\ 0 & 0 & 1 \\ 0 & -1 & 0 \end{bmatrix}$. *Then, by Algorithm 4,*
the feedback gain matrix L *of (2.2) is obtained as* $L = \begin{bmatrix} 0.3333 & 0 & -1.3333 \end{bmatrix}^T$. *By*
Theorem 7, the agents under this protocol can reach consensus with respect
to any communication graph containing a spanning tree and for any positive
coupling gain. One such graph with 6 nodes is shown in FIGURE 2.4. The
corresponding Laplacian matrix is

$$\mathcal{L} = \begin{bmatrix} 3 & 0 & 0 & -1 & -1 & -1 \\ -1 & 1 & 0 & 0 & 0 & 0 \\ -1 & -1 & 2 & 0 & 0 & 0 \\ -1 & 0 & 0 & 1 & 0 & 0 \\ 0 & 0 & 0 & -1 & 1 & 0 \\ 0 & 0 & 0 & 0 & -1 & 1 \end{bmatrix},$$

whose nonzero eigenvalues $1, 1.3376 \pm \iota 0.5623, 2, 3.3247$. *Select* $c = 1$ *for sim-*
plicity. It can be verified that the convergence rate in this case equals 0.0303.

Next, the protocol (2.21) is redesigned to achieve consensus with a specified
convergence rate larger than 1. The feedback gain F *is chosen to be the same as*
above. Solving LMI (2.32) with $\alpha = 1$ *by using SeDuMi toolbox [158] gives* $L =$
$\begin{bmatrix} -8.5763 & -15.2128 & -5.6107 \end{bmatrix}^T$. *For the graph in FIGURE 2.4, the threshold value*
for the coupling gain is $c_{th} = 1$ *by (2.19). Select* $c = 1$, *the same as before.*
The consensus errors $x_i - x_1$, $i = 2, \cdots, 6$, *for the graph in FIGURE 2.4,*
under the protocols generated by Algorithms 4 and 6 with $\alpha = 1$, *are depicted*
in FIGURE 2.5(a) and FIGURE 2.5(b), respectively. It can be observed that
the consensus process of the former case is indeed much slower than the latter.

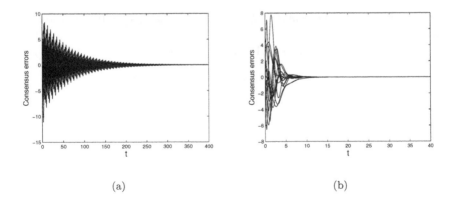

(a)   (b)

FIGURE 2.5: The consensus errors: (a) Algorithm 4, (b) Algorithm 6 with
$\alpha = 1$.

2.3.2   Full-Order Observer-Type Protocol II

In this subsection, we will propose another alternative dynamic consensus protocol. Based on the relative output information of neighboring agents, we propose the following full-order observer-type protocol:

$$
\dot{\bar{v}}_i = (A + \bar{L}C)\bar{v}_i + \sum_{j=1}^{N} a_{ij}[cB\bar{F}(\bar{v}_i - \bar{v}_j) + \bar{L}(y_i - y_j)],
$$

$$
u_i = c\bar{F}\bar{v}_i, \quad i = 1, \cdots, N,
$$

$$(2.33)$$

where $\bar{v}_i \in \mathbf{R}^n$, $c > 0$ denotes the coupling gain, $\bar{L} \in \mathbf{R}^{n \times q}$ and $\bar{F} \in \mathbf{R}^{p \times n}$ are the feedback gain matrices.

Different from the dynamic protocol (2.21) where the protocol states v_i take roles of intermediate variables, the states $\bar{v}_i$ of (2.33) can be the estimates of the relative states $\sum_{j=1}^{N} a_{ij}(x_i - x_j)$. Let $e_i = \bar{v}_i - \sum_{j=1}^{N} a_{ij}(x_i - x_j)$, $i = 1, \cdots, N$. Then, we can get from (2.33) that e_i, $i = 1, \cdots, N$, satisfy the following dynamics:

$$
\dot{e}_i = (A + \bar{L}C)e_i, \quad i = 1, \cdots, N.
$$

Clearly, if $\bar{L}$ is chosen such that $A + \bar{L}C$ is Hurwtiz, then $\bar{v}_i$ asymptotically converge to $\sum_{j=1}^{N} a_{ij}(x_i - x_j)$. Let $e = [e_1^T, \cdots, e_N^T]^T$ and $x = [x_1^T, \cdots, x_N^T]^T$. Then, we can get from (2.1) and (2.33) that the closed-loop network dynamics can be written in terms of e and x as

$$
\begin{bmatrix} \dot{x} \\ \dot{e} \end{bmatrix} = \begin{bmatrix} A + c\mathcal{L} \otimes B\bar{F} & cB\bar{F} \\ 0 & A + \bar{L}C \end{bmatrix} \begin{bmatrix} x \\ e \end{bmatrix}.
$$

$$(2.34)$$

By following similar steps in showing Theorem 6, we obtain the following result.

Theorem 9 *Suppose that Assumption 2.1 holds. Then, the observer-type protocol (2.33) solves the consensus problem for the agents in (2.1) if all the matrices $A + \bar{L}C$ and $A + c\lambda_i B\bar{F}$, $i = 2, \cdots, N$, are Hurwitz.*

Note that the design of $\bar{F}$ and $\bar{L}$ in Theorem 9 is actually dual to the design of L and F in Theorem 6. Algorithms 4 and 5 in the previous subsection can be modified to determine the parameters c, $\bar{F}$, and $\bar{L}$ of the dynamic protocol (2.33). The details are omitted here for conciseness.

2.3.3   Reduced-Order Observer-Based Protocol

The observer-type protocol (2.21) in the last subsection is of full order, i.e., whose dimension is equal to that of a single agent. Note that this full-order protocol possesses a certain degree of redundancy, which stems from the fact that while the observer constructs an "estimate" of the entire state, part of the state information is already reflected in the system outputs. In this subsection, we will design a reduced-order protocol so as to eliminate this redundancy and thereby can considerably reduce the dimension of the protocol especially for the case where the agents are multi-input multi-output (MIMO) systems.

Based on the local output information of neighboring agents, we introduce here a reduced-order observer-based consensus protocol as

$$\dot{v}_i = Fv_i + Gy_i + TBu_i,$$
$$u_i = cK[Q_1 \sum_{j=1}^{N} a_{ij}(y_i - y_j) + Q_2 \sum_{j=1}^{N} a_{ij}(v_i - v_j)], \ i = 1, \cdots, N, \quad (2.35)$$

where $v_i \in \mathbf{R}^{n-q}$ is the protocol state, $c > 0$ is the coupling gain, a_{ij} is the (i,j)-th entry of the adjacency matrix of $\mathcal{G}$, $F \in \mathbf{R}^{(n-q)\times(n-q)}$ is Hurwitz and has no eigenvalues in common with those of A, $G \in \mathbf{R}^{(n-q)\times q}$, $T \in \mathbf{R}^{(n-q)\times n}$ is the unique solution to the following Sylvester equation:

$$TA - FT = GC, \quad (2.36)$$

which further satisfies that $[{}^C_T]$ is nonsingular, $Q_1 \in \mathbf{R}^{n\times q}$ and $Q_2 \in \mathbf{R}^{n\times(n-q)}$ are given by $[\,Q_1 \ Q_2\,] = [{}^C_T]^{-1}$, and $K \in \mathbf{R}^{p\times n}$ is the feedback gain matrix.

Remark 13 *Different from the full-order observer-type protocols (2.21) and (2.33) which rely on the relative output information of neighboring agents, the above reduced-order protocol (2.35) requires the absolute measures of the agents' outputs, which might not be available in some circumstances, e.g., in deep-space formation flying [155].*

Let $\zeta_i = [x_i^T, v_i^T]^T$ and $\zeta = [\zeta_1^T, \cdots, \zeta_N^T]^T$. Then, the closed-loop network dynamics resulting from (2.1) and (2.35) can be written as

$$\dot{\zeta} = (I_N \otimes \mathcal{M} + c\mathcal{L} \otimes \mathcal{R})\zeta, \quad (2.37)$$

where $\mathcal{L} \in \mathbf{R}^{N\times N}$ is the Laplacian matrix of $\mathcal{G}$, and

$$\mathcal{M} = \begin{bmatrix} A & 0 \\ GC & F \end{bmatrix}, \quad \mathcal{R} = \begin{bmatrix} BKQ_1C & BKQ_2 \\ TBKQ_1C & TBKQ_2 \end{bmatrix}.$$

Next, an algorithm is presented to select the parameters in the consensus protocol (2.35).

Algorithm 7 *Supposing that (A, B, C) is stabilizable and observable and Assumption 2.1 holds, the protocol (2.35) can be constructed as follows:*

1) Choose a Hurwitz matrix F having no eigenvalues in common with those of A. Select G such that (F, G) is controllable.

2) Solve (2.36) to get a solution T, which satisfies that $[{}^C_T]$ is nonsingular. Then, compute matrices Q_1 and Q_2 by $[\,Q_1 \ Q_2\,] = [{}^C_T]^{-1}$.

3) *Solve the LMI (2.18) to get one solution $P > 0$. Then, choose the matrix $K = -B^T P^{-1}$.*

4) *Select the coupling gains $c \geq c_{th}$, with c_{th} given as in (2.19).*

Remark 14 *By Theorem 8.M6 in [23], a necessary condition for the matrix T to the unique solution to (2.36) and further to satisfy that $\begin{bmatrix} C \\ T \end{bmatrix}$ is nonsingular is that (F, G) is controllable, (A, C) is observable, and F and A have no common eigenvalues. In the case where the agent in (2.1) is single-input single-output (SISO), this condition is also sufficient. Under such a condition, it is shown for the general MIMO case that the probability for $\begin{bmatrix} C \\ T \end{bmatrix}$ to be nonsingular is 1 [23]. If $\begin{bmatrix} C \\ T \end{bmatrix}$ is singular in step 2), we need to go back to step 1) and repeat the process. As shown in Proposition 1, a necessary and sufficient condition for the existence of a positive-definite solution to the LMI (2.18) is that (A, B) is stabilizable. Therefore, a sufficient condition for Algorithm 7 to successfully construct a protocol (2.35) is that (A, B, C) is stabilizable and observable.*

Theorem 10 *Assuming that Assumption 2.1 holds, the reduced-order protocol (2.35) solves the consensus problem for the agents described by (2.1) if the matrices F and $A + c\lambda_i BK$, $i = 2, \cdots, N$ are Hurwitz. One such protocol solving the consensus problem is constructed by Algorithm 7. Then, the state trajectories of (2.37) satisfy*

$$
x_i(t) \rightarrow \varpi(t) \triangleq (r^T \otimes e^{At}) \begin{bmatrix} x_1(0) \\ \vdots \\ x_N(0) \end{bmatrix},
$$
(2.38)
$$
v_i(t) \rightarrow T\varpi(t), \ i = 1, \cdots, N, \ as \ t \rightarrow \infty,
$$

where $r \in \mathbf{R}^N$ is a nonnegative vector such that $r^T \mathcal{L} = 0$ and $r^T \mathbf{1} = 1$.

Proof 9 *By following similar steps as in the proof of Theorem 6, we can reduce the consensus problem of (2.37) to the simultaneous asymptotically stability problem of the following $N - 1$ systems:*

$$
\dot{\tilde{\zeta}}_i = \mathcal{M} + c\lambda_i \mathcal{R} \tilde{\zeta}_i, \quad i = 2, \cdots, N,
$$

where λ_i, $i = 2, \cdots, N$, are the nonzero eigenvalues of $\mathcal{L}$. Multiplying the left and right sides of the matrix $\mathcal{M} + c\lambda_i \mathcal{R}$ by $Q = \begin{bmatrix} I & 0 \\ -T & I \end{bmatrix}$ and Q^{-1}, respectively, and in virtue of (2.36) and step 2) of Algorithm 7, we get

$$
Q(\mathcal{M} + c\lambda_i \mathcal{R})Q^{-1} = \begin{bmatrix} A + c\lambda_i BK & c\lambda_i BKQ_2 \\ 0 & F \end{bmatrix}.
$$

Therefore, the consensus problem of (2.37) is reached if F and $A + c\lambda_i BK$, $i = 2, \cdots, N$ are Hurwitz. As shown in Proposition 1, $A + c\lambda_i BK$, $i = 2, \cdots, N$, with c and K designed as in Algorithm 7, are Hurwitz. The rest of the proof can be completed by following similar steps as in the proof of Theorem 2, which is thus omitted here.

For the case where the state matrix A is neutrally stable, Algorithm 7 can be modified by refereeing to Algorithm 1. Moreover, Algorithm 7 can also be modified to construct the protocol (2.35) to reach consensus with a prescribed convergence rate. We omit the details here for conciseness.

2.4   Extensions to Switching Communication Graphs

In the previous sections, the communication graph is assumed to be fixed throughout the whole process. As mentioned in Example 5, the communication graph may change with time in many practical situations due to various reasons, such as communication constraints and link variations. The objective of this section is to extend the results in the previous sections to the case with switching communication graphs. In this section, we restrict our attention to the case with undirected communication graphs.

Denote by $\mathscr{G}_N$ the set of all possible undirected connected graphs with N nodes. Let $\sigma(t) : [0, \infty) \to \mathscr{P}$ be a piecewise constant switching signal with switching times $t_0, t_1, \cdots$, and $\mathscr{P}$ be the index set associated with the elements of $\mathscr{G}_N$, which is clearly finite. The communication graph at time t is denoted by $\mathcal{G}_{\sigma(t)}$.

For illustration and conciseness, here we consider only the static consensus protocol (2.2). Extensions of the observer-type protocols can be similarly done. For the case with switching undirected communication graphs, (2.2) becomes

$$u_i = cK \sum_{j=1}^{N} a_{ij}(t)(x_i - x_j), \qquad (2.39)$$

where $a_{ij}(t)$ is the (i, j)-th entry of the adjacency matrix associated with $\mathcal{G}_{\sigma(t)}$ and the rest of the variables are the same as in (2.2). Using (2.39) for (2.1), we can obtain the closed-loop network dynamics as

$$\dot{x} = (I_N \otimes A + c\mathcal{L}_{\sigma(t)} \otimes BK)x, \qquad (2.40)$$

where $x = [x_1^T, \cdots, x_N^T]^T$ and $\mathcal{L}_{\sigma(t)}$ is the Laplacian matrix associated with $\mathcal{G}_{\sigma(t)}$.

Let $\varsigma = [(I_N - \frac{1}{N}\mathbf{1}\mathbf{1}^T) \otimes I_n]x$. Then, as shown in Section 2.2, the consensus problem of (2.40) is reduced to the asymptotical stability problem of ς, which satisfies

$$\dot{\varsigma} = (I_N \otimes A + c\mathcal{L}_{\sigma(t)} \otimes BK)\varsigma. \qquad (2.41)$$

Algorithm 2 will be modified as below to design the consensus protocol (2.39).

Algorithm 8 *Assuming that (A, B, C) that is stabilizable and detectable, and* *that $\mathcal{G}_{\sigma(t)} \in \mathscr{G}_N$, a protocol (2.40) can be constructed according to the following steps:*

1) *step 1) in Algorithm 2.*

2) *Select the coupling gain $c \geq \frac{1}{\lambda_2^{min}}$, where $\lambda_2^{min} \triangleq \min_{\mathcal{G}_{\sigma(t)} \in \mathscr{G}_N}\{\lambda_2(\mathcal{L}_{\sigma(t)})\}$ denotes the minimum of the smallest nonzero eigenvalues of $\mathcal{L}_{\sigma(t)}$ for $\mathcal{G}_{\sigma(t)} \in \mathscr{G}_N$.*

The following theorem shows that the protocol (2.40) does solve the consensus problem.

Theorem 11 *For arbitrary switching communication graphs $\mathcal{G}_{\sigma(t)}$ belonging to $\mathscr{G}_N$, the N agents described by (2.1) reach consensus under the protocol (2.40) designed by Algorithm 8. Further, the final consensus value is given by*

$$x_i(t) \to \chi(t) \triangleq (\frac{1}{N}\mathbf{1}^T \otimes e^{At}) \begin{bmatrix} x_1(0) \\ \vdots \\ x_N(0) \end{bmatrix}, \quad as\ t \to \infty. \qquad (2.42)$$

Proof 10 *Consider a common Lyapunov function candidate as follows:*

$$V_1 = \frac{1}{2}\varsigma^T(I_N \otimes P^{-1})\varsigma,$$

where $P > 0$ is a solution to the LMI (2.18). The time derivative of V_1 along the trajectory of (2.41) can be obtained as

$$\dot{V}_1 = \varsigma^T(I_N \otimes P^{-1}A - c\mathcal{L}_{\sigma(t)} \otimes P^{-1}BBP^{-1})\varsigma. \qquad (2.43)$$

Because $\mathcal{G}_{\sigma(t)}$ is connected and $(\mathbf{1}^T \otimes I)\varsigma = 0$, it follows from Lemma 2 that $\varsigma^T(\mathcal{L}_{\sigma(t)} \otimes I)\varsigma \geq \lambda_2^{min}\varsigma^T\varsigma$. Since $P^{-1}BBP^{-1} \geq 0$, we can further get that

$$\varsigma^T(\mathcal{L}_{\sigma(t)} \otimes P^{-1}BBP^{-1})\varsigma \geq \lambda_2^{min}\varsigma^T(I_N \otimes P^{-1}BBP^{-1})\varsigma. \qquad (2.44)$$

Then, it follows from (2.43) and (2.44) that

$$\dot{V}_1 \leq \frac{1}{2}\varsigma^T(I_N \otimes (P^{-1}A + A^TP^{-1} - c\lambda_2^{min}P^{-1}BBP^{-1})\varsigma$$
$$< 0,$$

where we have used that fact that $P^{-1}A + A^TP^{-1} - c\lambda_2^{min}P^{-1}BBP^{-1} < 0$, which follows readily from (2.18). Therefore, we obtain that the system (2.41) is asymptotically stable, i.e., the consensus problem is solved. The assertion that $x_i(t) \to \psi(t)$ as $t \to \infty$ can be shown by following similar lines in the proof of Theorem 2, which is omitted here.

Remark 15 *An interesting observation is that even though the undirected communication graphs are switching according to some time sequence, the final consensus value is still equal to the case with a fixed undirected graph. This actually can be explained by noting the fact that $\chi(t)$ depends on the left eigenvector $\mathbf{1}$ of the Laplacian matrix corresponding to the zero eigenvalue, which remain the same for all undirected communication graphs.*

2.5   Extension to Formation Control

In this section, the consensus protocols proposed in the preceding sections will be modified to solve formation control problem of multi-agent systems.

Let $\tilde{H} = (h_1, h_2, \cdots, h_N) \in \mathbf{R}^{n \times N}$ describe a constant formation structure of the agent network in a reference coordinate frame, where $h_i \in \mathbf{R}^n$ is the formation variable corresponding to agent i. Then, variable $h_i - h_j$ can be used to denote the relative formation vector between agents i and j, which is independent of the reference coordinate. For the agents (2.1), the static consensus protocol can be modified to get a distributed formation protocol, described by

$$u_i = cK \sum_{j=1}^{N} a_{ij}(x_i - x_j - h_i + h_j), \tag{2.45}$$

where the variables are the same as those in (2.2). It should be noted that (2.45) reduces to the consensus protocol (2.2), when $h_i - h_j = 0$, $\forall i, j = 1, \cdots, N$.

Definition 9 *The agents (2.1) under the protocol (2.45) achieve a given formation $\tilde{H} = (h_1, h_2, \cdots, h_N)$, if*

$$\lim_{t \to \infty} \|(x_i(t) - h_i) - (x_j(t) - h_j)\| \to 0, \quad \forall i, j = 1, \cdots, N. \tag{2.46}$$

Theorem 12 *For graph $\mathcal{G}$ satisfying Assumption 2.1, the agents (2.1) reach the formation $\tilde{H}$ under the protocol (2.45) if all the matrices $A + c\lambda_i BK$, $i = 2, \cdots, N$, are Hurwitz, and $Ah_i = 0$, $\forall i = 1, \cdots, N$, where λ_i, $i = 2, \cdots, N$, are the nonzero eigenvalues of the Laplacian matrix $\mathcal{L}$.*

Proof 11 *Let $\tilde{x}_i = x_i - h_i$, $i = 1, \cdots, N$. Then, it follows from (2.1) and (2.45) that*

$$\dot{\tilde{x}}_i = A\tilde{x}_i + c \sum_{j=1}^{N} \mathcal{L}_{ij} BK\tilde{x}_j + Ah_i, \quad i = 1, \cdots, N. \tag{2.47}$$

Note that the formation $\tilde{H}$ is achieved if the system (2.47) reaches consensus.

By following similar steps in the proof of Theorem 1, it is easy to see that the formation $\tilde{H}$ is achieved under the protocol (2.45) if all the matrices $A + c\lambda_i BK$, $i = 2, \cdots, N$, are Hurwitz, and $Ah_i = 0$, $\forall i = 1, \cdots, N$.

Remark 16 *Note that not all kinds of formation structure can be achieved for the agents (2.1) by using protocol (2.45). The achievable formation structures have to satisfy the constraints $Ah_i = 0$, $\forall i = 1, \cdots, N$. Note that h_i can be replaced by $h_i - h_1$, $i = 2, \cdots, N$, in order to be independent of the reference coordinate, by simply choosing h_1 corresponding to agent 1 as the origin. The formation protocol (2.45) satisfying Theorem 12 can be constructed by using Algorithms 1 and 2.*

The other consensus protocols discussed in the previous sections can also be modified to solve the formation control problem. For instance, a distributed formation protocol corresponding to the observer-type consensus protocol (2.1) can be described as follows:

$$\dot{v}_i = (A + BF)v_i + cL\sum_{j=1}^{N} a_{ij}[C(v_i - v_j) - (y_i - y_j - C(h_i - h_j))], \tag{2.48}$$

$$u_i = Fv_i,$$

where the variables are defined as in (2.21). Similar results can be accordingly obtained. The only difference is that the constraints $Ah_i = 0$, $\forall i = 1, \cdots, N$ need to be also satisfied.

Example 7 *Consider a network of six double integrators, described by*

$$\dot{x}_i = v_i,$$
$$\dot{\tilde{v}}_i = u_i,$$
$$y_i = x_i, \quad i = 1, \cdots, 6,$$

where $x_i \in \mathbf{R}^2$, $\tilde{v}_i \in \mathbf{R}^2$, $y_i \in \mathbf{R}^2$, and $u_i \in \mathbf{R}^2$ are the position, the velocity, the measured output, and the acceleration input of agent i, respectively.

The objective is to design a dynamic protocol (2.48) such that the agents will evolve to form a regular hexagon with edge length 4. In this case, choose $h_1 = [0\,0\,0\,0]^T$, $h_2 = [4\,0\,0\,0]^T$, $h_3 = [6\,2\sqrt{3}\,0\,0]^T$, $h_4 = [4\,4\sqrt{3}\,0\,0]^T$, $h_5 = [0\,4\sqrt{3}\,0\,0]^T$, $h_6 = [-2\,2\sqrt{3}\,0\,0]^T$. Take $F = [-1.5\,-2.5] \otimes I_2$ in (2.48) such that matrix $A + BF$ has eigenvalues -1 and -1.5. By solving LMI (2.32) with $\alpha = 1$, one obtains $L = [-3.6606\,-4.8221]^T \otimes I_2$. The six agents under protocol (2.48) with F, L given as above, and $c = 1$ will form a regular hexagon with a convergence rate larger than 1 for the communication topology given in FIGURE 2.4. The state trajectories of the six agents are depicted in FIGURE 2.6.

In the following, we will consider the formation keeping problem for satellites moving in the low earth orbit. The translational dynamics of satellites

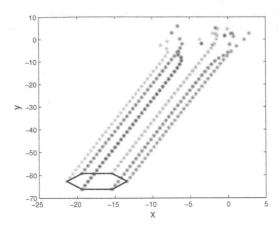

FIGURE 2.6: The six agents form a hexagon.

in the low earth orbit, different from those of deep-space satellites, cannot be modeled as double integrators. In order to simplify the analysis, assume that a virtual reference satellite is moving in a circular orbit of radius R_0. The relative dynamics of the other satellites with respect to the virtual satellite will be linearized in the following coordinate system, where the origin is on the mass center of the virtual satellite, the x-axis is along the velocity vector, the y-axis is aligned with the position vector, and the z-axis completes the right-hand coordinate system. The linearized equations of the relative dynamics of the i-th satellite with respect to the virtual satellite are given by the Hill's equations [69]:

$$\ddot{\tilde{x}}_i - 2\omega_0\dot{\tilde{y}}_i = u_{x_i},$$
$$\ddot{\tilde{y}}_i + 2\omega_0\dot{\tilde{x}}_i - 3\omega_0^2\tilde{y}_i = u_{y_i}, \qquad (2.49)$$
$$\ddot{\tilde{z}}_i + \omega_0^2\tilde{z}_i = u_{z_i},$$

where $\tilde{x}_i$, $\tilde{y}_i$, $\tilde{z}_i$ are the position components of the i-th satellite in the rotating coordinate, u_{x_i}, u_{y_i}, u_{z_i} are control inputs, and ω_0 denotes the angular rate of the virtual satellite. The main assumption inherent in Hill's equations is that the distance between the i-th satellite and the virtual satellite is very small in comparison to the orbital radius R_0.

Denote the position vector by $r_i = [\tilde{x}_i, \tilde{y}_i, \tilde{z}_i]^T$ and the control vector by $u_i = [u_{x_i}, u_{y_i}, u_{z_i}]^T$. Then, (2.49) can be rewritten as

$$\begin{bmatrix} \dot{r}_i \\ \ddot{r}_i \end{bmatrix} = \begin{bmatrix} 0 & I_3 \\ A_1 & A_2 \end{bmatrix} \begin{bmatrix} r_i \\ \dot{r}_i \end{bmatrix} + \begin{bmatrix} 0 \\ I_3 \end{bmatrix} u_i, \qquad (2.50)$$

where

$$A_1 = \begin{bmatrix} 0 & 0 & 0 \\ 0 & 3\omega_0^2 & 0 \\ 0 & 0 & -\omega_0^2 \end{bmatrix}, \quad A_2 = \begin{bmatrix} 0 & 2\omega_0 & 0 \\ -2\omega_0 & 0 & 0 \\ 0 & 0 & 0 \end{bmatrix}.$$

The satellites are said to achieve formation keeping, if their velocities converge to the same and their positions maintain a prescribed separation, i.e., $r_i - h_i \to r_j - h_j$, $\dot{r}_i \to \dot{r}_j$, as $t \to \infty$, where $h_i - h_j \in \mathbf{R}^3$ denotes the desired constant separation between satellite i and satellite j.

Represent the communication topology among the N satellites by a directed graph $\mathcal{G}$. Assume that measurements of both relative positions and relative velocities between neighboring satellites are available. The control input to satellite i is proposed here as

$$u_i = -A_1 h_i + c \sum_{j=1}^{N} a_{ij} \left[F_1 (r_i - h_i - r_j + h_j) + F_2 (\dot{r}_i - \dot{r}_j) \right], \qquad (2.51)$$

where $c > 0$, F_1 and $F_2 \in \mathbf{R}^{3 \times 3}$ are the feedback gain matrices to be determined. If satellite k is a leader, i.e., it does not receive information from any other satellite, then the term $A_1 h_k$ is set to zero. With (2.51), the equation (2.50) can be reformulated as

$$\begin{bmatrix} \dot{r}_i \\ \ddot{r}_i \end{bmatrix} = \begin{bmatrix} 0 & I_3 \\ A_1 & A_2 \end{bmatrix} \begin{bmatrix} r_i - h_i \\ \dot{r}_i \end{bmatrix} + c \sum_{j=1}^{N} a_{ij} \begin{bmatrix} 0 & 0 \\ F_1 & F_2 \end{bmatrix} \begin{bmatrix} r_i - h_i - r_j + h_j \\ \dot{r}_i - \dot{r}_j \end{bmatrix}. \quad (2.52)$$

The following result is a direct consequence of Theorem 12.

Corollary 2 *Assume that graph $\mathcal{G}$ has a directed spanning tree. Then, the protocol (2.51) solves the formation keeping problem if and only if the matrices $\begin{bmatrix} 0 & I_3 \\ A_1 & A_2 \end{bmatrix} + c\lambda_i \begin{bmatrix} 0 & 0 \\ F_1 & F_2 \end{bmatrix}$ are Hurwitz for $i = 2, \cdots, N$, where λ_i, $i = 2, \cdots, N$, denote the nonzero eigenvalues of the Laplacian matrix of $\mathcal{G}$.*

The feedback gain matrices F_1 and F_2 satisfying Corollary 2 can be designed by following Algorithm 2. Since the system (2.50) is stabilizable, they always exist.

Example 8 *Consider the formation keeping of four satellites with respect to a virtual satellite which moves in a circular orbit at rate $\omega_0 = 0.001$. Assume that the communication topology is given by FIGURE 2.7, from which it is observed that satellite 1 plays the leader role. The nonzero eigenvalues of the Laplacian matrix in this case are $1, 1, 2$. Select c in (2.51) to be 1 for simplicity. We can solve the formation keeping problem in the following steps: i) Select properly the initial state of the leading satellite 1 such that it moves in a spatial circular relative orbit with respect to the virtual satellite; ii) Design the consensus protocol in the form of (2.51) such that the four satellites together maintain a desired formation flying situation.*

Select the initial state of satellite 1 as $\tilde{x}_0 = 0$, $\tilde{y}_0 = 500$, $\tilde{z}_0 = 866$, $\dot{\tilde{x}}_0 = 1$, $\dot{\tilde{y}}_0 = 0$, $\dot{\tilde{z}}_0 = 0$. In such a case, satellite 1 maintains a circular relative orbit with respect to the virtual satellite, of radius $r = 1000m$, and with the tangent angle to the orbital place $\alpha = 30 \deg$. Suppose that the four satellites will

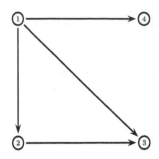

FIGURE 2.7: The communication topology.

maintain a square shape with the separation of 500m in a plane tangent to the orbit of the virtual satellite by an angle $\theta = 60$ deg. Let $h_1 = (100, 100, 0)$, $h_2 = (-100, 100, 0)$, $h_3 = (100, 0, 173.21)$, and $h_4 = (-100, 10, 173.21)$. The feedback gain matrices F_1 and F_2 in (2.51) can be obtained via Algorithm 2. The simulation result is presented in FIGURE 2.8.

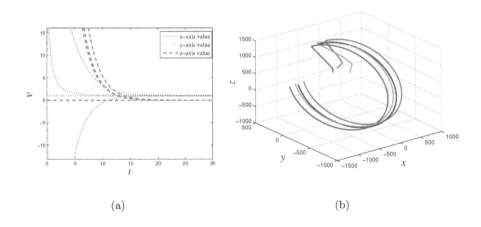

(a)   (b)

FIGURE 2.8: (a) The relative velocity; (b) the relative positions, $t \in [0, 5000]s$.

2.6 Notes

The materials of this chapter are mainly based on [83, 86, 91, 198]. The full-order observer-type consensus protocol (2.33) is adopted from [197], which is also proposed in [169].

For more results about consensus of general continuous-time linear multi-agent systems, please refer to [106, 116, 143, 146, 171, 173, 178, 186, 195] and the references therein. In [106, 116, 171, 173], the authors considered the case where the relative states of neighboring agents are available. Distributed dynamic consensus protocols based on the local output information were proposed in [143, 146, 195, 198]. Specially, [195] presented other types of observer-type protocols except of those in this chapter. The observed-based consensus protocol in [143] requires the absolute output of each agent but is applicable to jointly connected switching communication graphs. A low gain technique was used in [146, 198] to design an output feedback consensus protocol, which are based on only the relative outputs of neighboring agents. It is worth mentioning that the convergence rate of reaching consensus under the protocols using the low gain technique is generally much lower and a restriction on the state matrix A is required in [146, 198]. Reduced-order consensus protocols in [199] do not require the absolute output information, which however are based on both the relative outputs and inputs of neighboring agents. Necessary and sufficient conditions were derived in [186] for achieving continuous-time consensus over Markovian switching topologies. The consensus problem of linear multi-agent systems with uniform constant communication delay was concerned in [178], where an upper bound for delay tolerance was obtained which explicitly depends on the agent dynamics and the network topology.

2.8 Notes

3

Consensus Control of Linear Multi-Agent Systems: Discrete-Time Case

CONTENTS

3.1   Problem Statement 54
3.2   State Feedback Consensus Protocols 54
    3.2.1   Consensus Condition 55
    3.2.2   Discrete-Time Consensus Region 57
    3.2.3   Consensus Protocol Design 59
        3.2.3.1   The Special Case with Neutrally Stable Agents .. 60
        3.2.3.2   The General Case 62
3.3   Observer-Type Consensus Protocols 65
    3.3.1   Full-Order Observer-Type Protocol I 65
    3.3.2   Full-Order Observer-Type Protocol II 67
    3.3.3   Reduced-Order Observer-Based Protocol 68
3.4   Application to Formation Control 69
3.5   Discussions 71
3.6   Notes 72

In this chapter, we consider the consensus of discrete-time linear multi-agent systems with directed communication topologies, which is the discrete-time counterpart of the continuous-time consensus in the preceding chapter. The consensus problem of discrete-time multi-agent systems is formulated in Section 3.1. The communication graph among the agents is assumed to contain a directed spanning tree throughout this chapter. Different from the previous chapter where the Laplacian matrix is used to represent the communication graph, the stochastic matrix will be used instead in this chapter.

Based on the relative states of neighboring agents, a distributed static consensus protocol is proposed in Section 3.1. The consensus problem of the discrete-time multi-agent system under the proposed static protocol is cast into the stability of a set of matrices with the same dimension as a single agent. The final consensus value reached by the agents is also derived. Inspired by the continuous-time consensus region in Chapter 2, the notion of discrete-time consensus region is introduced and analyzed. It is pointed out via numerical examples that the consensus protocol should have a reasonably large bounded consensus region so as to be robust to variations of the communication topology. For the special case where the state matrix of each agent is neutrally stable, it is shown that there exists a static protocol with a bounded consensus region in the form of an open unit disk if each agent is stabilizable, implying that such static protocol can achieve consensus with respect to all communication topologies containing a directed spanning tree. For the general case where the state matrix might be unstable, an algorithm is proposed to construct a protocol with the origin-centered disk of radius δ $(0 < \delta < 1)$ as its consensus region. It is worth noting that δ has to further satisfy a constraint relying on the unstable eigenvalues of the state matrix for the case where each agent has a least one eigenvalue outside the unit circle, which shows that the consensus problem of the discrete-time multi-agent systems is generally more difficult to solve, compared to the continuous-time case in the previous chapter.

Based on the relative outputs of neighboring agents, distributed full-order and reduced-order observer-type consensus protocols are proposed in Section 3.3. The Separation Principle of the traditional observer-based controllers still holds in the discrete-time multi-agent setting presented in this paper. Algorithms are presented to design these observer-type protocols. In Section 3.4, the consensus protocols are modified to solve formation control problem for discrete-time multi-agent systems. The main differences of the discrete-time consensus in this chapter and the continuous-time consensus in the previous chapter are further highlighted in Section 3.5.

3.1   Problem Statement

Consider a network of N identical agents with general discrete-time linear dynamics, where the dynamics of the i-th agent are described by

$$
\begin{aligned}
x_i^+ &= Ax_i + Bu_i, \\
y_i &= Cx_i, \quad i = 1, \cdots, N,
\end{aligned} \tag{3.1}
$$

where $x_i = x_i(k) \in \mathbf{R}^n$ is the state, $x_i^+ = x_i(k+1)$ is the state at the next time instant, $u_i \in \mathbf{R}^p$ is the control input, $y_i \in \mathbf{R}^q$ is the measured output, and A, B, C are constant matrices with compatible dimensions.

The communication topology among the N agents is represented by a directed graph $\mathcal{G}$, which is assumed to satisfy the following assumption throughout this chapter.

Assumption 3.1 *The communication graph $\mathcal{G}$ contains a directed spanning tree.*

The objective of this chapter is to solve the consensus problem for the N agents described by (3.1), i.e., to design distributed consensus protocols using only local information to ensure that the N agents described by (3.1) can achieve consensus in the sense of $\lim_{k \to \infty} \|x_i(k) - x_j(k)\| = 0$, $\forall i, j = 1, \cdots, N$.

3.2   State Feedback Consensus Protocols

In this section, we consider the case where each agent can have access to its own state and the states from its neighbors. A distributed static consensus protocol in this case can be proposed as

$$u_i = K \sum_{j=1}^{N} d_{ij}(x_i - x_j), \tag{3.2}$$

where $K \in \mathbf{R}^{p \times n}$ is the feedback gain matrix and d_{ij} is the (i, j)-th entry of the row-stochastic matrix $\mathcal{D}$ associated with the graph $\mathcal{G}$. Note that different from the Laplacian matrix used in the last chapter, the stochastic matrix will be used to characterize the communication graph in this chapter.

Let $x = [x_1^T, \cdots, x_N^T]^T$. Using (3.2) for (3.1), it is not difficult to get that the closed-loop network dynamics can be written as

$$x^+ = [I_N \otimes A + (I_N - \mathcal{D}) \otimes BK]x. \tag{3.3}$$

By Lemma 7, 1 is an eigenvalue of $\mathcal{D}$. Let $r \in \mathbf{R}^N$ be the left eigenvector of $I_N - \mathcal{D}$ associated with the eigenvalue 0, satisfying $r^T \mathbf{1} = 1$. Introduce $\xi \in \mathbf{R}^{Nn \times Nn}$ by

$$\xi = [(I_N - \mathbf{1}r^T) \otimes I_n]x, \tag{3.4}$$

which satisfies $(r^T \otimes I_n)\xi = 0$. By the definition of r, it is not difficult to see that 0 is a simple eigenvalue of $I_N - \mathbf{1}r^T$ with $\mathbf{1}$ as its right eigenvector and 1 is another eigenvalue with multiplicity $N - 1$. Thus, it follows from (3.4) that $\xi = 0$ if and only if $x_1 = x_2 = \cdots = x_N$, i.e., the consensus problem of (3.3) can be cast into the Schur stability of vector ξ, which evolves according to the following dynamics:

$$\xi^+ = [I_N \otimes A + (I - \mathcal{D}) \otimes BK]\xi. \tag{3.5}$$

3.2.1   Consensus Condition

The following theorem presents a necessary and sufficient condition for solving the consensus problem of (3.3).

Theorem 13 *Supposing Assumption 3.1 holds, the agents in (3.1) reach consensus under the static protocol (3.2) if and only if all the matrices $A + (1 - \hat{\lambda}_i)BK$, $i = 2, \cdots, N$, are Schur stable, where $\hat{\lambda}_i$, $i = 2, \cdots, N$, denote the eigenvalues of $\mathcal{D}$ located in the open unit disk.*

Proof 12 *Because the graph $\mathcal{G}$ satisfies Assumption 3.1, it follows from Lemma 7 that 0 is a simple eigenvalue of $I_N - \mathcal{D}$ and the other eigenvalues lie in* *the open unit disk centered at $1 + \iota 0$ in the complex plane. Let $Y_1 \in \mathbf{C}^{N \times (N-1)}$, $Y_2 \in \mathbf{C}^{(N-1) \times N}$, $T \in \mathbf{R}^{N \times N}$, and upper-triangular $\Delta \in \mathbf{C}^{(N-1) \times (N-1)}$ be such that*

$$T = \begin{bmatrix} 1 & Y_1 \end{bmatrix}, \ T^{-1} = \begin{bmatrix} r^T \\ Y_2 \end{bmatrix}, \ T^{-1}(I_N - \mathcal{D})T = J = \begin{bmatrix} 0 & 0 \\ 0 & \Delta \end{bmatrix}, \qquad (3.6)$$

where the diagonal entries of Δ are the nonzero eigenvalues of $I_N - \mathcal{D}$. Let $\zeta = (T^{-1} \otimes I_n)\xi$ with $\zeta = [\zeta_1^T, \cdots, \zeta_N^T]^T$. Then, (3.5) can be rewritten in terms of ζ as

$$\zeta^+ = [I_N \otimes A + (I - J) \otimes BK]\zeta, \qquad (3.7)$$

where $\zeta_1 = (r^T \otimes I_n)\xi \equiv 0$, which follows directly from the definition of ξ in (3.4). Note that the elements of the state matrix of (3.7) are either block diagonal or block upper-triangular. Hence, ζ_i, $i = 2, \cdots, N$, converge asymptotically to zero if and only if the $N - 1$ subsystems along the diagonal, i.e.,

$$\zeta_i^+ = [A + (1 - \hat{\lambda}_i)BK]\zeta_i, \quad i = 2, \cdots, N, \qquad (3.8)$$

are Schur stable. Therefore, the Schur stability of the matrices $A + (1 - \hat{\lambda}_i)BK$, $i = 2, \cdots, N$, is equivalent to that the state ζ of (3.7) converges asymptotically to zero, implying that consensus is achieved.

Remark 17 *The above theorem is a discrete-time counterpart of Theorem 1 in the last chapter. As mentioned earlier, the main difference is that the Laplacian matrix is used in the last chapter while the stochastic matrix is used here to characterize the communication topology. The effects of the communication topology on consensus are characterized by the eigenvalues of the stochastic matrix $\mathcal{D}$ except the eigenvalue one.*

Theorem 14 *Consider the multi-agent network (3.3) whose communication topology $\mathcal{G}$ satisfying Assumption 3.1. If the protocol (3.2) satisfies Theorem 13, then*

$$x_i(k+1) \to \varpi(k) \triangleq (r^T \otimes A^k) \begin{bmatrix} x_1(0) \\ \vdots \\ x_N(0) \end{bmatrix}, \ i = 1, \cdots, N, \ as \ k \to \infty,$$

$$(3.9)$$

where $r \in \mathbf{R}^N$ satisfies $r^T(I_N - \mathcal{D}) = 0$ and $r^T \mathbf{1} = 1$.

Proof 13 *The solution of (3.3) can be obtained as*

$$x(k+1) = (I_N \otimes A + (I_N - \mathcal{D}) \otimes BK)^k x(0)$$
$$= (T \otimes I)(I_N \otimes A + J \otimes BK)^k (T^{-1} \otimes I)x(0)$$
$$= (T \otimes I) \begin{bmatrix} A^k & 0 \\ 0 & (I_{N-1} \otimes A + \Delta \otimes BK)^k \end{bmatrix} (T^{-1} \otimes I)x(0),$$

where matrices T, J and Δ are defined in (3.6). By Theorem 13, $I_{N-1} \otimes A + \Delta \otimes BK$ is Schur stable. Thus,

$$x(k+1) \to (\mathbf{1} \otimes I) A^k (r^T \otimes I) x(0)$$
$$= (\mathbf{1} r^T) \otimes A^k x(0) \quad \text{as } k \to \infty,$$

implying that

$$x_i(k) \to \varpi(k) \quad \text{as } k \to \infty, \; i = 1, \cdots, N.$$

Remark 18 *Some observations on the final consensus value $\varpi(k)$ in (3.9) can be concluded as follows. If the state matrix A in (3.1) is Schur stable, then $\varpi(k) \to 0$ as $k \to \infty$, i.e., the discrete-time consensus problem becomes trivial. If A is unstable (having eigenvalues located outside the open unit circle), then $\varpi(k)$ will tend to infinity exponentially. On the other hand, if A has eigenvalues in the closed unit circle, then the agents in (3.1) may reach consensus nontrivially.*

3.2.2   Discrete-Time Consensus Region

To facilitate the design of the consensus protocol and elucidate the robustness of the protocol with respect to the communication topology, we introduce the following discrete-time consensus region.

Definition 10 *The region $\mathcal{S}$ of the parameter $\sigma \in \mathbf{C}$, such that matrix $A + (1 - \sigma)BK$ is Schur stable, is called the discrete-time consensus region of the network (3.3).*

The above definition presents a discrete-time counterpart of the continuous-time consensus region introduced in the previous chapter. The following result is a direct consequence of Theorem 13.

Corollary 3 *The agents in (3.1) reach consensus under the static protocol (3.2) if $\hat{\lambda}_i \in \mathcal{S}$, $i = 2, \cdots, N$, where $\hat{\lambda}_i$, $i = 2, \cdots, N$, are the eigenvalues of $\mathcal{D}$ located in the open unit disk.*

For an undirected communication graph, the consensus region of the network (3.3) is a bounded interval or a union of several intervals on the real axis. However, for a directed graph where the eigenvalues of $\mathcal{D}$ are generally complex numbers, the consensus region $\mathcal{S}$ is either a bounded region or a set of several disconnected regions in the complex plane. Due to the fact that the eigenvalues of the row-stochastic matrix $\mathcal{D}$ lie in the unit disk, unbounded consensus regions, desirable for consensus in the continuous-time setting as in the previous chapter, generally do not exist for the discrete-time consensus considered here.

The following example has a disconnected consensus region.

Example 9 *The agent dynamics and the consensus protocol are given by (3.1) and (3.2), respectively, with*

$$A = \begin{bmatrix} 0 & 1 \\ -1 & 1.02 \end{bmatrix}, \quad B = \begin{bmatrix} 1 & 0 \\ 0 & 1 \end{bmatrix}, \quad K = \begin{bmatrix} 0 & -1 \\ 1 & 0 \end{bmatrix}.$$

For simplicity in illustration, assume that the communication graph $\mathcal{G}$ is undirected here. Then, the consensus region is a set of intervals on the real axis. The characteristic equation of $A + (1 - \sigma)BK$ is

$$\det(zI - A - (1 - \sigma)BK) = z^2 - 1.02z + \sigma^2 = 0. \qquad (3.10)$$

Applying bilinear transformation $z = \frac{s+1}{s-1}$ to (3.10) gives

$$(\sigma^2 - 0.02)s^2 + (1 - \sigma^2)s + 2.02 + \sigma^2 = 0. \qquad (3.11)$$

It is well known that, under the bilinear transformation, (3.10) has all roots within the unit disk if and only if the roots of (3.11) lie in the open left-half plane (LHP). According to the Hurwitz criterion [23], (3.11) has all roots in the open LHP if and only if $0.02 < \sigma^2 < 1$. Therefore, the consensus region in this case is $\mathcal{S} = (-1, -0.1414) \cup (0.1414, 1)$, a union of two disconnected intervals, which can be obtained from the plot of the eigenvalues of $A + (1 - \sigma)BK$ with respect to σ as depicted in FIGURE 3.1.

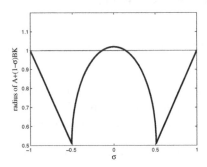

FIGURE 3.1: A disconnected discrete-time consensus region.

For the communication graph shown in FIGURE 3.1, the corresponding row-stochastic matrix is

$$\mathcal{D} = \begin{bmatrix} 0.3 & 0.2 & 0.2 & 0.2 & 0 & 0.1 \\ 0.2 & 0.6 & 0.2 & 0 & 0 & 0 \\ 0.2 & 0.2 & 0.6 & 0 & 0 & 0 \\ 0.2 & 0 & 0 & 0.4 & 0.4 & 0 \\ 0 & 0 & 0 & 0.4 & 0.2 & 0.4 \\ 0.1 & 0 & 0 & 0 & 0.4 & 0.5 \end{bmatrix},$$

whose eigenvalues, other than 1, are $-0.2935, 0.164, 0.4, 0.4624, 0.868$, which

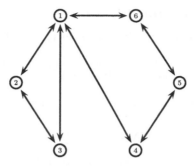

FIGURE 3.2: An undirected communication topology.

all belong to S. Thus, it follows from Corollary 3 that the network (3.3) with graph given in FIGURE 3.2 can achieve consensus.

Let us see how modifications of the communication topology affect the consensus. Consider the following two simple cases:

1) An edge is added between nodes 1 and 5, thus more information exchange will exist inside the network. Then, the row-stochastic matrix D becomes

$$
\begin{bmatrix}
0.2 & 0.2 & 0.2 & 0.2 & 0.1 & 0.1 \\
0.2 & 0.6 & 0.2 & 0 & 0 & 0 \\
0.2 & 0.2 & 0.6 & 0 & 0 & 0 \\
0.2 & 0 & 0 & 0.4 & 0.4 & 0 \\
0.1 & 0 & 0 & 0.4 & 0.2 & 0.3 \\
0.1 & 0 & 0 & 0 & 0.4 & 0.5
\end{bmatrix},
$$

whose eigenvalues, in addition to 1, are $-0.2346, 0.0352, 0.4, 0.4634, 0.836$. *Clearly, the eigenvalue 0.0352 does not belong to S, i.e., consensus can not be achieved in this case.*

2) The edge between nodes 5 and 6 is removed. The row-stochastic matrix D becomes

$$
\begin{bmatrix}
0.3 & 0.2 & 0.2 & 0.2 & 0 & 0.1 \\
0.2 & 0.6 & 0.2 & 0 & 0 & 0 \\
0.2 & 0.2 & 0.6 & 0 & 0 & 0 \\
0.2 & 0 & 0 & 0.4 & 0.4 & 0 \\
0.1 & 0 & 0 & 0.4 & 0.6 & 0 \\
0.1 & 0 & 0 & 0 & 0 & 0.9
\end{bmatrix},
$$

whose eigenvalues, other than 1, are $-0.0315, 0.2587, 0.4, 0.8676, 0.9052$. *In this case, the eigenvalue* -0.0315 *does not belong to S, i.e., consensus can not be achieved either.*

These sample cases imply that, for disconnected consensus regions, consensus can be quite fragile to the variations of the network's communication topology. Hence, the consensus protocol should be designed to have a sufficiently large bounded consensus region in order to be robust with respect to the communication topology, which is the topic of the following subsection.

3.2.3   Consensus Protocol Design

3.2.3.1   The Special Case with Neutrally Stable Agents

First we consider a special case where the matrix A is neutrally stable. Before moving forward, we introduce the following lemma.

Lemma 25 ([200]) *For a matrix $Q = Q^H \in \mathbf{C}^{n \times n}$, consider the following Lyapunov equation:*

$$A^H X A - X + Q = 0.$$

If $X > 0$, $Q \geq 0$, and (Q, V) is observable, then the matrix A is Schur stable.

Proposition 4 *For matrices $Q \in \mathbf{R}^{n \times n}$, $V \in \mathbf{R}^{m \times n}$, $\sigma \in \mathbf{C}$, where Q is orthogonal, $VV^T = I$, and (Q, V) is observable, if $|\sigma| < 1$, then the matrix $Q - (1 - \sigma)QV^TV$ is Schur stable.*

Proof 14 *Observe that*

$$
\begin{aligned}
[Q &- (1 - \sigma)QV^TV]^H[Q - (1 - \sigma)QV^TV] - I \\
&= Q^TQ - (1 - \sigma)Q^TQV^TV - (1 - \bar{\sigma})V^TVQ^TQ \\
&\quad + |1 - \sigma|^2V^TVQ^TQV^TV - I \\
&= (-2\mathrm{Re}(1 - \sigma) + |1 - \sigma|^2)V^TV \\
&= (|\sigma|^2 - 1)V^TV.
\end{aligned}
\tag{3.12}
$$

Since (Q, V) is observable, it is easy to verify that $(Q - (1 - \sigma)QV^TV, V^TV)$ is also observable. Then, by Lemma 25, (3.12) implies that $Q - (1 - \sigma)QV^TV$ is Schur stable for any $|\sigma| < 1$.

Next, an algorithm for protocol (3.1) is presented, which will be used later.

Algorithm 9 *Given that A is neutrally stable and that (A, B) is stabilizable, the protocol (3.1) can be constructed as follows:*

1) Choose $U \in \mathbf{R}^{n_1 \times n}$ and $W \in \mathbf{R}^{(n-n_1) \times n}$ such that [1]

$$
\begin{bmatrix} U \\ W \end{bmatrix} A \begin{bmatrix} U \\ W \end{bmatrix}^{-1} = \begin{bmatrix} M & 0 \\ 0 & X \end{bmatrix}
\tag{3.13}
$$

where $M \in \mathbf{R}^{n_1 \times n_1}$ is orthogonal and $X \in \mathbf{R}^{(n-n_1) \times (n-n_1)}$ is Schur stable (assume that without loss of generality U^TB has full column rank).

2) Choose $V \in \mathbf{R}^{p \times n_1}$ such that $VV^T = I_p$ and range(V^T) = range(UB).

3) Define $K = -(VUB)^{-1}VMU$.

---

[1] Matrices U and W can be derived by transforming matrix A into the real Jordan canonical form [58].

Theorem 15 *Suppose that A is neutrally stable and that (A, B) is stabiliz-able. The protocol (3.2) constructed via Algorithm 9 has the open unit disk as its bounded consensus region. Thus, such a protocol solves the consensus problem for (3.1) with respect to Γ_N (which denotes the set of all the communication topologies containing a directed spanning tree).*

Proof 15 *Let the related variables be defined as in Algorithm 9. By step 2) in Algorithm 9, it is easy to see that $V^T V$ is an orthogonal projection onto $\text{range}(V^T) = \text{range}(UB)$. By the definition of orthogonal projection (Definition 4 in Chapter 1), we can get that $V^T V U B = U B$. By the fact that $\text{range}(V^T) = \text{range}(UB)$, it is not difficult to get that $B^T U^T V^T$ is invertible. Thus, we have $V = (B^T U^T V^T)^{-1} B^T U^T$ and $UBK = -V^T V M U$. Also, the stablizability of (A, B) implies that (M, V) is controllable. Let $U^\dagger \in \mathbf{R}^{n \times n_1}$ and $W^\dagger \in \mathbf{R}^{n \times (n-n_1)}$ be such that $[U^\dagger \; W^\dagger] = [\begin{smallmatrix} U \\ W \end{smallmatrix}]^{-1}$, where $U U^\dagger = I$, $W W^\dagger = I$, $W U^\dagger = 0$, and $U W^\dagger = 0$. It can be verified by some algebraic manipulations that*

$$\begin{bmatrix} U \\ W \end{bmatrix} [A + (1-\sigma)BK] \begin{bmatrix} U \\ W \end{bmatrix}^{-1}$$

$$= \begin{bmatrix} M + (1-\sigma)UBKU^\dagger & (1-\sigma)UBKW^\dagger \\ (1-\sigma)WBKU^\dagger & X + (1-\sigma)WBKW^\dagger \end{bmatrix} \qquad (3.14)$$

$$= \begin{bmatrix} M - (1-\sigma)V^T V M & -(1-\sigma)U^\dagger BKW \\ 0 & X \end{bmatrix}.$$

By Lemma 25, matrix $M - (1-\sigma)V^T V M$ is Schur stable for any $|\sigma| < 1$. Hence, by Proposition 4, (3.14) implies that matrix $A + (1-\sigma)BK$ with K given by Algorithm 9 is Schur stable for any $|\sigma| < 1$, i.e., the protocol (3.2) constructed via Algorithm 9 has a bounded consensus region in the form of the open unit disk. Since the eigenvalues of any communication topology containing a spanning tree lie in the open unit disk, except eigenvalue 1, it follows from Corollary 3 that this protocol solves the consensus problem with respect to Γ_N.

Remark 19 *It should be mentioned that Algorithm 9 and the proof of Theorem 15 are partly inspired by [172]. Here we use a different yet comparatively much simpler method to derive Theorem 15.*

Example 10 *Consider a network of agents described by (3.1), with*

$$A = \begin{bmatrix} 0.2 & 0.6 & 0 \\ -1.4 & 0.8 & 0 \\ 0.7 & 0.2 & -0.5 \end{bmatrix}, \quad B = \begin{bmatrix} 0 \\ 1 \\ 0 \end{bmatrix}.$$

The eigenvalues of matrix A are -0.5, $0.5 \pm \iota 0.866$, thus A is neutrally stable. By transforming A into its real Jordan canonical form, we can get that the matrices

$$U = \begin{bmatrix} 0.1709 & -0.4935 & 0 \\ 0.7977 & 0 & 0 \end{bmatrix}, \quad W = \begin{bmatrix} -0.0570 & -0.2961 & 1 \end{bmatrix}$$

satisfy (3.13) with $M = \begin{bmatrix} 0.5 & 0.866 \\ -0.866 & 0.5 \end{bmatrix}$ *and* $X = -0.5$. *Thus,* $UB = \begin{bmatrix} -0.4935 & 0 \end{bmatrix}^T$. *Thus, we can take* $V = \begin{bmatrix} 1 & 0 \end{bmatrix}$ *such that* $VV^T = 1$ *and* range(V^T) = range(UB). *Then, by Algorithm 9, we obtain that* $K = \begin{bmatrix} 1.5732 & -0.5 & 0 \end{bmatrix}$. *In light of Theorem 15, the agents considered in this example will reach consensus under the protocol (3.2) with* K *given as above, with respect to all the communication topologies containing a directed spanning tree.*

3.2.3.2 The General Case

In this subsection, we consider the general case without assuming that A is neutrally stable. Before moving forward, we introduce the following modified algebraic Riccati equation (MARE) [152, 144]:

$$P = A^T PA - (1 - \delta^2)A^T PB(B^T PB + I)^{-1}B^T PA + Q, \qquad (3.15)$$

where $P \geq 0$, $Q > 0$, and $\delta \in \mathbf{R}$. For $\delta = 0$, the MARE (3.15) is reduced to the commonly-used discrete-time Riccati equation discussed in, e.g., [200].

The following lemma concerns the existence of solutions for the MARE.

Lemma 26 ([152, 144]) *Let* (A, B) *be stablizable. Then, the following statements hold.*

a) *Suppose that the matrix* A *has no eigenvalues with magnitude larger than 1, Then, the MARE (3.15) has a unique positive-definite solution* P *for any* $0 < \delta < 1$.

b) *For the case where* A *has a least one eigenvalue with magnitude larger than 1 and the rank of* B *is one, the MARE (3.15) has a unique positive-definite solution* P, *if* $0 < \delta < \frac{1}{\prod_i |\lambda_i^u(A)|}$, *where* $\lambda_i^u(A)$ *denote the unstable eigenvalues of* A.

c) *If the MARE (3.15) has a unique positive-definite solution* P, *then* $P = \lim_{k \to \infty} P_k$ *for any initial condition* $P_0 \geq 0$, *where* P_k *satisfies*

$$P(k+1) = A^T P(k)A - \delta A^T P(k)B(B^T P(k)B + I)^{-1}B^T P(k)A + Q.$$

Theorem 16 *Let* (A, B) *be stabilizable. The feedback gain matrix* K *of (3.2) is given by* $K = -(B^T PB + I)^{-1}B^T PA$, *where* $P > 0$ *is the unique solution of the MARE (3.15). Then, the protocol (3.2) designed as above has a bounded consensus region in the form of an origin-centered disk of radius* δ, *i.e., this protocol solves the consensus problem for networks with agents (3.1) with respect to* $\Gamma_{\leq \delta}$ *(which denotes the set of all directed graphs containing a directed spanning tree, whose non-one eigenvalues lie in the disk of radius* δ *centered at the origin), where* δ *satisfies* $0 < \delta < 1$ *for the case where* A *has no eigenvalues with magnitude larger than 1 and satisfies* $0 < \delta < \frac{1}{\prod_i |\lambda_i^u(A)|}$ *for the case where* A *has a least one eigenvalue outside the unit circle and* B *is of rank one.*

Proof 16 *Observe that*

$$
[A + (1 - \delta)BK]^H P[A + (1 - \delta)BK] - P
$$
$$
= A^T PA - 2\mathrm{Re}(1 - \delta)A^T PB(B^T PB + I)^{-1}B^T PA - P
$$
$$
+ |1 - \delta|^2 A^T PB(B^T PB + I)^{-1}B^T PB(B^T PB + I)^{-1}B^T PA
$$
$$
= A^T PA + [-2\mathrm{Re}(1 - \delta) + |1 - \delta|^2]A^T PB(B^T PB + I)^{-1}B^T PA - P
$$
$$
+ |1 - \delta|^2 A^T PB(B^T PB + I)^{-1}[-I + B^T PB(B^T PB + I)^{-1}]B^T PA
$$
$$
= A^T PA + (|\delta|^2 - 1)A^T PB(B^T PB + I)^{-1}B^T PA - P
$$
$$
- |1 - \delta|^2 A^T PB(B^T PB + I)^{-2}B^T PA
$$
$$
\leq A^T PA - (1 - |\delta|^2)A^T PB(B^T PB + I)^{-1}B^T PA - P
$$
$$
= -Q < 0,
$$

(3.16)

where the identity $-I + B^T PB(B^T PB + I)^{-1} = -(B^T PB + I)^{-1}$ *has been applied. Then, the assertion follows directly from (3.16), Lemma 26, and the discrete-time Lyapunov inequality.*

Remark 20 *Note that* $\Gamma_{\leq \delta}$ *for the general case is a subset of* Γ_N *for the special case where A is neutrally stable. This is consistent with the intuition that unstable behaviors are more difficult to synchronize than the neutrally stable ones. By Lemma 26, it follows that a sufficient and necessary condition for the existence of the consensus protocol (3.2) is that (A, B) is stabilizable for the case where A has no eigenvalues with magnitude larger than 1. In contrast, δ has to further satisfy $\delta < \frac{1}{\prod_i |\lambda_i^u(A)|}$ for the case where A has at least eigenvalue outside the unit circle and B is of rank one. This implies that contrary to the continuous-time case in the previous chapter, both the eigenvalues of the communication graph and the unstable eigenvalues of the agent dynamics are critical for the design of the consensus protocol. In other words, the consensus problem of discrete-time multi-agent systems in this chapter is generally more challenging to solve.*

Remark 21 *For the case where B is of full column rank, the feedback gain matrix K of (3.2) can be chosen as $K = -(B^T PB)^{-1}B^T PA$, where $P > 0$ is the unique solution to this simplified MARE: $P = A^T PA - (1 - \delta^2)A^T PB(B^T PB)^{-1}B^T PA + Q$.*

Example 11 *Let the agents in (3.1) be discrete-time double integrators, with*

$$
x_i = \begin{bmatrix} x_{i1} \\ x_{i2} \end{bmatrix}, \quad A = \begin{bmatrix} 1 & 1 \\ 0 & 1 \end{bmatrix}, \quad B = \begin{bmatrix} 0 \\ 1 \end{bmatrix}.
$$

Because the spectral radius of A is 1, we know from Lemma 26 that the MARE 3.15 is solvable for any $0 < \delta < 1$. Solving the MARE (3.15) with $\delta = 0.95$ and $Q = 3I$ gives $P = 10^4 \times \begin{bmatrix} 0.0062 & 0.0602 \\ 0.0602 & 1.1780 \end{bmatrix}$. Then we can obtain

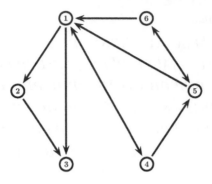

FIGURE 3.3: A directed communication topology.

that $K = [\,-0.0511\ -1.0510\,]$. *It follows from Theorem 16 that the agents (3.1)*
reach consensus under the protocol (3.2) with K given as above with respect
to $\Gamma_{\leq 0.95}$. Assume that the communication topology $\mathcal{G}$ is given as in FIGURE
3.3, and the corresponding row-stochastic matrix is

$$
\mathcal{D} = \begin{bmatrix}
0.4 & 0 & 0 & 0.1 & 0.3 & 0.2 \\
0.5 & 0.5 & 0 & 0 & 0 & 0 \\
0.3 & 0.2 & 0.5 & 0 & 0 & 0 \\
0.5 & 0 & 0 & 0.5 & 0 & 0 \\
0 & 0 & 0 & 0.4 & 0.4 & 0.2 \\
0 & 0 & 0 & 0 & 0.3 & 0.7
\end{bmatrix},
$$

whose eigenvalues, other than 1, are $\lambda_i = 0.5, 0.5565, 0.2217 \pm \iota 0.2531$. Clearly,
$|\lambda_i| < 0.95$, *for $i = 2, \cdots, 6$. FIGURE 3.4 depicts the state trajectories of*
the discrete-time double integrators, which shows that consensus is actually
achieved.

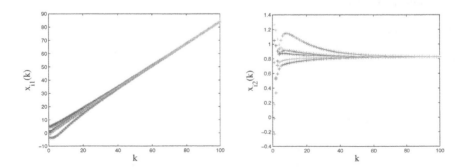

FIGURE 3.4: The state trajectories of a double-integrator network.

3.3 Observer-Type Consensus Protocols

In the previous section, the consensus protocol (3.2) rely on the relative states of neighboring agents. However, in many circumstances the state information might not be available. In this section, we will consider the consensus problem of (3.1) for the case where each agent can have access to the local output information, i.e., its own output and the outputs from its neighbors.

3.3.1 Full-Order Observer-Type Protocol I

Based on the relative output information of neighboring agents, we propose a full-order observer-type consensus protocol as follows:

$$v_i^+ = (A + BF)v_i + L \sum_{j=1}^{N} d_{ij}[C(v_i - v_j) - (y_i - y_j)],$$

$$u_i = Fv_i,$$

(3.17)

where $v_i \in \mathbf{R}^n$ is the protocol state, $i = 1, \cdots, N$, $L \in \mathbf{R}^{q \times n}$ and $F \in \mathbf{R}^{p \times n}$ are the feedback gain matrices to be determined. In (3.17), the term $\sum_{j=1}^{N} d_{ij}C(v_i - v_j)$ denotes the information exchanges between the protocol of agent i and those of its neighboring agents.

Let $z_i = [x_i^T, v_i^T]^T$ and $z = [z_1^T, \cdots, z_N^T]^T$. Then, the closed-loop system resulting from (3.1) and (3.17) can be written as

$$z^+ = [I_N \otimes \mathcal{A} + (I_N - \mathcal{D}) \otimes \mathcal{H}]z,$$

(3.18)

where

$$\mathcal{A} = \begin{bmatrix} A & BF \\ 0 & A + BF \end{bmatrix}, \quad \mathcal{H} = \begin{bmatrix} 0 & 0 \\ -LC & LC \end{bmatrix}.$$

Theorem 17 *For any $\mathcal{G}$ satisfying Assumption 3.1, the agents in (3.1) reach consensus under the observer-type protocol (3.17) if all the matrices $A + BK$ and $A + (1 - \hat{\lambda}_i)LC$, $i = 2, \cdots, N$, are Schur stable, where $\hat{\lambda}_i$, $i = 2, \cdots, N$, denote the eigenvalues of $\mathcal{D}$ located in the open unit disk. Furthermore, the final consensus value is given by*

$$x_i(k + 1) \to \varpi(k), \quad v_i(k + 1) \to 0, \quad i = 1, \cdots, N, \quad as \ k \to \infty,$$

(3.19)

where $\varpi(k)$ is defined as in (3.9).

Proof 17 *By following similar steps in the proof of Theorem 13, we can derive that the consensus problem of (3.18) can be reduced to the asymptotical stability problem of the following systems:*

$$\zeta_i^+ = [\mathcal{A} + (1 - \hat{\lambda}_i)\mathcal{H}]\zeta_i, \quad i = 2, \cdots, N.$$

(3.20)

It is easy to verify that matrices $A + \hat{\lambda}_i \mathcal{H}$ are similar to $\begin{bmatrix} A + (1 - \hat{\lambda}_i)LC & 0 \\ -(1 - \hat{\lambda}_i)LC & A + BF \end{bmatrix}$, $i = 2, \cdots, N$. Therefore, the Schur stability of the matrices $A + BK$ and $A + (1 - \hat{\lambda}_i)LC$, $i = 2, \cdots, N$, is equivalent to the asymptotical stability of (3.20), implying the solvability of the consensus problem of (3.18). The final consensus value of (3.18) can also be derived by following the steps in the proof of Theorem 14, which is thus omitted here for brevity.

For the observer-type protocol (3.17), the consensus region can be similarly defined as in Definition 10. Assuming that $A + BF$ is Schur stable, the region of the parameter $\sigma \in \mathbf{C}$ such that $A + (1 - \sigma)LC$ is Schur stable, is the discrete-time consensus region of the network (3.18).

In the following, we shall design the observer-type protocol (3.17). It is noted that the design of $A + (1 - \hat{\lambda}_i)LC$ is dual to the design of $A + (1 - \hat{\lambda}_i)BK$, the latter of which has been discussed in the preceding section. First, for the special case where A is neutrally stable, we can modify Algorithm 9 to get the following.

Algorithm 10 *Given that A is neutrally stable and that (A, B, C) is stabilizable and detectable, the protocol (3.17) can be constructed as follows:*

1) Select F be such that $A + BF$ is Schur stable.

2) Choose $\hat{U} \in \mathbf{R}^{n \times n_1}$ and $\hat{W} \in \mathbf{R}^{n \times (n - n_1)}$, satisfying

$$[U \ \ W]^{-1} A [U \ \ W] = \begin{bmatrix} \hat{M} & 0 \\ 0 & \hat{X} \end{bmatrix},$$

where $\hat{M} \in \mathbf{R}^{n_1 \times n_1}$ is orthogonal and $\hat{X} \in \mathbf{R}^{(n - n_1) \times (n - n_1)}$ is Schur stable.

3) Choose $\hat{V} \in \mathbf{R}^{q \times n_1}$ such that $VV^T = I_q$ and $\mathrm{range}(\hat{V}^T) = \mathrm{range}(\hat{U}^T C^T)$.

4) Define $L = -\hat{U}\hat{M}\hat{V}^T (C\hat{U}\hat{V}^T)^{-1}$.

Theorem 18 *Suppose that matrix A is neutrally stable and that (A, B, C) is stabilizable and detectable. The protocol (3.17) constructed via Algorithm 10 has the open unit disk as its bounded consensus region. Thus, such a protocol solves the consensus problem for (3.1) with respect to Γ_N.*

Proof 18 *The proof can be completed by following similar steps in proving Theorem 15.*

Next, consider the general case without assumptions on A.

Theorem 19 *Assume that (A, B, C) is stabilizable and detectable. The feedback gain matrices of (3.17) is designed as F satisfying that $A + BF$ is Schur stable and $L = -APC^T(CPC^T + I)^{-1}$, where $P > 0$ is the unique solution of the following MARE:*

$$P = APA^T - (1 - \delta^2)APC^T(CPC^T + I)^{-1}CPA^T + Q,$$

where $Q > 0$ and $\delta \in \mathbf{R}$. *Then, the protocol (3.17) designed as above has a bounded consensus region in the form of an origin-centered disk of radius δ, i.e., this protocol solves the consensus problem for networks with agents (3.1) with respect to $\Gamma_{\leq\delta}$, where δ satisfies $0 < \delta < 1$ for the case where A has no eigenvalues with magnitude larger than 1 and satisfies $0 < \delta < \frac{1}{\prod_i |\lambda_i^u(A)|}$ for the case where A has a least one eigenvalue outside the unit circle and C is of rank one.*

Proof 19 *The proof of this theorem can be completed by following similar steps as in the proof of Theorem 16.*

3.3.2 Full-Order Observer-Type Protocol II

In this subsection, we will propose another alternative dynamic consensus protocol. Based on the relative output information of neighboring agents, we propose the following full-order observer-type protocol:

$$\bar{v}_i^+ = (A + \bar{L}C)\bar{v}_i + \sum_{j=1}^{N} d_{ij}[B\bar{F}(v_i - v_j) + \bar{L}(y_i - y_j)],$$

$$u_i = \bar{F}v_i, \quad i = 1, \cdots, N,$$

(3.21)

where $\bar{v}_i \in \mathbf{R}^n$, $\bar{L} \in \mathbf{R}^{n \times q}$ and $\bar{F} \in \mathbf{R}^{p \times n}$ are the feedback gain matrices.

Let $e_i = \bar{v}_i - \sum_{j=1}^{N} a_{ij}(x_i - x_j)$, $i = 1, \cdots, N$. Then, we can get from (3.21) that e_i satisfy the following dynamics:

$$e_i^+ = (A + \bar{L}C)e_i, \quad i = 1, \cdots, N.$$

Clearly, if $\bar{L}$ is chosen such that $A + \bar{L}C$ is Schur stable, then $\bar{v}_i$ asymptotically converge to $\sum_{j=1}^{N} a_{ij}(x_i - x_j)$. Let $e = [e_1^T, \cdots, e_N^T]^T$ and $x = [x_1^T, \cdots, x_N^T]^T$. Then, we can get from (3.1) and (3.21) that the closed-loop network dynamics can be written in terms of e and x as

$$\begin{bmatrix} x^+ \\ e^+ \end{bmatrix} = \begin{bmatrix} A + c\mathcal{L} \otimes B\bar{F} & cB\bar{F} \\ 0 & A + \bar{L}C \end{bmatrix} \begin{bmatrix} x \\ e \end{bmatrix}.$$

By following similar steps in showing Theorem 17, we obtain the following result.

Theorem 20 *For any $\mathcal{G}$ satisfying Assumption 3.1, the agents in (3.1) reach consensus under the observer-type protocol (3.21) if all the matrices $A + \bar{L}C$ and $A + (1 - \hat{\lambda}_i)B\bar{F}$, $i = 2, \cdots, N$, are Schur stable, where $\hat{\lambda}_i$, $i = 2, \cdots, N$, denote the eigenvalues of $\mathcal{D}$ located in the open unit disk.*

Note that the design of $\bar{F}$ and $\bar{L}$ in Theorem 20 is actually dual to the design of L and F in Theorem 17. The algorithms in the previous subsection can be modified to determine the parameters $\bar{F}$ and $\bar{L}$ of the dynamic protocol (3.21). The details are omitted here for conciseness.

3.3.3   Reduced-Order Observer-Based Protocol

The observer-type protocol (3.17) in the last subsection is of full order, which however possesses a certain degree of redundancy, which stems from the fact that while the observer constructs an "estimate" of the entire state, part of the state information is already reflected in the system outputs. In this subsection, we will design a reduced-order protocol so as to eliminate this redundancy.

Based on the relative output information of neighboring agents, we introduce here a reduced-order observer-based consensus protocol as

$$\hat{v}_i^+ = F\hat{v}_i + Gy_i + TBu_i,$$
$$u_i = KQ_1 \sum_{j=1}^{N} d_{ij}(y_i - y_j) + KQ_2 \sum_{j=1}^{N} d_{ij}(\hat{v}_i - \hat{v}_j), \ i = 1, \cdots, N, \tag{3.22}$$

where $\hat{v}_i \in \mathbf{R}^{n-q}$ is the protocol state, $F \in \mathbf{R}^{(n-q)\times(n-q)}$ is Schur stable and has no eigenvalues in common with those of A, $G \in \mathbf{R}^{(n-q)\times q}$, $T \in \mathbf{R}^{(n-q)\times n}$ is the unique solution to (2.36), satisfying that $[\begin{smallmatrix} C \\ T \end{smallmatrix}]$ is nonsingular, $[Q_1 \ Q_2] = [\begin{smallmatrix} C \\ T \end{smallmatrix}]^{-1}$, $K \in \mathbf{R}^{p\times n}$ is the feedback gain matrix to be designed, and d_{ij} is the (i,j)-th entry of the row-stochastic matrix $\mathcal{D}$ associated with the graph $\mathcal{G}$.

Let $\hat{z}_i = [x_i^T, \hat{v}_i^T]^T$ and $\hat{z} = [\hat{z}_1^T, \cdots, \hat{z}_N^T]^T$. Then, the collective network dynamics can be written as

$$\hat{z}^+ = [I_N \otimes \mathcal{M} + (I_N - \mathcal{D}) \otimes \mathcal{R}]\hat{z}, \tag{3.23}$$

where

$$\mathcal{M} = \begin{bmatrix} A & 0 \\ GC & F \end{bmatrix}, \quad \mathcal{R} = \begin{bmatrix} BKQ_1C & BKQ_2 \\ TBKQ_1C & TBKQ_2 \end{bmatrix}.$$

Theorem 21 *For any $\mathcal{G}$ satisfying Assumption 3.1, the agents in (3.1) reach consensus under the observer-type protocol (3.17) if the matrices F and $A + (1-\hat{\lambda}_i)BK$, $i = 2, \cdots, N$, are Schur stable, where $\hat{\lambda}_i$, $i = 2, \cdots, N$, denote the eigenvalues of $\mathcal{D}$ located in the open unit disk. Moreover, the final consensus value is given by*

$$x_i(k+1) \to \varpi(k), \quad v_i(k+1) \to T\varpi(k), \ i = 1, \cdots, N, \quad as \ k \to \infty,$$

where $\varpi(k)$ is defined as in (3.9).

Proof 20 *By following similar steps as in the proof of Theorem 13, we can reduce the consensus problem of (2.37) to the simultaneous asymptotically stability problem of the following $N - 1$ systems:*

$$\tilde{\zeta}_i^+ = [\mathcal{M} + (1 - \hat{\lambda}_i)\mathcal{R}]\tilde{\zeta}_i, \quad i = 2, \cdots, N, \tag{3.24}$$

where $\hat{\lambda}_i$, $i = 2, \cdots, N$, denote the eigenvalues of $\mathcal{D}$ located in the open unit disk. It is not difficult to verify that $\mathcal{M} + (1 - \hat{\lambda}_i)\mathcal{R}$ is similar to

$\begin{bmatrix} A+c\lambda_i BK & c\lambda_i BKQ_2 \\ 0 & F \end{bmatrix}$. *Therefore, the consensus problem of (2.37) is reached if F and $A + c\lambda_i BK$, $i = 2, \cdots , N$ are Schur stable. The rest of the proof can be completed by following similar steps as in the proof of Theorem 14. The details are omitted here for brevity.*

The feedback gain matrix K satisfying Theorem (21) can be constructed by using Algorithms 9 or Theorem 19.

3.4 Application to Formation Control

Similarly to the continuous-time case in Chapter 2, the consensus protocols proposed in the previous sections can also be modified to solve formation control problem of discrete-time multi-agent systems. For illustration, we only consider the observer-type protocol (3.17).

Let $\widetilde{H} = (h_1, \cdots , h_N) \in \mathbf{R}^{n \times N}$ describe a constant formation structure of the agent network in a reference coordinate frame, where $h_i \in \mathbf{R}^n$ is the formation variable corresponding to agent i. For example, $h_1 = [0\ 0]^T$, $h_2 = [0\ 1]^T$, $h_3 = [1\ 0]^T$, and $h_4 = [1\ 1]^T$ represent a unit square. Variable $h_i - h_j$ denotes the relative formation vector between agents i and j, which is independent of the reference coordinate.

Given a formation structure $\widetilde{H}$, the observer-type protocol (3.17) can be modified into the following form:

$$v_i^+ = (A + BF)v_i + \sum_{j=1}^{N} d_{ij} L[C(v_i - v_j) - (y_i - y_j - C(h_i - h_j))], \tag{3.25}$$

$$u_i = Fv_i.$$

It should be noted that (3.25) reduces to the consensus protocol (3.17), when $h_i - h_j = 0$, $\forall i, j = 1, \cdots , N$.

Definition 11 *The agents (3.1) under the protocol (3.25) achieve a given formation $\widetilde{H} = (h_1, \cdots , h_N)$ if*

$$\|x_i(k) - h_i - x_j(k) + h_j\| \to 0, \ as \ k \to \infty, \ \forall\ i, j = 1, \cdots , N.$$

Theorem 22 *For any $\mathcal{G}$ satisfying Assumption 3.1, the agents in (3.1) reach the formation $\widetilde{H}$ under the protocol (3.25) if all the matrices $A + BF$ and $A + (1 - \hat{\lambda}_i)LC$, $i = 2, \cdots , N$, are Schur stable, and $(A - I)h_i = 0$, $\forall i, j = 1, \cdots , N$, where $\hat{\lambda}_i$, $i = 2, \cdots , N$, denote the eigenvalues of $\mathcal{D}$ located in the open unit disk.*

Proof 21 *Let $\tilde{z}_i = \begin{bmatrix} x_i - h_i \\ v_i \end{bmatrix}$, $i = 1, \cdots, N$. Then, it follows from (3.1) and (3.25) that*

$$\tilde{z}^+ = [I_N \otimes \mathcal{A} + (I_N - \mathcal{D}) \otimes \mathcal{H}]\tilde{z} + \begin{bmatrix} A - I \\ 0 \end{bmatrix} \otimes \widetilde{H}, \qquad (3.26)$$

where the matrices $\mathcal{A}$ and $\mathcal{H}$ are defined in (3.18). Note that the formation $\widetilde{H}$ is achieved if the system (3.26) reaches consensus. By following similar steps in the proof of Theorem 17, it is easy to see that the formation $\widetilde{H}$ is achieved under the protocol (3.25) if the matrices $A + BF$ and $A + (1 - \hat{\lambda}_i)LC$, $i = 2, \cdots, N$, are Schur stable, and $(A - I)h_i = 0$, $\forall i = 1, \cdots, N$.

Remark 22 *The achievable formation structures have to satisfy the constraints $(A - I)h_i = 0$, $\forall i = 1, \cdots, N$. Note that h_i can be replaced by $h_i - h_1$, $i = 2, \cdots, N$, in order to be independent of the reference coordinate, by simply choosing h_1 corresponding to agent 1 as the origin. The formation protocol (3.25) for a given achievable formation structure can be constructed by using Algorithm 10 and Theorem 19.*

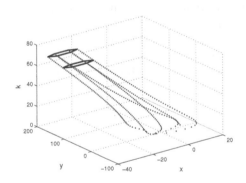

FIGURE 3.5: The agents form a regular hexagon.

Example 12 *Consider a network of six double integrators, described by*

$$x_i^+ = x_i + \tilde{v}_i,$$
$$\tilde{v}_i^+ = \tilde{v}_i + u_i,$$
$$y_i = x_i, \quad i = 1, \cdots, 6,$$

where $x_i \in \mathbf{R}^2$, $\tilde{v}_i \in \mathbf{R}^2$, $y_i \in \mathbf{R}^2$, and $u_i \in \mathbf{R}^2$ are the position, the velocity, the measured output, and the acceleration input of agent i, respectively.

The objective is to design a protocol (3.25) such that the agents will evolve to a regular hexagon with edge length 8. In this case, choose $h_1 = \begin{bmatrix} 0 & 0 & 0 & 0 \end{bmatrix}^T$,

$h_2 = \begin{bmatrix} 8\ 0\ 0\ 0 \end{bmatrix}^T$, $h_3 = \begin{bmatrix} 12\ 4\sqrt{3}\ 0\ 0 \end{bmatrix}^T$, $h_4 = \begin{bmatrix} 8\ 8\sqrt{3}\ 0\ 0 \end{bmatrix}^T$, $h_5 = \begin{bmatrix} 0\ 8\sqrt{3}\ 0\ 0 \end{bmatrix}^T$, $h_6 = \begin{bmatrix} -4\ 4\sqrt{3}\ 0\ 0 \end{bmatrix}^T$. *Choose* $F = -\begin{bmatrix} 0.5\ 1.5 \end{bmatrix} \otimes I_2$ *so that* $A + BF$ *is Schur stable. Solving equation (3.15) with* $\delta = 0.95$ *gives* $P = 10^4 \times \begin{bmatrix} 1.1780 & 0.0602 \\ 0.0602 & 0.0062 \end{bmatrix} \otimes I_2$. *By Theorem 19, we obtain* $L = -\begin{bmatrix} 1.051\ 0.051 \end{bmatrix}^T \otimes I_2$. *Then, the agents with such a protocol (3.25) will form a regular hexagon with respect to* $\Gamma_{\le 0.95}$. *The state trajectories of the six agents are depicted in FIGURE 3.5 for the communication topology given in FIGURE 3.3.*

3.5 Discussions

This chapter has addressed the consensus control problem for multi-agent systems with general discrete-time linear agent dynamics. The results obtained in this chapter can be regarded as extensions of those results in the preceding chapter in the discrete-time setting.

Regarding the continuous-time consensus in Chapter 2 and the discrete-time consensus in this chapter, there exist some important differences, which are summarized as follows.

i) For the continuous-time multi-agent systems in Chapter 2, the existence of consensus protocol relies only on the stabilizability or controllability of each agent if the given communication topology contains a directed spanning tree. In contrast, for the discrete-time multi-agent systems in this chapter, the existence condition for the consensus protocols additionally depends on the unstable eigenvalues of the state matrix of each agent, implying that the consensus problem of discrete-time multi-agent systems is usually much harder to tackle than that of continuous-time multi-agent systems.

ii) The consensus region for the continuous-time multi-agent systems in Chapter 2 is the open right half plane or a subset of the open right half plane. On the contrary, the consensus region for the discrete-time multi-agent systems in this chapter is the unit disk centered in the origin or a subset of the unit disk. The continuous-time consensus region is related to the discrete-time consensus region by a bilinear transformation, similarly to the relationship between the continuous-time and discrete-time stability of linear time-invariant systems.

iii) Laplacian matrices are utilized to depict the communication topology for the continuous-time multi-agent systems in Chapter 2. Differently, stochastic matrices are chosen for the discrete-time multi-agent systems in this chapter. Because the discrete-time consensus region is the unit disk centered in the origin or a subset of the unit disk, both largest and smallest nonzero eigenvalues of the Laplacian matrix (for undirected graphs) are required for the design of the consensus protocols when adopting Laplacian matrices as in related works [185, 50, 55]. The advantage of using stochastic matrices in this chapter is that only the eigenvalue with the largest modulus matters, which facilitates the consensus protocol design.

3.6   Notes

The materials of this chapter are mainly based on [82, 91]. For further results on consensus of general linear discrete-time multi-agent systems, see, e.g., [50, 55, 164, 172, 185]. Specially, distributed consensus protocols were designed in [172] for discrete-time multi-agent systems with each node being neutrally stable. Reference [185], similar to this chapter, also used modified algebraic Riccati equation to design consensus protocol and further considered the effect of a finite communication data rate. By introducing a properly designed dynamic filter into the local control protocols, the consensusability condition in [185] was further relaxed in [50]. Methods based on the H_∞ and H_2 type Riccati inequalities were given in [55] to design the consensus protocols. The effect of random link failures and random switching topology on discrete-time consensus was investigated in [186]. The consensus problem of discrete-time multi-agent systems with a directed topology and communication delay was considered in [164], where a consensus protocol based on the networked predictive control scheme was proposed to overcome the effect of delay.

4

H_∞ and H_2 Consensus Control of Linear Multi-Agent Systems

CONTENTS

4.1	H_∞ Consensus on Undirected Graphs		74
	4.1.1	Problem Formulation and Consensus Condition	74
	4.1.2	H_∞ Consensus Region	77
	4.1.3	H_∞ Performance Limit and Protocol Synthesis	80
4.2	H_2 Consensus on Undirected Graphs		83
4.3	H_∞ Consensus on Directed Graphs		85
	4.3.1	Leader-Follower Graphs	85
	4.3.2	Strongly Connected Directed Graphs	87
4.4	Notes		92

In the previous chapters, we have considered the consensus problems for multi-agent systems with linear node dynamics. In many circumstances, the agent dynamics might be subject to external disturbances, for which case the consensus protocols should maintain certain required disturbance attenuation performance. In this chapter, we consider the distributed H_∞ and H_2 consensus control problems for linear multi-agent systems in order to examine the disturbance attenuation performance of the consensus protocols with respect to external disturbances.

In Section 4.1, the H_∞ consensus problem for undirected communication graphs is formulated. A distributed consensus protocol based on the relative states of neighboring agents is proposed. The distributed H_∞ consensus problem of a linear multi-agent network is converted to the H_∞ control problem of a set of independent systems of the same dimension as a single agent. The notion of H_∞ consensus region is then defined and discussed. The H_∞ consensus region can be regarded as an extension of the notion of consensus region introduced in Chapter 2 in order to evaluate the performance of the multi-agent network subject to external disturbances. It is be pointed out via several examples that the H_∞ consensus regions can serve as a measure for the robustness of the consensus protocol with respect to the variations of the communication graph. A necessary and sufficient condition for the existence of a protocol yielding an unbounded H_∞ consensus region is derived. A multi-step procedure for constructing such a protocol is further presented, which maintains a favorable decoupling property. Such a design procedure involves no conservatism and is convenient to implement, especially when the agent number is large. The exact H_∞ performance limit of the consensus of the agents under the protocol is derived in this section, which is equal to the minimal H_∞ norm of an isolated agent achieved by using a state feedback controller, independent of the communication graph as long as it is connected.

The distributed H_2 consensus problem for undirected communication graphs is similarly addressed in Section 4.2. The consensus protocol is then designed to achieve consensus with a given H_2 performance and the H_2 performance limit of the consensus is also derived. It should be highlighted that the H_2 consensus condition and the H_2 performance limit, different from the H_∞ consensus case, depend on the number of agents in the network.

The distributed H_∞ consensus problems for leader-follower and strongly connected communication graphs are further studied in Section 4.3, which extend the results in Section 4.1. It is worth noting that for the H_∞ consensus problem with strongly connected graphs, the design of the consensus protocol generally requires the knowledge of the left eigenvector of the Laplacian matrix associated with the zero eigenvalue, implying the H_∞ consensus problem for the directed graph case is much harder than the undirected graph case.

4.1 H_∞ Consensus on Undirected Graphs

4.1.1 Problem Formulation and Consensus Condition

Consider a network of N identical linear agents subject to external disturbances. The dynamics of the i-th agent are described by

$$\dot{x}_i = Ax_i + Bu_i + D\omega_i, \quad i = 1, \cdots, N, \tag{4.1}$$

where $x_i \in \mathbf{R}^n$ is the state of the i-th agent, $u_i \in \mathbf{R}^p$ is the control input, $\omega_i \in \mathcal{L}_2^{m_1}[0, \infty)$ is the external disturbance, $\mathcal{L}_2^{m_1}[0, \infty)$ denotes the space of m_1-dimensional square integrable vector functions over $[0, \infty)$, and A, B, and D are constant matrices with compatible dimensions.

In this section, the communication graph among the N agents is represented by an undirected graph $\mathcal{G}$. It is assumed that at each time instant each agent knows the relative states of its neighboring agent with respect to itself. Based on the relative states of neighboring agents, the following distributed consensus protocol is proposed:

$$u_i = cK \sum_{j=1}^{N} a_{ij}(x_i - x_j), \quad i = 1, \cdots, N, \tag{4.2}$$

where $c > 0$ denotes the coupling gain, $K \in \mathbf{R}^{p \times n}$ is the feedback gain matrix, and a_{ij} denotes the (i, j)-th entry of the adjacency matrix associated with the graph $\mathcal{G}$.

The objective here is to find an appropriate protocol (4.2) for the agents in (4.1) to reach consensus and meanwhile maintain a desirable performance with respect to external disturbances ω_i. To this end, define the performance variable z_i, $i = 1, \cdots, N$, as the average of the weighted relative states of the agents, described by

$$z_i = \frac{1}{N} \sum_{i=1}^{N} C(x_i - x_j), \quad i = 1, \cdots, N, \tag{4.3}$$

where $z_i \in \mathbf{R}^{m_2}$ and $C \in \mathbf{R}^{m_2 \times n}$ is a given constant matrix.

Let $x = [x_1^T, \cdots, x_N^T]^T$, $\omega = [\omega_1^T, \cdots, \omega_N^T]^T$, and $z = [z_1^T, \cdots, z_N^T]^T$. Then, the agent network resulting from (4.1), (4.2), and (4.3) can be written as

$$\begin{aligned}
\dot{x} &= (I_N \otimes A + c\mathcal{L} \otimes BK)x + (I_N \otimes D)\omega, \\
z &= (M \otimes C)x,
\end{aligned} \tag{4.4}$$

where $\mathcal{L}$ is the Laplacian matrix associated with the graph $\mathcal{G}$ and $M \triangleq I_N - \frac{1}{N}\mathbf{1}\mathbf{1}^T$. Denote by $T_{\omega z}$ the transfer function matrix from ω to z of the network (4.4).

The distributed H_∞ consensus problem for the network (4.4) is first defined.

Definition 12 *Given the agents in (4.1) and an allowable $\gamma > 0$, the protocol (4.2) is said to solve the distributed suboptimal H_∞ consensus problem if*

i) *the network (4.4) with $w_i = 0$ can reach consensus in the sense of $\lim_{t \to \infty} \|x_i - x_j\| = 0$, $\forall i, j = 1, \cdots, N$;*

ii) *$\|T_{\omega z}\|_\infty < \gamma$, where $\|T_{\omega z}\|_\infty$ is the H_∞ norm of $T_{\omega z}$, defined by $\|T_{\omega z}(s)\|_\infty = \sup_{w \in \mathbf{R}} \bar{\sigma}(T_{\omega z}(jw))$ [200].*

The H_∞ performance limit of the consensus of the network (4.4) is the minimal $\|T_{\omega z}(s)\|_\infty$ of the network (4.4) achieved by using the protocol (4.2).

The following presents a necessary and sufficient condition for solving the H_∞ consensus problem of (4.4).

Theorem 23 *Assume that the communication graph $\mathcal{G}$ is connected. For a given $\gamma > 0$, there exists a consensus protocol (4.2) solving the suboptimal H_∞ consensus problem if and only if the following $N-1$ systems are simultaneously asymptotically stable and the H_∞ norms of their transfer function matrices are all less than γ:*

$$\begin{aligned}
\dot{\hat{x}}_i &= (A + c\lambda_i BK)\hat{x}_i + D\hat{\omega}_i, \\
\hat{z}_i &= C\hat{x}_i, \quad i = 2, \cdots, N,
\end{aligned} \tag{4.5}$$

where λ_i, $i = 2, \cdots, N$, are the nonzero eigenvalues of $\mathcal{L}$.

Proof 22 *By using the fact that* **1** *is the right eigenvector and also the left eigenvector of* $\mathcal{L}$ *associated with eigenvalue 0, we have*

$$M\mathcal{L} = \mathcal{L} = \mathcal{L}M. \tag{4.6}$$

Let $\xi = (M \otimes I_n)x$. *Then, by invoking (4.6), it follows from (4.4) that* ξ *evolves according to the following dynamics:*

$$\begin{aligned} \dot{\xi} &= (I_N \otimes A + c\mathcal{L} \otimes BK)\xi + (M \otimes D)\omega, \\ z &= (I_N \otimes C)\xi, \end{aligned} \tag{4.7}$$

whose transfer function matrix from ω *to* z *is denoted by* $\tilde{T}_{\omega z}$. *By the definition of* M, *it is easy to see that 0 is a simple eigenvalue of matrix* M *with* **1** *as its right eigenvector and 1 is another eigenvalue with multiplicity* $N - 1$. *Then it follows that* $\xi = 0$ *if and only if* $x_1 = x_2 = \cdots = x_N$. *Because (4.6) holds, there exists a unitary matrix* $U \in \mathbf{R}^{N \times N}$ *such that both* $U^T MU$ *and* $U^T \mathcal{L} U$ *are diagonal [58]. Since* $\mathcal{L}$ *and* M *have* **1** *as the same right and left eigenvectors associated with eigenvalue 0, we can choose* $U = [\frac{1}{\sqrt{N}} \ Y]$, $U^T = \begin{bmatrix} \frac{1^T}{\sqrt{N}} \\ W \end{bmatrix}$, *with* $Y \in \mathbf{R}^{N \times (N-1)}$ *and* $W \in \mathbf{R}^{(N-1) \times N}$, *such that*

$$\begin{aligned} U^T MU &= \Psi \triangleq \mathrm{diag}(0, 1, \cdots, 1), \\ U^T \mathcal{L} U &= \Lambda \triangleq \mathrm{diag}(0, \lambda_2, \cdots, \lambda_N). \end{aligned} \tag{4.8}$$

In virtue of (4.8), we can verify that

$$\begin{aligned} \tilde{T}_{\omega z} &= (I_N \otimes C)(sI - I_N \otimes A - c\mathcal{L} \otimes BK)^{-1}(M \otimes D) \\ &= (U \otimes C)(sI - I_N \otimes A - c\Lambda \otimes BK)^{-1}(\Psi U^T \otimes D) \\ &= (U \otimes I_n)\mathrm{diag}(0, \tilde{T}_{\hat{\omega}_2 \hat{z}_2}, \cdots, \tilde{T}_{\hat{\omega}_N \hat{z}_N})(U^T \otimes I_n) \\ &= (U\Psi \otimes C)(sI - I_N \otimes A - c\Lambda \otimes BK)^{-1}(U^T \otimes D) \\ &= (M \otimes C)(sI - I_N \otimes A - c\mathcal{L} \otimes BK)^{-1}(I_N \otimes D) \\ &= T_{\omega z}, \end{aligned} \tag{4.9}$$

where $\tilde{T}_{\hat{\omega}_i \hat{z}_i} = C(sI - A - c\lambda_i BK)^{-1}D$, $i = 2, \cdots, N$, *identical to the transfer function matrices of (4.5). Then, it follows that the distributed* H_∞ *consensus problem is solved if and only if (4.7) is asymptotically stable and* $\|\tilde{T}_{\omega z}\|_\infty < \gamma$.

Introduce the following transformations:

$$\xi = (U \otimes I_n)\tilde{\xi}, \quad \omega = (U \otimes I_{m_1})\hat{\omega}, \quad z = (U \otimes I_{m_2})\hat{z}, \tag{4.10}$$

where $\tilde{\xi} = [\tilde{\xi}_1^T, \cdots, \tilde{\xi}_N^T]^T$, $\hat{\omega} = [\hat{\omega}_1^T, \cdots, \hat{\omega}_N^T]^T$, *and* $\hat{z} = [\hat{z}_1^T, \cdots, \hat{z}_N^T]^T$. *Then, substituting (4.10) into (4.7) yields*

$$\begin{aligned} \dot{\tilde{\xi}} &= (I_N \otimes A + c\Lambda \otimes BK)\tilde{\xi} + (\Psi \otimes D)\hat{\omega}, \\ \hat{z} &= (I_N \otimes C)\tilde{\xi}, \end{aligned} \tag{4.11}$$

which clearly is composed of the following N independent systems:

$$\begin{aligned}
\dot{\tilde{\xi}}_1 &= A\tilde{\xi}_1, \\
\hat{z}_1 &= C\tilde{\xi}_1, \\
\dot{\tilde{\xi}}_i &= (A + c\lambda_i BK)\tilde{\xi}_i + D\hat{\omega}_i, \\
\hat{z}_i &= C\tilde{\xi}_i, \quad i = 2, \cdots, N.
\end{aligned} \tag{4.12}$$

Denote by $\tilde{T}_{\hat{\omega}\hat{z}}$ the transfer function matrices of (4.11). Then, it follows from (4.12), (4.11), and (4.10) that

$$\tilde{T}_{\hat{\omega}\hat{z}} = \mathrm{diag}(0, \tilde{T}_{\hat{\omega}_2\hat{z}_2}, \cdots, \tilde{T}_{\hat{\omega}_N\hat{z}_N}) = (U^T \otimes I_{m_1})\tilde{T}_{\omega z}(U \otimes I_{m_2}), \tag{4.13}$$

which implies that

$$\|\tilde{T}_{\hat{\omega}\hat{z}}\|_\infty = \max_{i=2,\cdots,N}\|\tilde{T}_{\hat{\omega}_i\hat{z}_i}\|_\infty = \|\tilde{T}_{\omega z}\|_\infty. \tag{4.14}$$

As to $\tilde{\xi}_1$, we have

$$\tilde{\xi}_1 = (\frac{\mathbf{1}^T}{\sqrt{N}} \otimes I_n)\xi = (\frac{\mathbf{1}^T}{\sqrt{N}} \otimes I_n)Mx \equiv 0.$$

Therefore, the suboptimal H_∞ consensus problem for the network (4.4) is solved if and only if the $N-1$ systems in (4.5) are simultaneously asymptotically stable and $\|\tilde{T}_{\hat{\omega}_i\hat{z}_i}\|_\infty < \gamma$, $i = 2, \cdots, N$.

Remark 23 The usefulness of this theorem lies in that it converts the distributed H_∞ consensus problem of the high-dimensional multi-agent network (4.4) into the H_∞ control problems of a set of independent systems having the same dimensions as a single agent in (4.1), thereby significantly reducing the computational complexity. A unique feature of protocol (4.2) proposed here is that by introducing a constant scalar $c > 0$, called the coupling gain, the notions of H_∞ and H_2 consensus regions can be brought forward, as detailed in the following subsections.

4.1.2 H_∞ Consensus Region

Given a protocol in the form of (4.2), the distributed H_∞ consensus problem of the network (4.4) can be recast into analyzing the following system:

$$\begin{aligned}
\dot{\zeta} &= (A + \sigma BK)\zeta + D\omega_i, \\
z_i &= C\zeta,
\end{aligned} \tag{4.15}$$

where $\zeta \in \mathbf{R}^n$ and $\sigma \in \mathbf{R}$, with σ depending on c. The transfer function of system (4.15) is denoted by $\widehat{T}_{\omega_i z_i}$. Clearly, the stability and H_∞ performance of the system (4.15) depends on the scalar parameter σ.

The notion of H_∞ consensus region is defined as follows.

Definition 13 *The region S_γ of the parameter $\sigma \subset \mathbf{R}^+$, such that system (4.15) is asymptotically stable and $\|\widehat{T}_{w_i z_i}\|_\infty < \gamma$, is called the H_∞ consensus region with performance index γ of the network (4.4).*

The H_∞ consensus region can be regarded as an extension of the consensus regions introduced in Chapter 2, used to evaluate the performance of a multi-agent network subject to external disturbances. According to Theorem 23, we have

Corollary 4 *For a given $\gamma > 0$, the protocol (4.2) solves the suboptimal H_∞ consensus problem for the agents (4.1) if and only if $c\lambda_i \in S_\gamma$, for $i = 2, \cdots, N$.*

For a protocol of the form (4.2), its H_∞ consensus region with index γ, if it exists, is an interval or a union of several intervals on the real axis, where the intervals themselves can be either bounded or unbounded. The H_∞ consensus region can serve as a measure for the robustness of the consensus protocol (4.2) with respect to the variations of the communication topology, as illustrated by the following example.

Example 13 *(bounded H_∞ consensus region) The agent dynamics and the protocol are given by (4.1) and (4.2), respectively, with*

$$A = \begin{bmatrix} -2 & 2 \\ -1 & 1 \end{bmatrix}, \quad B = \begin{bmatrix} 1 \\ -1 \end{bmatrix}, \quad D = \begin{bmatrix} 1 \\ 0.6 \end{bmatrix}$$
$$C = \begin{bmatrix} 1 & 0.8 \end{bmatrix}, \quad K = \begin{bmatrix} 1 & 0.2 \end{bmatrix}.$$

The H_∞ performance of (4.15) with respect to the parameter σ is depicted in FIGURE 4.1 (a). It can be observed that $S_{\gamma > 1.683}$, i.e., the H_∞ consensus region with index γ larger than the minimal value 1.683, is a bounded interval of σ in $\mathbf{R}$; for example, $S_{\gamma = 1.782}$ is $[0.0526, 0.203]$.

For illustration, let the communication graph $\mathcal{G}$ be given as in FIGURE 4.1 (b), with Laplacian matrix

$$\mathcal{L} = \begin{bmatrix} 4 & -1 & -1 & -1 & -1 & 0 \\ -1 & 3 & -1 & 0 & 0 & -1 \\ -1 & -1 & 2 & 0 & 0 & 0 \\ -1 & 0 & 0 & 2 & -1 & 0 \\ -1 & 0 & 0 & -1 & 3 & -1 \\ -1 & 0 & 0 & 0 & -1 & 2 \end{bmatrix},$$

whose nonzero eigenvalues are $1.382, 1.6972, 3.618, 4, 5.3028$. Thus, the protocol (4.2) given as above solves the suboptimal H_∞ consensus problem with $\gamma = 1.782$ for the graph given in FIGURE 4.1(b) if and only if the coupling gain c lies within the set $[0.0381, 0.0383]$.

Let us see how modifications of the communication graph affect the H_∞ consensus problem by considering the following simple cases.

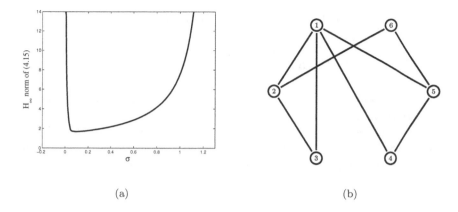

FIGURE 4.1: (a) The H_∞ consensus region; (b) the communication graph.

- *An edge between nodes 3 and 6 is added, i.e., more information exchanges exist inside the network. The minimal and maximal eigenvalues of the resulting Laplacian matrix are 1.4384 and 5.5616, respectively. Thus, it can be verified that the protocol (4.2) fails to solve the suboptimal H_∞ consensus problem with $\gamma = 1.782$ for any $c \in \mathbf{R}^+$.*

- *The edge between nodes 2 and 3 is removed, i.e., less information links in the network. The minimal and maximal eigenvalues of the resulting Laplacian matrix become 0.8817 and 5.2688, respectively. The protocol (4.2) in this case also fails to solve the H_∞ consensus problem with $\gamma = 1.782$.*

- *The edge between nodes 2 and 6 is removed and an edge between nodes 3 and 4 is added. In this case, the number of communication links in the network remains unchanged. The minimal and maximal eigenvalues of the resulting Laplacian matrix are 0.7639 and 5.2361, respectively. Once again, the protocol fails to solve the suboptimal H_∞ problem with $\gamma = 1.782$.*

These sample cases imply that for a bounded H_∞ consensus region, the distributed protocol (4.2), if not well designed, can be quite fragile to variations of the network's communication graph. In other words, it is desirable for the H_∞ consensus region to be large enough in order to ensure the protocol maintain a desired robustness margin with respect to the communication graph. This is similar to the consensus region introduced in Chapter 2.

The second example has a disconnected H_∞ consensus region.

Example 14 *The agent dynamics and the protocol are given by (4.1) and*

(4.2), respectively, with

$$A = \begin{bmatrix} -0.4 & -19.5998 \\ 4.333 & 0.4 \end{bmatrix}, \quad B = \begin{bmatrix} 1.5 & 0.3 \\ 0 & 1 \end{bmatrix}, \quad D = \begin{bmatrix} 1 \\ 0 \end{bmatrix}$$

$$C = \begin{bmatrix} 0 & 1 \end{bmatrix}, \quad K = \begin{bmatrix} 0 & 2 \\ -0.8 & 0 \end{bmatrix}.$$

The H_∞ consensus region of (4.15) with respect to parameter σ is depicted in FIGURE 4.2, which is composed of two disjoint subregions, of which one is bounded and the other is unbounded. It can be verified that $S_{\gamma=2} = [0.962, 5.1845] \cup [6.928, \infty)$. For the communication graph given in FIGURE 4.1(b), the protocol (4.2) solves the H_∞ consensus problem with $\gamma = 2$ if and only if $c \in [0.6961, 0.9777] \cup [5.013, \infty)$.

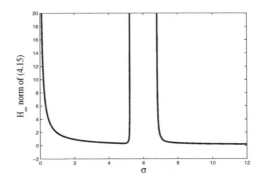

FIGURE 4.2: The disconnected H_∞ consensus region.

4.1.3 H_∞ Performance Limit and Protocol Synthesis

It was shown in the last subsection that the distributed protocol (4.2) should have a large enough H_∞ consensus region to be robust with respect to the communication graph. One convenient and desirable choice is to design the protocol with an unbounded H_∞ consensus region.

Lemma 27 (Bounded Real Lemma, [200]) *Let $\gamma > 0$ and $G(s) = C(sI - A)^{-1}B$. Then, the following two statements are equivalent:*
1) A is stable and $\|G(s)\|_\infty < \gamma$.
2) there exists an $X > 0$ such that

$$AX + XA^T + XC^TCX + \frac{1}{\gamma^2}BB^T < 0.$$

In the following, a necessary and sufficient condition is derived for the existence of a protocol (4.2) having an unbounded H_∞ consensus region.

Theorem 24 *For a given $\gamma > 0$, there exists a protocol (4.2) having an unbounded H_∞ consensus region $\mathcal{S}_\gamma \triangleq [\tau, \infty)$ if and only if there exist a matrix $P > 0$ and a scalar $\tau > 0$ satisfying the following linear matrix inequality (LMI):*

$$\begin{bmatrix} AP + PA^T - \tau BB^T & D & PC^T \\ D^T & -\gamma^2 I & 0 \\ CP & 0 & -I \end{bmatrix} < 0. \tag{4.16}$$

Proof 23 *(Necessity) According to Definition 12, if the network (4.4) has an unbounded H_∞ consensus region, then $A + \sigma BK$ is Hurwitz and $\|C(sI - A - \sigma BK)^{-1}D\|_\infty < \gamma$ for some matrix K and scalar σ. Since K is to be designed, without loss of generality, choose $\sigma = 1$. By Lemma 27, there exists a matrix K such that $A + BK$ is Hurwitz and $\|C(sI - A - BK)^{-1}D\|_\infty < \gamma$ if and only if there exists a matrix $P > 0$ such that*

$$(A + BK)P + P(A + BK)^T + \frac{1}{\gamma^2}DD^T + PC^T CP < 0.$$

Let $Y = KP$. Then, the above inequality becomes

$$AP + PA^T + BY + Y^T B^T + \frac{1}{\gamma^2}DD^T + PC^T CP < 0.$$

By Lemma 20 (Finsler's Lemma), there exists a matrix Y satisfying the above inequality if and only if there exists a scalar $\tau > 0$ such that

$$AP + PA^T - \tau BB^T + \frac{1}{\gamma^2}DD^T + PC^T CP < 0, \tag{4.17}$$

which, in virtue of Lemma 19 (the Schur Complement Lemma), is equivalent to

$$\begin{bmatrix} AP + PA^T - \tau BB^T & D & PC^T \\ D^T & -\gamma^2 I & 0 \\ CP & 0 & -I \end{bmatrix} < 0. \tag{4.18}$$

(Sufficiency) If (4.18) holds for some matrix $P > 0$ and scalar $\tau > 0$, then (4.17) holds also. Take $K = -\frac{1}{2}B^T P^{-1}$. Then, for $c\lambda_i \geq \tau$, $i = 2, \cdots, N$, it follows from (4.17) that

$$\begin{aligned} (A + c\lambda_i BK)P &+ P(A + c\lambda_i BK)^T + \frac{1}{\gamma^2}DD^T + PC^T CP \\ &= AP + PA^T - c\lambda_i BB^T + \frac{1}{\gamma^2}DD^T + PC^T CP < 0, \end{aligned} \tag{4.19}$$

implying that $\|C(sI - A - c\lambda_i BK)^{-1}D\|_\infty < \gamma$, $i = 2, \cdots, N$, i.e., the protocol (4.2) with K given as above has an unbounded H_∞ consensus region $[\tau, \infty)$.

The exact H_∞ performance limit of the network (4.4) under the protocol (4.2) is now obtained as a consequence.

Corollary 5 *The H_∞ performance limit $\gamma_{\min}$ of the consensus of the network (4.4) under the protocol (4.2) is given by the following optimization problem:*

$$\begin{aligned} &\text{minimize} \quad \gamma \\ &\text{subject to} \quad LMI\ (4.16),\ \text{with}\ P > 0,\ \tau > 0,\ \gamma > 0. \end{aligned} \tag{4.20}$$

Remark 24 *It should be noted that the H_∞ performance limit $\gamma_{\min}$ of the network (4.4) under the consensus protocol (4.2) is actually equal to the minimal H_∞ norm of a single agent (4.1) by using a state feedback controller of the form $u_i = Fx_i$, independent of the communication graph $\mathcal{G}$ as long as it is connected. Moreover, the minimum $\gamma_{\min}$ achieved by solving LMI (4.20) generally corresponds to a high-gain controller (4.2), as depicted in the example below. Choosing a γ a little bigger than $\gamma_{\min}$ would help keep the gain smaller.*

A procedure for constructing the protocol (4.2) is now presented.

Algorithm 11 *For any $\gamma \geq \gamma_{\min}$, where $\gamma_{\min}$ is given by (4.20), the protocol (4.2) solving the distributed H_∞ consensus problem can be constructed as follows:*

1) *Solve LMI (4.16) for a solution $P > 0$ and $\tau > 0$. Then, choose the feedback gain matrix $K = -\frac{1}{2}B^T P^{-1}$.*

2) *Select the coupling gain c not less than the threshold value $c_{th} = \dfrac{\tau}{\min\limits_{i=2,\cdots,N}\lambda_i}$, where λ_i, $i = 2, \cdots, N$, are the nonzero eigenvalues of $\mathcal{L}$.*

Remark 25 *Similar to the design algorithms based on the consensus region in Chapter 2, the above protocol design procedure based on the H_∞ consensus region also has a favorable decoupling feature. Specifically, step 1) deals only with the agent dynamics and the feedback gain matrix K of the protocol (4.2), leaving the communication graph of the network to be handled in step 2) by adjusting the coupling gain c. This feature is very desirable for the case where the agent number N is large, for which the eigenvalues of the corresponding Laplacian matrix $\mathcal{L}$ are hard to determine or even troublesome to estimate. Here, we need to choose the coupling gain to be large enough, which of course involves conservatism.*

Now, Example 13 in the above subsection is revisited.

Example 15 *The agent dynamics (4.1) remain the same as in Example 13, while matrix K will be redesigned via Algorithm 11. Solving LMI (4.16) with $\gamma = 1$ by using toolboxes YALMIP [103] and SeDuMi [158] gives feasible solutions $P = \left[\begin{smallmatrix} 1.3049 & 0.0369 \\ 0.0369 & 0.1384 \end{smallmatrix}\right]$ and $\tau = 1.7106$. Thus, the feedback gain matrix of (4.2) is chosen as $K = \left[\,-0.4890 \;\; 3.7443\,\right]$. Different from Example 13, the protocol (4.2) with this matrix K has an unbounded H_∞ consensus region with index $\gamma = 1$ in the form of $[1.7106, \infty)$. For the graph in FIGURE 4.1(b),* *the protocol (4.2) with K chosen here solves the suboptimal H_∞ consensus problem with $\gamma = 1$, if the coupling gain c is not less than the threshold value $c_{th} = 1.2378$.*

By solving the optimization problem (4.20), the H_∞ performance limit of the consensus of the network (4.4) under the protocol (4.2) can be further obtained as $\gamma_{\min} = 0.0535$. The corresponding optimal feedback gain matrix of (4.2) is obtained as $K = [\,48.6183\ 48.6450\,]$, and the scalar τ in (4.16) is $\tau = 9.5328 \times 10^6$. For the graph in FIGURE 4.1 (b), the threshold c_{th} corresponding to $\gamma_{\min}$ is $c_{th} = 6.8978 \times 10^6$ in this numerical example.

4.2 H_2 Consensus on Undirected Graphs

In this section, the H_2 performance of the network (4.4) will be discussed. For a stable transfer function $F(s)$, its H_2 norm is defined as $\|F\|_2 = \sqrt{\frac{1}{2\pi} \int_{-\infty}^{\infty} \operatorname{tr}(F^*(j\omega)F(j\omega))d\omega}$ [200].

Similar to the H_∞ case, the distributed H_2 consensus problem is stated as follows.

Definition 14 *Given the agents in (4.1) and an allowable $\gamma > 0$, the consensus protocol (4.2) is said to solve the distributed suboptimal H_2 consensus problem if*

i) the network (4.4) with $w_i = 0$ can reach consensus in the sense of $\lim_{t\to\infty} \|x_i - x_j\| = 0$, $\forall i, j = 1, \cdots, N$;

ii) $\|T_{wz}\|_2 < \gamma$.

The H_2 performance limit of the consensus of the network (4.4) is the minimal $\|T_{wz}(s)\|_2$ of the network (4.4) achieved by using the consensus protocol (4.2).

Theorem 25 *For a given $\tilde{\gamma} > 0$, there exists a protocol (4.2) solving the suboptimal H_2 consensus problem for the agents in (4.1) if and only if the $N - 1$ systems in (4.5) are simultaneously asymptotically stable and $\sqrt{\sum_{i=1}^{N} \|\widetilde{T}_{\hat{\omega}_i \hat{z}_i}\|_2^2} < \tilde{\gamma}$, where $\widetilde{T}_{\hat{\omega}_i \hat{z}_i}$, $i = 2, \cdots, N$, are the transfer function matrices of the systems in (4.5).*

Proof 24 *It can be shown by following similar steps in proving Theorem 23.*

The H_2 consensus region is defined as follows.

Definition 15 *The region $\widetilde{S}_{\tilde{\gamma}}$ of the parameter $\sigma \subset \mathbf{R}^+$, such that system (4.15) is asymptotically stable, with $\|\widetilde{T}_{\omega_i z_i}\|_2 < \frac{\tilde{\gamma}}{\sqrt{N}}$, is called the H_2 consensus region with performance index $\tilde{\gamma}$ of the network (4.4).*

According to Theorem 25, we have

Corollary 6 *The consensus protocol (4.2) solves the suboptimal H_2 consensus problem for the agents in (4.1), if $c\lambda_i \in \tilde{S}_{\tilde{\gamma}}$, for $i = 2, \cdots, N$.*

Remark 26 *Contrary to the H_∞ case, the H_2 consensus region is related to the number of agents in the network and using H_2 consensus region to characterize the H_2 performance of the network (4.4) involves certain conservatism. This is essentially due to the inherent difference between the H_2 and H_∞ norms, and due to the fact that, in the H_2 case, the performance index $\tilde{\gamma}$ couples the $N-1$ systems in (4.5), which thereby is more difficult to analyze.*

The H_2 consensus region analysis can be discussed similarly to Section 4.1, therefore is omitted here for brevity. The synthesis issue is somewhat different, hence is further discussed below.

Lemma 28 ([200]) *Let $\tilde{\gamma} > 0$ and $G(s) = C(sI-A)^{-1}B$. Then, the following two statements are equivalent.*
  1) A is stable and $\|G(s)\|_2 < \tilde{\gamma}$.
  2) there exists an $X > 0$ such that

$$AX + XA^T + BB^T < 0, \quad \mathrm{tr}(CXC^T) < \tilde{\gamma}^2.$$

Theorem 26 *For a given $\tilde{\gamma} > 0$, there exists a distributed protocol (4.2) having an unbounded H_2 consensus region $\tilde{S}_{\tilde{\gamma}} \triangleq [\tilde{\tau}, \infty)$ if and only if there exist a matrix $Q > 0$ and a scalar $\tilde{\tau}$ such that*

$$AQ + QA^T - \tilde{\tau}BB^T + DD^T < 0, \quad \mathrm{tr}(CQC^T) < \frac{\tilde{\gamma}^2}{N}. \qquad (4.21)$$

Proof 25 *(Necessity) Similar to the proof of Theorem 24, if the network (4.4) has an unbounded H_2 consensus region, then there exists a matrix K such that $A + BK$ is Hurwitz and $\|C(sI - A - BK)^{-1}D\|_2 < \frac{\tilde{\gamma}}{\sqrt{N}}$, which is equivalent to that there exists a matrix $Q > 0$ such that*

$$(A + BK)Q + Q(A + BK)^T + DD^T < 0, \quad \mathrm{tr}(CQC^T) < \frac{\tilde{\gamma}^2}{N}.$$

Let $V = KQ$. Then, the above inequality becomes

$$AQ + QA^T + BV + V^TB^T + DD^T < 0, \quad \mathrm{tr}(CQC^T) < \frac{\tilde{\gamma}^2}{N},$$

which, by Lemma 20, is equivalent to that there exist a matrix $Q > 0$ and a scalar $\tilde{\tau} > 0$ such that (4.21) holds.
  (Sufficiency) Take $K = -\frac{1}{2}B^TQ^{-1}$. For $c\lambda_i \geq \tilde{\tau}$, $i = 2, \cdots, N$, we have

$$(A + c\lambda_i BK)Q + Q(A + c\lambda_i BK)^T + DD^T$$
$$= AQ + QA^T - c\lambda_i BB^T + DD^T < 0,$$

$$\mathrm{tr}(CQC^T) < \frac{\tilde{\gamma}^2}{N},$$

which together imply that $\|C(sI-A-c\lambda_i BK)^{-1}D\|_2 < \frac{\tilde{\gamma}}{\sqrt{N}}, i = 2, \cdots, N$, i.e., the protocol (4.2) with K chosen as above has an unbounded H_2 performance region $[\tilde{\tau}, \infty)$.

Corollary 7 *The H_2 performance limit $\tilde{\gamma}_{\min}$ of the network (4.4) under the consensus protocol (4.2) is given by the optimization problem:*

$$\begin{aligned} \text{minimize} \quad & \tilde{\gamma} \\ \text{subject to} \quad & LMI\ (4.21),\ with\ Q > 0,\ \tilde{\tau} > 0,\ \tilde{\gamma} > 0. \end{aligned} \tag{4.22}$$

Proof 26 *Solving the optimization problem (4.22) gives solutions $\tilde{\gamma}_{\min}$ and $\tilde{\tau}_{\min}$. Select c such that $c\lambda_i \geq \tilde{\tau}_{\min}$, for $i = 1, 2, \cdots, N$. Then, the protocol (4.2) with K given as in the proof of Theorem 26 yields $\|T_{\hat{\omega}_i \hat{z}_i}\|_2 = \frac{\tilde{\gamma}_{\min}}{\sqrt{N}}$, for $i = 1, \cdots, N$, which by Theorem 25 implies that $\tilde{\gamma}_{\min}$ is the H_2 performance limit of the network (4.4).*

Remark 27 *The H_2 performance limit of the consensus of the network (4.4) under the protocol (4.2) is related to two factors: the minimal H_2 norm of a single agent (4.1) by using the state feedback controller $u_i = Fx_i$ and the number of agents in the network. Contrary to the H_∞ case, the H_2 performance limit of the network (4.4) scales with the size of the network.*

Algorithm 12 *For any $\tilde{\gamma} \geq \tilde{\gamma}_{\min}$, where $\tilde{\gamma}_{\min}$ is given by (4.22), the protocol (4.2) solving the distributed H_2 consensus problem can be constructed as follows:*

1) *Solve LMI (4.21) to get solutions $Q > 0$ and $\tilde{\tau} > 0$. Then, choose the feedback gain matrix $K = -\frac{1}{2}B^T Q^{-1}$.*

2) *Select the coupling gain $c > \tilde{c}_{th}$, with $\tilde{c}_{th} = \dfrac{\tilde{\tau}}{\min\limits_{i=2,\cdots,N} \lambda_i}$, where $\lambda_i, i = 2, \cdots, N$, are the nonzero eigenvalues of $\mathcal{L}$.*

4.3 H_∞ Consensus on Directed Graphs

In the above sections, the communication graph is assumed to undirected. In this section, we extend to consider the case where the multi-agent system maintains a directed communication graph.

4.3.1 Leader-Follower Graphs

In this subsection, we consider the case where the communication graph $\mathcal{G}$ is in the leader-follower form. Without loss of generality, assume that the

agents indexed by $2\cdots, N$, are the followers, whose dynamics are given by
(4.1), and the agent indexed by 1 is the leader, whose dynamics are described
by $\dot{x}_1 = Ax_1$, where $x_1 \in \mathbf{R}^n$. It is assumed that the leader receives no
information from any follower and the state of the leader is available to only
a subset of the followers. The communication graph $\mathcal{G}$ satisfies the following
assumption.

Assumption 4.1 $\mathcal{G}$ *contains a directed spanning tree with the leader as the*
root and the subgraph associated with the N followers is undirected.

Denote by $\mathcal{L}$ the Laplacian matrix associated with $\mathcal{G}$. Because the leader
has no neighbors, $\mathcal{L}$ can be partitioned as $\mathcal{L} = \begin{bmatrix} 0 & 0_{1\times(N-1)} \\ \mathcal{L}_2 & \mathcal{L}_1 \end{bmatrix}$, where $\mathcal{L}_2 \in$
$\mathbf{R}^{(N-1)\times 1}$ and $\mathcal{L}_1 \in \mathbf{R}^{(N-1)\times(N-1)}$. By Lemma 1 and Assumption 4.1, it is
clear that $\mathcal{L}_1 > 0$.

Since the followers are desired to follow the leader, it is natural to define
the performance variables $z_i \in \mathbf{R}^{m_2}$ as

$$z_i = C_2(x_i - x_1), \quad i = 2, \cdots, N. \tag{4.23}$$

In this case, the following distributed consensus protocol is proposed:

$$u_i = cK \sum_{j=1}^{N} a_{ij}(x_i - x_j), \quad i = 2, \cdots, N, \tag{4.24}$$

where c, a_{ij}, and K are defined as in (4.2).

Let $\varepsilon = [x_2^T - x_1^T, \cdots, x_N^T - x_1^T]^T$, $\bar{\omega} = [\omega_2^T, \cdots, \omega_N^T]^T$, and $\bar{z} = [z_2^T, \cdots, z_N^T]^T$. Then, the agent network can be written as

$$\begin{aligned} \dot{\varepsilon} &= (I_{N-1} \otimes A + c\mathcal{L}_1 \otimes BK)\varepsilon + (I_{N-1} \otimes D)\bar{\omega}, \\ \bar{z} &= (I_{N-1} \otimes C)\varepsilon. \end{aligned} \tag{4.25}$$

The transfer function matrix from $\bar{\omega}$ to $\bar{z}$ of (4.25) is denoted by $\overline{T}_{\bar{\omega}\bar{z}}$.

The leader-follower H_∞ consensus problem for the network (4.25) is stated
as follows.

Definition 16 *Given the agents in (4.1) and an allowable $\gamma > 0$, the protocol*
(4.24) is said to solve the distributed leader-follower H_∞ consensus problem if
i) the states of the network (4.25) with $w_i = 0$ satisfy $\lim_{t\to\infty} \|x_i - x_1\| = 0$,
$\forall i = 2, \cdots, N$; ii) $\|\overline{T}_{\bar{\omega}\bar{z}}\|_\infty < \gamma$.

Theorem 27 *Assume that the graph $\mathcal{G}$ satisfies Assumption 4.1. For a given*
$\gamma > 0$, there exists a protocol (4.24) solving the leader-follower H_∞ consensus
problem if and only if the following $N - 1$ systems are simultaneously asymp-
totically stable and the H_∞ norms of their transfer function matrices are all
less than γ:

$$\begin{aligned} \dot{\tilde{x}}_i &= (A + c\lambda_i BK)\tilde{x}_i + D\tilde{\omega}_i, \\ \tilde{z}_i &= C\tilde{x}_i, \quad i = 2, \cdots, N, \end{aligned} \tag{4.26}$$

where $0 < \lambda_2 \leq \cdots \leq \lambda_N$ are the nonzero eigenvalues of $\mathcal{L}$.

Proof 27 *It can be completed by following similar steps in the proof of Theorem 23.*

The consensus protocol (4.24) can be designed by using Algorithm 11 to satisfy Theorem 27. The H_2 consensus problem for leader-follower graphs satisfying Assumption 4.1 can be discussed by following the steps in Section 4.1, which is omitted here.

4.3.2 Strongly Connected Directed Graphs

In this subsection, we extend the results in Section 4.1 to consider the H_∞ consensus problem for the case of strongly connected directed communication graphs. The H_2 consensus problem for strongly connected directed graphs can be similarly done.

Similarly as in Section 4.1, the objective in this subsection is to design the consensus protocol (4.2) for the agents in (4.1) to reach consensus and meanwhile maintain a desirable disturbance rejection performance. Different from the performance variable in (4.3) for the case with undirected graphs, we define a new performance variables for the directed communication graphs in this subsection as

$$z_i = C(x_i - \sum_{j=1}^{N} r_j x_j), \quad i = 1, \cdots, N, \tag{4.27}$$

where $C \in \mathbf{R}^{m \times n}$ is a constant matrix, $r = [r_1, \cdots, r_N]^T$ denotes the positive left eigenvector of the Laplacian matrix $\mathcal{L}$ corresponding to the zero eigenvalue, satisfying $r^T \mathcal{L} = 0$ and $r^T \mathbf{1} = 1$. Clearly, (4.27) will reduce to (4.3) for strongly connected and balanced graphs.

Let $x = [x_1^T, \cdots, x_N^T]^T$, $\omega = [\omega_1^T, \cdots, \omega_N^T]^T$, and $z = [z_1^T, \cdots, z_N^T]^T$. Then, the agent network resulting from (4.1), (4.2), and (4.27) can be written as

$$\begin{aligned} \dot{x} &= (I_N \otimes A + c\mathcal{L} \otimes BK)x + (I_N \otimes D)\omega, \\ z &= (\widetilde{M} \otimes C)x, \end{aligned} \tag{4.28}$$

where $\widetilde{M} \triangleq I_N - \mathbf{1}_N r^T$. Denote by $T_{\omega z}$ the transfer function matrix of (4.28) from ω to z.

The distributed H_∞ consensus problem of the network (4.28) is defined in Definition 12.

Define $\zeta = (\widetilde{M} \otimes I)x$, where $x = [x_1^T, \ldots, x_N^T]^T$. By the definition of r^T, it is easy to check that 0 is a simple eigenvalue of $\widetilde{M}$ with $\mathbf{1}$ as a right eigenvector and 1 is the other eigenvalue with algebraic multiplicity $N-1$. Thus, it is easy to see that $\zeta = 0$ if and only if $x_1 = x_2 = \ldots = x_N$. The network (4.28) can be rewritten in terms of ζ as

$$\begin{aligned} \dot{\zeta} &= (I_N \otimes A + c\mathcal{L} \otimes BK)\zeta + (\widetilde{M} \otimes D)\omega, \\ z &= (I_N \otimes C)\zeta, \end{aligned} \tag{4.29}$$

whose transfer function matrix from w to z is denoted by $\widetilde{T}_{wz}$. Since $\widetilde{M}\mathcal{L} = \mathcal{L}\widetilde{M}$, by following the steps in the proof of Theorem 23, it is not difficult to show that $\widetilde{T}_{wz} = T_{wz}$. Therefore, the distributed H_∞ consensus problem with $\gamma > 0$ is solved if and only if (4.29) is asymptotically stable and $\widetilde{T}_{wz} < \gamma$.

The following algorithm is presented to construct the consensus protocol (4.2).

Algorithm 13 *For a scalar $\gamma > 0$, the protocol (4.2) solving the distributed H_∞ consensus problem can be constructed as follows:*

1) Solve the following LMI:

$$
\begin{bmatrix}
XA^T + AX - \tau BB^T & XC^T & D \\
CX & -r_{\min}I_n & 0 \\
D^T & 0 & -\frac{\gamma^2 r_{\min}}{r_{\max}}I_n
\end{bmatrix} < 0 \qquad (4.30)
$$

to get a matrix $X > 0$ and a scalar $\tau > 0$, where $r_{\min} = \min\limits_{i} r_i$ and $r_{\max} = \max\limits_{i} r_i$. Then, choose the feedback gain matrix $K = -\frac{1}{2}B^TX^{-1}$.

2) Select the coupling gain $c > \tilde{c}_{th}$, with $\tilde{c}_{th} = \frac{\tau}{a(\mathcal{L})}$, where $a(\mathcal{L})$ denotes the generalized algebraic connectivity of the communication graph $\mathcal{G}$ (see Lemma 4 in Chapter 1).

Remark 28 *In the last step, the generalized algebraic connectivity $a(\mathcal{L})$ is used to determine the required coupling gain c. For a given graph, its $a(\mathcal{L})$ can be obtained by using Lemma 8 in [189], which might be not easy to compute especially if the graph is of large scale. For the case where the communication graph is balanced and strongly connected, then $r^T = \mathbf{1}^T/N$ and $a(\mathcal{L}) = \lambda_2(\frac{1}{2}(\mathcal{L} + \mathcal{L}^T))$. In this case, the LMI (4.30) will reduce to the LMI (4.16) in Section 4.1. Further, step 2) of Algorithm 13 will be identical to step 2) of Algorithm 11 if the communication graph is undirected and connected.*

Theorem 28 *Assume that the communication graph $\mathcal{G}$ is strongly connected. For a given $\gamma > 0$, the consensus protocol (4.2) designed in Algorithm 13 can solve the suboptimal H_∞ consensus problem.*

Proof 28 *Consider the Lyapunov function candidate*

$$
V = \frac{1}{2}\zeta^T(R \otimes X^{-1})\zeta,
$$

where $R = \mathrm{diag}(r_1, r_2, \ldots, r_N)$. Differentiating V with respect to t along the trajectories of (4.29) and substituting $K = \frac{1}{2}B^TX^{-1}$ give

$$
\begin{aligned}
\dot{V} &= \zeta^T(R \otimes X^{-1}A - \frac{1}{2}cR\mathcal{L} \otimes X^{-1}BB^TX^{-1})\zeta + \zeta^T(R\widetilde{M} \otimes X^{-1}D)w \\
&= \tilde{\zeta}^T[R \otimes AX - \frac{c}{4}(R\mathcal{L} + \mathcal{L}^T R) \otimes BB^T]\tilde{\zeta} + \tilde{\zeta}^T(R\widetilde{M} \otimes D)w,
\end{aligned}
$$

$$(4.31)$$

where $\tilde\zeta = (I \otimes X^{-1})\zeta$.

Since $(r^T \otimes B^T)\tilde\zeta = (r^T \widetilde{M} \otimes B^T X^{-1})x = 0$, we can obtain from Lemma 4 that

$$\frac{1}{2}\tilde\zeta^T[(R\mathcal{L} + \mathcal{L}^T R) \otimes BB^T]\tilde\zeta \geq a(\mathcal{L})\tilde\zeta^T(I_N \otimes BB^T)\tilde\zeta, \qquad (4.32)$$

where $a(\mathcal{L}) > 0$. In light of (4.32), it follows from (4.31) and step 2) of Algorithm 13 that

$$\dot{V} \leq \frac{1}{2}\tilde\zeta^T[R \otimes (AX + XA^T - \tau BB^T)]\tilde\zeta + \tilde\zeta^T(R\widetilde{M} \otimes D)\omega. \qquad (4.33)$$

For the case of $\omega(t) \equiv 0$, it follows from the above analysis that

$$\dot{V} \leq \frac{1}{2}\tilde\zeta^T[R \otimes (AX + XA^T - \tau BB^T)]\tilde\zeta < 0,$$

where the last inequality follows from the LMI (4.30). This indicates that consensus of (4.29) with $\omega(t) \equiv 0$ is achieved. Thus, the first condition in Definition 12 is satisfied.

Next, analyze the H_∞ consensus performance of the multi-agent system with $\omega(t) \neq 0$. Observe that

$$\begin{aligned}
\dot{V} \leq{}& \frac{1}{2}\tilde\zeta^T[R \otimes (AX + XA^T - \tau BB^T)]\tilde\zeta + \tilde\zeta^T(R\widetilde{M} \otimes D)\omega \\
& + \frac{1}{2r_{\min}}\tilde\zeta^T(R \otimes XC^TCX)\tilde\zeta - \frac{1}{2}z^T z \\
& - \frac{\gamma^2}{2r_{\max}}\omega^T(R \otimes I_n)\omega + \frac{\gamma^2}{2}\omega^T\omega \\
={}& \frac{1}{2}\begin{bmatrix}\tilde\zeta \\ \omega\end{bmatrix}^T \Omega \begin{bmatrix}\tilde\zeta \\ \omega\end{bmatrix} - \frac{1}{2}z^T z + \frac{\gamma^2}{2}\omega^T\omega,
\end{aligned} \qquad (4.34)$$

where

$$\Omega = \begin{bmatrix} R \otimes \Pi & R\widetilde{M} \otimes D \\ \widetilde{M}^T R \otimes D^T & -\frac{\gamma^2}{r_{\max}}R \otimes I \end{bmatrix}, \qquad (4.35)$$

and $\Pi = AX + XA^T - \tau BB^T + \frac{1}{r_{\min}}XC^TCX$.

Noting that $\Omega < 0$ if and only if the following inequality holds,

$$R \otimes \Pi + \frac{r_{\max}}{\gamma^2}\widetilde{M}^T R\widetilde{M} \otimes DD^T < 0, \qquad (4.36)$$

where we have used the fact that $R\widetilde{M} = \widetilde{M}^T R$.

Further, we have

$$\begin{aligned}
R \otimes \Pi + \frac{r_{\max}}{\gamma^2}\widetilde{M}^T R\widetilde{M} \otimes DD^T &\leq R \otimes \Pi + \frac{r_{\max}}{\gamma^2}\lambda_{\max}(\widetilde{M}^T R\widetilde{M})I \otimes DD^T \\
&\leq R \otimes [\Pi + \frac{r_{\max}}{\gamma^2 r_{\min}}\lambda_{\max}(\widetilde{M}^T R\widetilde{M})DD^T].
\end{aligned} \qquad (4.37)$$

By using the Gershgorin's disc theorem [58], it is not difficult to get that $\lambda_{\max}(\widetilde{M}^T R\widetilde{M}) \leq 1$. *Then, it follows from (4.37) that*

$$R \otimes \Pi + \frac{r_{\max}}{\gamma^2}\widetilde{M}^T R\widetilde{M} \otimes DD^T \leq R \otimes (\Pi + \frac{r_{\max}}{\gamma^2 r_{\min}}DD^T) \tag{4.38}$$
$$< 0,$$

which implies that (4.36) holds, i.e., $\Omega < 0$. Note that the last inequality in (4.38) follows from the LMI (4.30).

Then, invoking (4.34) gives that

$$\dot{V} + \frac{1}{2}y^T y - \frac{\gamma^2}{2}\omega^T \omega < 0, \tag{4.39}$$

for all z and ω such that $|z|^2 + |\omega|^2 \neq 0$. Integrating inequality (4.39) over infinite horizon yields

$$\|y\|^2 - \gamma^2\|\omega\|^2 - z^T(0)(R \otimes P)z(0) < 0.$$

Recalling that $z(0) = 0$, we have $\|y\|^2 - \gamma^2\|\omega\|^2 < 0$, i.e., $\|\widetilde{T}_{\omega z}\|_\infty = \|T_{\omega z}\|_\infty < \gamma$. Thus, the second condition in Definition 12 is satisfied. Therefore, the H_∞ consensus problem is solved.

Remark 29 *Note that the Laplacian matrix $\mathcal{L}$ associated with a directed communication graph is generally not positive semi-definite. Especially, for a strongly connected graph without balanced conditions which is concerned herein, the matrix $\mathcal{L} + \mathcal{L}^T$ may not bet positive semi-definite either. Thus, this unfavorable feature will not facilitate the decomposition method employed as in Section 4.1. The key tools leading to Theorem 28 are the properly designed Lyapunov function and the generalized algebraic connectivity of strongly connected directed graphs. Compared to the H_∞ consensus problem for undirected communication graphs in Section 4.1, a distinct feature for the case with directed graphs is that now the LMI condition in Algorithm 13 is related to the positive left eigenvector of $\mathcal{L}$ associated with the zero eigenvalue.*

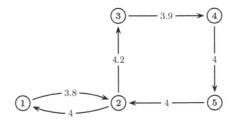

FIGURE 4.3: A strongly connected directed communication graph.

Example 16 *Consider a network consisting of five agents whose communication graph as shown in FIGURE 4.3, where the weights are indicated on the edges. The dynamics of the i-th agent is described by (4.1), with*

$$x_i = \begin{bmatrix} x_{i1} \\ x_{i2} \end{bmatrix}, A = \begin{bmatrix} -0.4 & -19.5998 \\ 4.333 & 0.4 \end{bmatrix}, B = \begin{bmatrix} 1.5 & 0.3 \\ 0 & 1 \end{bmatrix}, C = \begin{bmatrix} 1 & 0 \end{bmatrix}, D = \begin{bmatrix} 1 \\ 1 \end{bmatrix}.$$

The external disturbance $\omega = [2w, \ 5w, \ 4w, \ w, \ 1.5w]^T$, where $w(t)$ is a ten-period square wave starting at $t = 0$ with width 5 and height 1.

From FIGURE 4.3, it is easy to see that the network $\mathcal{G}$ is strongly connected. Some simple calculations give that both the minimum and maximum values of the left eigenvector r are equal to 0.4472. By using Lemma 8 in [189], we can get that $a(\mathcal{L}) = 2.2244$. Choose the H_∞ performance index $\gamma = 1$. Solving LMI (4.30) by using the LMI toolbox of MATLAB gives a feasible solution $P = \begin{bmatrix} 0.21 & 0.08 \\ 0.08 & 0.03 \end{bmatrix}$ and $\tau = 7.52$. Thus, by Algorithm 13, the feedback gain matrix of the protocol (4.2) is given as $K = \begin{bmatrix} -23.52 & 57.73 \\ 33.78 & -83.06 \end{bmatrix}$. Then, according to Theorem 28 and Algorithm 13, the protocol (4.2) with K chosen as above could solve H_∞ consensus with performance index $\gamma = 1$, if the coupling gain $c \geq 3.52$. For the case of $\omega = 0$, the state trajectories of the agents are respectively shown in FIGURE 4.4, from which it can be seen that consensus is indeed achieved. Furthermore, the trajectories of the performance variables z_i, $i = 1, \cdots, 5$, in the presence of disturbances under the zero initial conditions are shown in FIGURE 4.5.

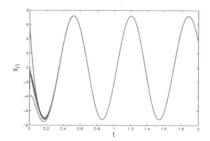 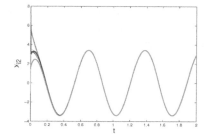

FIGURE 4.4: The state trajectory of the agents under (4.2) constructed via Algorithm 13.

4.4 Notes

The materials of Sections 4.1 and 4.2 are mainly adopted from [81, 85]. The results in Section 4.3 are mainly based on [177]. For further results on distributed H_∞ consensus and control problems of multi-agent systems, please

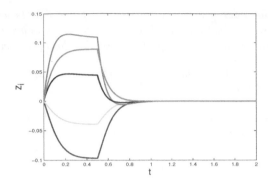

FIGURE 4.5: The performance variables under (4.2) constructed via Algorithm 13.

refer to [87, 98, 99, 102, 108, 197]. Specifically, the H_∞ consensus problem for multi-agent systems of first-order and second-order integrators with external disturbances and parameter uncertainties were considered in [98, 99]. A decomposition approach was proposed in [108] to solve the distributed H_2 and H_∞ control of identical coupled linear systems. An observer-type consensus protocol was provided by using dynamic output feedback approach in [197] to deal with the H_∞ consensus problem for the case where the disturbances satisfy the matching condition. Distributed H_∞ consensus of linear multi-agent systems with switching directed and balanced topologies was studied in [180]. Dynamic controllers were proposed for distributed H_∞ control problem in [87, 102], whose design needs to solve a set of linear matrix inequalities associated with all the nonzero eigenvalues of the Laplacian matrix.

5

Consensus Control of Linear Multi-Agent Systems Using Distributed Adaptive Protocols

CONTENTS

5.1   Distributed Relative-State Adaptive Consensus Protocols 94
    5.1.1   Consensus Using Edge-Based Adaptive Protocols 96
    5.1.2   Consensus Using Node-Based Adaptive Protocols 100
    5.1.3   Extensions to Switching Communication Graphs 101
5.2   Distributed Relative-Output Adaptive Consensus Protocols 103
    5.2.1   Consensus Using Edge-Based Adaptive Protocols 104
    5.2.2   Consensus Using Node-Based Adaptive Protocols 107
    5.2.3   Simulation Examples 109
5.3   Extensions to Leader-Follower Graphs 111
5.4   Robust Redesign of Distributed Adaptive Protocols 114
    5.4.1   Robust Edge-Based Adaptive Protocols 115
    5.4.2   Robust Node-Based Adaptive Protocols 119
    5.4.3   Simulation Examples 121
5.5   Distributed Adaptive Protocols for Graphs Containing Directed
    Spanning Trees 122
    5.5.1   Distributed Adaptive Consensus Protocols 123
    5.5.2   Robust Redesign in the Presence of External Disturbances 129
5.6   Notes 135

In this chapter, we continue Chapter 2 to further consider the consensus control problem for multi-agent systems with general continuous-time linear agent dynamics. The consensus protocols in Chapter 2 are *proposed* in a distributed fashion, using only the local information of each agent and its neighbors. However, these consensus protocols involves some design issues. To be specific, for the general case where the agents are not neutrally stable, the design of the consensus protocols as in Chapter 2 generally requires the knowledge of some eigenvalue information of the Laplacian matrix associated with the communication graph (the smallest nonzero eigenvalue of the Laplacian matrix for undirected graphs and the smallest real part of the nonzero eigenvalues of the Laplacian matrix for directed graphs). Similar situations take place in related works [106, 146, 195]. Note that the smallest real part of the nonzero eigenvalues of the Laplacian matrix is global information in the sense that each agent has to know the entire communication graph $\mathcal{G}$ to compute it. Therefore, although these consensus protocols are proposed and can be implemented in a distributed fashion, they cannot be *designed* by each agent in a distributed fashion. In other words, these consensus protocols are not *fully distributed*. As explained in Chapter 1, due to the large number of agents, limited sensing capability of sensors, and short wireless communication ranges, fully distributed consensus protocols, which can be designed and implemented by each agent using only local information of its own and neighbors, are more desirable and have been widely recognized. Fully distributed consensus protocols have the advantages of high robustness, strong adaptivity, and flexible scalability.

In this paper, we intend to remove the limitation of requiring global information of the communication graph and introduce some fully distributed consensus protocols. The main idea stems from the decoupling feature of the consensus region approach proposed in Chapter 2 for consensus protocol design. As shown in Algorithm 2 in Chapter 2, the feedback gain matrix of the consensus protocol (2.2) can be designed by using only the agent dynamics, and the coupling gain c is used to deal with the effect of the communication topology on consensus. The coupling gain c is essentially a uniform weight on the edges in the communication graph. Since the knowledge of global information of the communication graph is required for the selection of c, we intend to implement some adaptive control tools to dynamically update the weights on the communication graph in order to compensate for the lack of the global information, and thereby to present fully distributed consensus protocols.

In Sections 5.1 and 5.2, we propose distributed consensus protocols based on the relative state and output information, combined with an adaptive law for adjusting the coupling weights between neighboring agents. Two types of distributed adaptive consensus protocols are proposed, namely, the edge-based adaptive protocol which assigns a time-varying coupling weight to each edge in the communication graph and the node-based adaptive protocol which uses a time-varying coupling weight for each node. For the case with undirected communication graphs, these two classes of adaptive protocols are designed to ensure that consensus is reached in a fully distributed fashion for any undirected connected communication graph without using any global information about the communication graph. The case with switching communication graphs is also studied. It is shown that the edge-based adaptive consensus protocol is applicable to arbitrary switching connected graphs. Extensions of the obtained results to the case with a leader-follower communication graph are further discussed in Section 5.3.

The robustness of the proposed adaptive protocols is discussed in Section 5.4. The σ-modification technique in [63] is implemented to present robust adaptive protocols, which is shown to be able to guarantee the ultimate boundedness of both the consensus error and the adaptive weights in the presence of bounded external disturbances. The upper bounds for the consensus error are explicitly given.

Note that the aforementioned protocols are applicable to only undirected communication graphs or leader-follower graphs where the subgraphs among the followers are undirected. How to design fully distributed adaptive consensus protocols for the case with general directed graphs is quite challenging. The main difficulty lies in that the Laplacian matrices of directed graphs are generally asymmetric, which renders the construction of adaptive consensus protocols and the selection of appropriate Lyapunov function quite difficult. Based on the relative states of neighboring agents, a novel distributed adaptive consensus protocol is constructed in Section 5.5. It is shown via a integral Lyapunov function that the proposed adaptive protocol can achieve leader-follower consensus for any communication graph containing a directed spanning tree with the leader as the root node. A distinct feature of the adaptive protocol for directed graphs is that monotonically increasing functions are introduced to provide extra freedom for design. The adaptive protocol for directed graphs is further redesigned in order to be robust in the presence of matching external disturbances.

5.1 Distributed Relative-State Adaptive Consensus Protocols

Consider a group of N identical agents with general linear continuous-time dynamics. The dynamics of the i-th agent are described by

$$
\begin{aligned}
\dot{x}_i &= Ax_i + Bu_i, \\
y_i &= Cx_i, \quad i = 1, \cdots, N,
\end{aligned}
\tag{5.1}
$$

where $x_i \in \mathbf{R}^n$ is the state, $u_i \in \mathbf{R}^p$ is the control input, $y_i \in \mathbf{R}^q$ is the measured output, and A, B, C are constant matrices with compatible dimensions.

In this section, the communication graph among the agents is represented by an undirected graph $\mathcal{G}$. It is assumed that each agent knows the relative states, of its neighbors with respect to itself. Based on the relative information of neighboring agents, we propose two types of distributed adaptive consensus protocols, namely, the edge-based and node-based adaptive consensus protocols.

The edge-based adaptive consensus protocol dynamically updates the coupling weight for each edge (i.e., each communication link) and is given by

$$
\begin{aligned}
u_i &= K \sum_{j=1}^{N} c_{ij} a_{ij} (x_i - x_j), \\
\dot{c}_{ij} &= \kappa_{ij} a_{ij} (x_i - x_j)^T \Gamma (x_i - x_j), \quad i = 1, \cdots, N,
\end{aligned}
\tag{5.2}
$$

where a_{ij} denotes the (i,j)-th entry of the adjacency matrix associated with $\mathcal{G}$, $c_{ij}(t)$ denotes the time-varying coupling weight for the edge (i,j) with

$c_{ij}(0) = c_{ji}(0)$, $\kappa_{ij} = \kappa_{ji}$ are positive constants, and $K \in \mathbf{R}^{p \times n}$ and $\Gamma \in \mathbf{R}^{n \times n}$ are the feedback gain matrices.

The node-based adaptive consensus protocol assigns a time-varying coupling weight to each node (i.e., each agent) and is described by

$$
\begin{aligned}
u_i &= d_i K \sum_{j=1}^{N} a_{ij}(x_i - x_j), \\
\dot{d}_i &= \epsilon_i [\sum_{j=1}^{N} a_{ij}(x_i - x_j)]^T \Gamma [\sum_{j=1}^{N} a_{ij}(x_i - x_j)], \quad i = 1, \cdots, N,
\end{aligned}
\tag{5.3}
$$

where $d_i(t)$ denotes the coupling weight for agent i, ϵ_i are positive constants, and the rest of the variables are defined as in (5.2).

The adaptive protocols (5.2) and (5.3) are actually extensions of the static protocol (2.2) in Chapter 2 by dynamically updating the coupling weights of neighboring agents, which is equivalent to adaptively weighting the communication graph. The edge-based adaptive protocol (5.2) assigns different time-varying weights for different communication edges while the node-based adaptive protocol (5.3) assigns the same time-varying weight for all the ingoing edges of each node, as illustrated in FIGURE 5.1. For each time instant, the adaptive protocols (5.2) and (5.3) are in more general forms of the static protocol (2.2) with nonidentical constant weights on the communication graph.

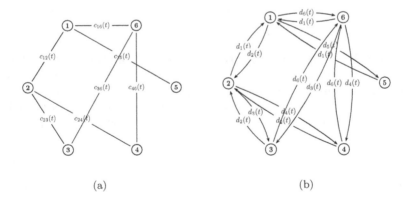

(a)   (b)

FIGURE 5.1: The adaptive consensus protocols (5.2) and (5.3) assign different time-varying weights on a given graph: (a) the edge-based adaptive protocol; (b) the node-based adaptive protocol.

In the adaptive protocols (5.2) and (5.3), it can be observed that the adaptive coupling weights c_{ij} and d_i are monotonically increasing and will converge to constant values when consensus is reached. An intuitive explanation for the reason why (5.2) and (5.3) may solve the consensus problem is that the coupling weights c_{ij} and d_i may be adaptively tuned such that all the nonzero eigenvalues of the Laplacian matrix of the weighted communication graph are located within the unbounded consensus region as in Theorem 4 of Chapter 2 (since the coupling weights (5.2) and (5.3) vary very slowly when consensus is nearly reached, the closed-loop network resulting from (2.1) and (5.2) or (5.3) may be regarded as a linear time-invariant system). However, how to design the feedback gain matrices of (5.2) and (5.3) and theoretically prove that they can solve the consensus problem for the N agents in (2.1) is far from being easy, which will be addressed in the following subsections.

5.1.1 Consensus Using Edge-Based Adaptive Protocols

In this subsection, we study the consensus problem of the agents in (5.1) under the edge-based adaptive protocol (5.2). Let $\xi_i = x_i - \frac{1}{N}\sum_{j=1}^{N} x_j$ and $\xi = [\xi_1^T, \cdots, \xi_N^T]^T$. Then, we get that

$$\xi = [(I_N - \frac{1}{N}\mathbf{1}\mathbf{1}^T) \otimes I_n]x. \tag{5.4}$$

It is easy to see that 0 is a simple eigenvalue of $I_N - \frac{1}{N}\mathbf{1}\mathbf{1}^T$ with $\mathbf{1}$ as the corresponding right eigenvector, and 1 is the other eigenvalue with multiplicity $N - 1$. Then, it follows that $\xi = 0$ if and only if $x_1 = \cdots = x_N$. Therefore, the consensus problem under the adaptive protocol (5.2) is solved if and only if ξ converges to zero. Hereafter we refer to ξ as the consensus error. Because $\kappa_{ij} = \kappa_{ji}$ and $c_{ij}(0) = c_{ji}(0)$, it follows from (5.2) that $c_{ij}(t) = c_{ji}(t)$, $\forall t \geq 0$. Using (5.2) for (5.1), it follows that ξ and c_{ij} satisfy the following dynamics:

$$\dot{\xi}_i = A\xi_i + \sum_{j=1}^{N} c_{ij}a_{ij}BK(\xi_i - \xi_j),$$

$$\dot{c}_{ij} = \kappa_{ij}a_{ij}(\xi_i - \xi_j)^T\Gamma(\xi_i - \xi_j), \quad i = 1, \cdots, N. \tag{5.5}$$

The following theorem presents a sufficient condition for designing (5.2) to solve the consensus problem.

Theorem 29 *For any given connected graph $\mathcal{G}$, the N agents described by (5.1) reach consensus under the edge-based adaptive protocol (5.2) with $K = -B^T P^{-1}$ and $\Gamma = P^{-1}BB^T P^{-1}$, where $P > 0$ is a solution to the following linear matrix inequality (LMI):*

$$AP + PA^T - 2BB^T < 0. \tag{5.6}$$

Moreover, each coupling weight c_{ij} converges to some finite steady-state value.

Proof 29 *Consider the Lyapunov function candidate*

$$V_1 = \frac{1}{2}\sum_{i=1}^{N} \xi_i^T P^{-1}\xi_i + \sum_{i=1}^{N}\sum_{j=1, j\neq i}^{N} \frac{(c_{ij} - \alpha)^2}{4\kappa_{ij}}, \tag{5.7}$$

where α is a positive constant to be determined later. The time derivative of V_1 along the trajectory of (5.5) is given by

$$
\begin{aligned}
\dot{V}_1 &= \sum_{i=1}^{N} \xi_i^T P^{-1} \dot{\xi}_i + \frac{1}{2} \sum_{i=1}^{N} \sum_{j=1, j \neq i}^{N} \frac{c_{ij} - \alpha}{\kappa_{ij}} \dot{c}_{ij} \\
&= \sum_{i=1}^{N} \xi_i^T P^{-1} [A\xi_i + \sum_{j=1}^{N} c_{ij} a_{ij} BK(\xi_i - \xi_j)] \\
&\quad + \frac{1}{2} \sum_{i=1}^{N} \sum_{j=1, j \neq i}^{N} (c_{ij} - \alpha) a_{ij} (\xi_i - \xi_j)^T \Gamma(\xi_i - \xi_j).
\end{aligned}
\tag{5.8}
$$

Because $c_{ij}(t) = c_{ji}(t)$, $\forall t \geq 0$, we have

$$
\begin{aligned}
&\sum_{i=1}^{N} \sum_{j=1, j \neq i}^{N} (c_{ij} - \alpha) a_{ij} (\xi_i - \xi_j)^T \Gamma(\xi_i - \xi_j) \\
&= 2 \sum_{i=1}^{N} \sum_{j=1}^{N} (c_{ij} - \alpha) a_{ij} \xi_i^T \Gamma(\xi_i - \xi_j).
\end{aligned}
\tag{5.9}
$$

Let $\tilde{\xi}_i = P^{-1} \xi_i$ and $\tilde{\xi} = [\tilde{\xi}_1^T, \cdots, \tilde{\xi}_N^T]^T$. Substituting $K = -B^T P^{-1}$ and $\Gamma = P^{-1} BB^T P^{-1}$ into (5.8), we can obtain

$$
\begin{aligned}
\dot{V}_1 &= \sum_{i=1}^{N} \xi_i^T P^{-1} A\xi_i - \alpha \sum_{i=1}^{N} \sum_{j=1}^{N} a_{ij} \xi_i^T P^{-1} BB^T P^{-1} (\xi_i - \xi_j) \\
&= \frac{1}{2} \sum_{i=1}^{N} \tilde{\xi}_i^T (AP + PA^T) \tilde{\xi}_i - \alpha \sum_{i=1}^{N} \sum_{j=1}^{N} \mathcal{L}_{ij} \tilde{\xi}_i^T BB^T \tilde{\xi}_j \\
&= \frac{1}{2} \tilde{\xi}^T [I_N \otimes (AP + PA^T) - 2\alpha \mathcal{L} \otimes BB^T] \tilde{\xi},
\end{aligned}
\tag{5.10}
$$

where $\mathcal{L}_{ij}$ denotes the (i, j)-th entry of the Laplacian matrix $\mathcal{L}$ associated with $\mathcal{G}$.

By the definitions of ξ and $\tilde{\xi}$, it is easy to see that $(\mathbf{1}^T \otimes I)\tilde{\xi} = (\mathbf{1}^T \otimes P^{-1})\xi = 0$. Because $\mathcal{G}$ is connected, it then follows from Lemma 2 that $\tilde{\xi}^T (\mathcal{L} \otimes I)\tilde{\xi} \geq \lambda_2(\mathcal{L})\tilde{\xi}^T \tilde{\xi}$, where $\lambda_2(\mathcal{L})$ is the smallest nonzero eigenvalue of $\mathcal{L}$. Therefore, it follows from (5.10) that

$$
\dot{V}_1 \leq \frac{1}{2} \tilde{\xi}^T [I_N \otimes (AP + PA^T - 2\alpha \lambda_2(\mathcal{L}) BB^T] \tilde{\xi},
\tag{5.11}
$$

By choosing α sufficiently large such that $\alpha \lambda_2(\mathcal{L}) \geq 1$, it follows from (5.6) that

$$
AP + PA^T - 2\alpha \lambda_2(\mathcal{L}) BB^T \leq AP + PA^T - 2BB^T < 0.
$$

Therefore, $\dot{V}_1 \leq 0$.

Since $\dot{V}_1 \leq 0$, $V_1(t)$ is bounded, and so is each c_{ij}. By noting $\Gamma \geq 0$, it can be seen from (5.5) that c_{ij} is monotonically increasing. Then, it follows that each coupling weight c_{ij} converges to some finite value. Let $S = \{\tilde{\xi}, c_{ij} | \dot{V}_1 = 0\}$. Note that $\dot{V}_1 \equiv 0$ implies that $\tilde{\xi}_i = 0$, $i = 1, \cdots, N$, which further implies that $\xi = 0$. Hence, by Lemma 11 (LaSalle's Invariance principle), it follows that $\xi(t) \to 0$ as $t \to \infty$. That is, the consensus problem is solved.

Remark 30 *Comparing the above theorem to Algorithm 2 in Chapter 2, it is easy to see that the feedback gain matrix K of the adaptive protocol (5.2) is identical to that of the static protocol (2.2). This is essentially thanks to the decoupling feature of the consensus region approach, where the feedback gain matrix is independent of the communication topology. As shown in Proposition 1 in Chapter 2, a sufficient condition for the existence of an adaptive protocol (5.2) satisfying Theorem 29 is that (A, B) is stabilizable. The consensus protocol (5.2) can also be designed by solving the algebraic Riccati equation: $A^T Q + QA + I - QBB^T Q = 0$, as in [171, 195]. In this case, the parameters in (5.2) can be chosen as $K = -B^T Q$ and $\Gamma = QBB^T Q$. The solvability of the above Ricatti equation is equal to that of the LMI (5.6).*

Remark 31 *Equation (5.2) presents an adaptive protocol, under which the agents in (5.1) can reach consensus for all connected communication topologies. In contrast to the static consensus protocol (2.2) in Chapter 2, the adaptive protocol (5.2) can be computed and implemented by each agent in a fully distributed way without using any global information of the communication graph. It should be mentioned that compared to the static protocol (2.2) in Chapter 2, one disadvantage of the adaptive protocol (5.2) is that it is nonlinear and has a higher dimension.*

The following theorem presents the explicit expression for the final consensus value.

Theorem 30 *Suppose that the communication graph $\mathcal{G}$ is connected and the adaptive protocol (5.2) satisfies Theorem 33. Then, the final consensus value achieved by the agents in (5.1) is given by*

$$\varpi(t) \triangleq \frac{1}{N} \sum_{i=1}^{N} e^{At} x_i(0). \tag{5.12}$$

Proof 30 *Using (5.2) for (5.1), it follows that*

$$\dot{x} = (I_N \otimes A + \mathcal{L}_c \otimes BK)x, \tag{5.13}$$

where $x = [x_1^T, \cdots, x_N^T]^T$ and $\mathcal{L}^c$ is defined as $\mathcal{L}_{ij}^c = -c_{ij}a_{ij}$, $i \neq j$, and $\mathcal{L}_{ii}^c = \sum_{j=1, j\neq i}^{N} c_{ij}a_{ij}$. It is easy to see that $\mathcal{L}^c$ is also a symmetric Laplacian matrix of the undirected graph $\mathcal{G}$, with a time-varying weight $c_{ij}a_{ij}$ for the *edge (i, j). For a connected graph $\mathcal{G}$, 0 is a simple eigenvalue of $\mathcal{L}^c$ with $\mathbf{1}$ as the corresponding eigenvector, i.e., $\mathbf{1}^T \mathcal{L}^c = 0$.*

Consider the quantity $H = (\mathbf{1}^T \otimes e^{-At})x$. By using (5.13), the time derivative of H is given by

$$
\begin{aligned}
\dot{H} &= -(\mathbf{1}^T \otimes e^{-At} A)x + (\mathbf{1}^T \otimes e^{-At})(I_N \otimes A + \mathcal{L}^c \otimes BK)x \\
&= (\mathbf{1}^T \mathcal{L}^c \otimes e^{-At} BK)x \equiv 0.
\end{aligned}
$$

which implies that H is an invariant quantity. Therefore,

$$(\mathbf{1}^T \otimes e^{-At})(\mathbf{1} \otimes \varpi(t)) = (\mathbf{1}^T \otimes I_n)x(0), \tag{5.14}$$

from which we get the final consensus value $\varpi(t)$ in (5.12).

Remark 32 *By comparing Theorem 30 to Theorem 1 in Chapter 2, it can be found that under either the adaptive protocol (5.2) or the static linear protocol (2.2), the final consensus value achieved by the agents remains the same, depending on the state matrix A, the Laplacian matrix $\mathcal{L}$ with unit weights, and the initial states of the agents.*

5.1.2   Consensus Using Node-Based Adaptive Protocols

This subsection considers the consensus problem of the agents in (5.1) under the node-based adaptive protocol (5.3). Let the consensus error ξ be defined as in (5.4). From (5.1) and (5.3), We can obtain that ξ and d_i satisfy the following dynamics:

$$
\begin{aligned}
\dot{\xi} &= [I_N \otimes A + ((I_N - \frac{1}{N}\mathbf{1}\mathbf{1}^T)D\mathcal{L}) \otimes BK]\xi, \\
\dot{d}_i &= \epsilon_i [\sum_{j=1}^{N} \mathcal{L}_{ij}\xi_j^T]\Gamma[\sum_{j=1}^{N} \mathcal{L}_{ij}\xi_j], \quad i = 1, \cdots, N,
\end{aligned} \tag{5.15}
$$

where $D(t) = \text{diag}(d_1(t), \cdots, d_N(t))$.

The following result presents a sufficient condition for designing (5.3).

Theorem 31 *Assume that the communication graph $\mathcal{G}$ is undirected and connected. Then, the N agents in (5.1) reach consensus under the node-based adaptive protocol (5.3) with K and Γ given as in Theorem 29. Moreover, each coupling weight d_i converges to some finite steady-state value.*

Proof 31 *Consider the Lyapunov function candidate*

$$V_2 = \frac{1}{2}\xi^T(\mathcal{L} \otimes P^{-1})\xi + \sum_{i=1}^{N} \frac{(d_i - \beta)^2}{2\epsilon_i}, \tag{5.16}$$

where β is a positive constant to be determined later. For a connected graph $\mathcal{G}$, *it follows from Lemma 2 and the definition of ξ in (5.4) that $\xi^T(\mathcal{L} \otimes P^{-1})\xi \geq \lambda_2 \xi^T(I_N \otimes P^{-1})\xi$. Therefore, it is easy to see that $\Omega_c = \{\xi, d_i | V_2 \leq c\}$ is compact for any positive c.*

Following similar steps to those in the proof of Theorem 29, we can obtain the time derivative of V_2 along the trajectory of (5.15) as

$$\dot{V}_2 = \xi^T(\mathcal{L} \otimes P^{-1})\dot{\xi} + \sum_{i=1}^{N} \frac{d_i - \beta}{\tau_i}\dot{d}_i$$

$$= \xi^T[\mathcal{L} \otimes P^{-1}A + (\mathcal{L}D\mathcal{L}) \otimes P^{-1}BK]\xi \qquad (5.17)$$

$$+ \sum_{i=1}^{N}(d_i - \beta)(\sum_{j=1}^{N}\mathcal{L}_{ij}\xi_j^T)\Gamma(\sum_{j=1}^{N}\mathcal{L}_{ij}\xi_j),$$

where we have used the fact that $\mathcal{L}\mathbf{1} = 0$. By noting that $K = -B^T P^{-1}$, it is easy to get that

$$\xi^T(\mathcal{L}D\mathcal{L} \otimes P^{-1}BK)\xi = -\sum_{i=1}^{N}d_i(\sum_{j=1}^{N}\mathcal{L}_{ij}\xi_j^T)P^{-1}BB^T P^{-1}(\sum_{j=1}^{N}\mathcal{L}_{ij}\xi_j). \quad (5.18)$$

Substituting (5.18) into (5.17) yields

$$\dot{V}_2 = \frac{1}{2}\xi^T[\mathcal{L} \otimes (P^{-1}A + A^T P^{-1}) - 2\beta\mathcal{L}^2 \otimes P^{-1}BB^T P^{-1}]\xi. \qquad (5.19)$$

Because $\mathcal{G}$ is connected, it follows from Lemma 2 that zero is a simple eigenvalue of $\mathcal{L}$ and all the other eigenvalues are positive. Let $U \in \mathbf{R}^{N \times N}$ be such a unitary matrix that $U^T \mathcal{L}U = \Lambda \triangleq \operatorname{diag}(0, \lambda_2, \cdots, \lambda_N)$. Because the right and left eigenvectors of $\mathcal{L}$ corresponding to the zero eigenvalue are equal to $\mathbf{1}$, we can choose $U = [\frac{1}{\sqrt{N}} \ Y_1]$ and $U^T = \begin{bmatrix} \frac{\mathbf{1}^T}{\sqrt{N}} \\ Y_2 \end{bmatrix}$, with $Y_1 \in \mathbf{R}^{N \times (N-1)}$ and $Y_2 \in \mathbf{R}^{(N-1) \times N}$. Let $\bar{\xi} \triangleq [\bar{\xi}_1^T, \cdots, \bar{\xi}_N^T]^T = (U^T \otimes P^{-1})\xi$. By the definitions of ξ and $\bar{\xi}$, it is easy to see that

$$\bar{\xi}_1 = (\frac{\mathbf{1}^T}{\sqrt{N}} \otimes P^{-1})\xi = 0. \qquad (5.20)$$

Then, we have

$$\dot{V}_2 = \frac{1}{2}\bar{\xi}^T[\Lambda \otimes (AP + PA^T) - 2\beta\Lambda^2 \otimes BB^T]\bar{\xi}$$

$$= \frac{1}{2}\sum_{i=2}^{N}\lambda_i \bar{\xi}_i^T(AP + PA^T - 2\beta\lambda_i BB^T)\bar{\xi}_i. \qquad (5.21)$$

By choosing β sufficiently large such that $\beta\lambda_i \geq 1$, $i = 2, \cdots, N$, it follows from (5.6) and (5.21) that $\dot{V}_2 \leq 0$. Since $\dot{V}_2 \leq 0$, $V_2(t)$ is bounded and so is each d_i. By noting $\Gamma \geq 0$, it can be seen from (5.15) that d_i is monotonically *increasing. Then, it follows that each coupling weight d_i converges to some finite value. Note that $\dot{V}_2 \equiv 0$ implies that $\bar{\xi}_i = 0$, $i = 2, \cdots, N$, which, by noticing that $\bar{\xi}_1 \equiv 0$ in (5.20), further implies that $\xi = 0$. Hence, by Lemma 11 (LaSalle's Invariance principle), it follows that $\bar{\xi}(t) \to 0$ and thereby $\xi(t) \to 0$ as $t \to \infty$.*

Remark 33 *Similar to the edge-based adaptive protocol (5.2), the node-based adaptive protocol (5.3) can also be computed in a fully distributed fashion. In the following, some comparisons between (5.2) and (5.3) are now briefly discussed. The dimension of the edge-based adaptive protocol (5.2) is proportional to the number of edges in the communication graph. Since the number of edges is usually larger than the number of nodes in a connected graph, the dimension of the edge-based adaptive protocol (5.2) is generally higher than that of the node-based protocol (5.3). On the other hand, the edge-based adaptive protocol (5.2) is applicable to the case with switching communication graphs, which will be shown in the following subsection. Moreover, the final consensus value under the node-based adaptive protocol (5.3), different from the case of the edge-based adaptive protocol (5.2), is not easy to derive, due to the asymmetry of $D(t)\mathcal{L}$.*

5.1.3   Extensions to Switching Communication Graphs

In the last subsections, the communication graph is assumed to be fixed throughout the whole process. However, the communication graph may change with time in many practical situations due to various reasons, such as communication constraints and link variations. In this subsection, the consensus problem under the edge-based adaptive protocol (5.2) with switching communication graphs will be considered.

Denote by $\mathscr{G}_N$ the set of all possible undirected connected graphs with N nodes. Let $\sigma(t) : [0, \infty) \to \mathscr{P}$ be a piecewise constant switching signal with switching times $t_0, t_1, \cdots$, and $\mathscr{P}$ be the index set associated with the elements of $\mathscr{G}_N$, which is clearly finite. The communication graph at time t is denoted by $\mathcal{G}_{\sigma(t)}$. Accordingly, (5.2) becomes

$$u_i = K \sum_{j=1}^{N} c_{ij} a_{ij}(t)(x_i - x_j),$$

$$\dot{c}_{ij} = \kappa_{ij} a_{ij}(t)(x_i - x_j)^T \Gamma(x_i - x_j), \quad i = 1, \cdots, N, \tag{5.22}$$

where $a_{ij}(t)$ is the (i, j)-th entry of the adjacency matrix associated with $\mathcal{G}_{\sigma(t)}$ and the rest of the variables are the same as in (5.2).

Theorem 32 *For arbitrary switching communication graphs $\mathcal{G}_{\sigma(t)}$ belonging to $\mathscr{G}_N$, the N agents in (5.1) reach consensus under the edge-based protocol (5.22) with K and Γ given as in Theorem 29. Besides, the coupling weights c_{ij} converge to some finite values.*

Proof 32 *Let ξ_i and ξ be defined as in (5.4). By following similar steps to those in the proof of Theorem 29, the consensus problem of the agents (5.1) under the protocol (5.22) is solved if ξ converges to zero. Clearly, ξ_i and c_{ij} satisfy the following dynamics:*

$$\dot{\xi}_i = A\xi_i + \sum_{j=1}^{N} c_{ij}a_{ij}(t)BK(\xi_i - \xi_j),$$

$$\dot{c}_{ij} = \kappa_{ij}a_{ij}(t)(\xi_i - \xi_j)^T\Gamma(\xi_i - \xi_j), \quad i = 1, \cdots, N. \tag{5.23}$$

Take a common Lyapunov function candidate

$$V_3 = \frac{1}{2}\sum_{i=1}^{N} \xi_i^T P^{-1}\xi_i + \sum_{i=1}^{N}\sum_{j=1,j\neq i}^{N} \frac{\tilde{c}_{ij}^2}{4\varepsilon_{ij}},$$

where $\tilde{c}_{ij} = c_{ij} - \delta$ and δ is a positive scalar. Following similar steps as in the proof of Theorem 29, we can obtain the time derivative of V_3 along (5.23) as

$$\dot{V}_3 = \frac{1}{2}\tilde{\xi}^T\left[I_N \otimes (AP + PA^T) - 2\delta\mathcal{L}_{\sigma(t)} \otimes BB^T\right]\tilde{\xi}, \tag{5.24}$$

where $\mathcal{L}_{\sigma(t)}$ is the Laplacian matrix associated with $\mathcal{G}_{\sigma(t)}$ and the rest of the variables are the same as in (5.10). Since $\mathcal{G}_{\sigma(t)}$ is connected and $(\mathbf{1}^T \otimes I)\tilde{\xi} = 0$, it follows from Lemma 2 that $\tilde{\xi}^T(\mathcal{L}_{\sigma(t)} \otimes I)\tilde{\xi} \geq \lambda_2^{\min}\tilde{\xi}^T\tilde{\xi}$, where $\lambda_2^{\min} \triangleq \min_{\mathcal{G}_{\sigma(t)}\in\mathscr{G}_N}\{\lambda_2(\mathcal{L}_{\sigma(t)})\}$ denotes the minimum of the smallest nonzero eigenvalues of $\mathcal{L}_{\sigma(t)}$ for all $\mathcal{G}_{\sigma(t)} \in \mathscr{G}_N$. Therefore, we can get from (5.24) that

$$\dot{V}_3 \leq W(\tilde{\xi}) \triangleq \frac{1}{2}\tilde{\xi}^T[I_N \otimes (AP + PA^T - 2\delta\lambda_2^{\min}BB^T)]\tilde{\xi}.$$

As shown in the proof of Theorem 29, by choosing $\delta > 0$ sufficiently large such that $\delta\lambda_2^{\min} \geq 1$, we have $AP+PA^T-2\delta\lambda_2^{\min}BB^T < 0$. Therefore, $\dot{V}_3 \leq 0$. Note that V_3 is positive definite and radically unbounded. By Lemma 14 (LaSalle-Yoshizawa theorem), it follows that $\lim_{t\to\infty} W(\tilde{\xi}) = 0$, implying that $\tilde{\xi}(t) \to 0$, as $t \to \infty$, which further implies that $\xi(t) \to 0$, as $t \to \infty$. The convergence of c_{ij} can be similarly shown as in the proof of Theorem 29.

Remark 34 *Theorem 32 shows that the edge-based adaptive consensus protocol (5.2) given by Theorem 29 is applicable to arbitrary switching communication graphs which are connected at any time instant and the final consensus value remains the same as in (5.12). Because the Lyapunov function in (5.16) for the node-based adaptive protocol (5.3) is explicitly related with the communication graph, it cannot be taken as a feasible common Lyapunov function in the case of switching topologies.*

5.2 Distributed Relative-Output Adaptive Consensus Protocols

In the previous section, the adaptive consensus protocols (5.2) and (5.3) depend on the relative state information of neighboring agents, which however might not be available in some circumstances. In this section, we extend to consider the case where that each agent knows the relative outputs, rather than the relative states, of its neighbors with respect to itself. Based on the relative output information of neighboring agents, distributed edge-based and node-based adaptive dynamic consensus protocols are proposed in the following.

The distributed edge-based adaptive observer-type consensus protocol is given by

$$
\begin{aligned}
\dot{v}_i &= (A + BF)v_i + L\sum_{j=1}^{N} c_{ij}a_{ij}[C(v_i - v_j) - (y_i - y_j)], \\
\dot{c}_{ij} &= \varepsilon_{ij}a_{ij}\begin{bmatrix} y_i - y_j \\ C(v_i - v_j) \end{bmatrix}^T \Gamma \begin{bmatrix} y_i - y_j \\ C(v_i - v_j) \end{bmatrix}, \\
u_i &= Fv_i, \quad i = 1, \cdots, N,
\end{aligned}
\tag{5.25}
$$

where $v_i \in \mathbf{R}^n$ is the protocol state, $i = 1, \cdots, N$, a_{ij} is the (i,j)-th entry of the adjacency matrix associated with $\mathcal{G}$, $c_{ij}(t)$ denotes the time-varying coupling weight for the edge (i,j) with $c_{ij}(0) = c_{ji}(0)$, $\varepsilon_{ij} = \varepsilon_{ji}$ are positive constants, and $F \in \mathbf{R}^{p \times n}$, $L \in \mathbf{R}^{n \times q}$, and $\Gamma \in \mathbf{R}^{2q \times 2q}$ are the feedback gain matrices.

The node-based adaptive observer-type consensus protocol is described by

$$
\begin{aligned}
\dot{\tilde{v}}_i &= (A + BF)\tilde{v}_i + d_i L\sum_{j=1}^{N} a_{ij}[C(\tilde{v}_i - \tilde{v}_j) - (y_i - y_j)], \\
\dot{d}_i &= \tau_i \left(\sum_{j=1}^{N} a_{ij}\begin{bmatrix} y_i - y_j \\ C(\tilde{v}_i - \tilde{v}_j) \end{bmatrix} \right)^T \Gamma \left(\sum_{j=1}^{N} a_{ij}\begin{bmatrix} y_i - y_j \\ C(\tilde{v}_i - \tilde{v}_j) \end{bmatrix} \right), \\
u_i &= F\tilde{v}_i, \quad i = 1, \cdots, N,
\end{aligned}
\tag{5.26}
$$

where $d_i(t)$ denotes the coupling weight for agent i, τ_i are positive constants, and the rest of the variables are defined as in (5.25).

Note that the terms $\sum_{j=1}^{N} c_{ij}a_{ij}C(v_i - v_j)$ in (5.25) and $\sum_{j=1}^{N} a_{ij}C(\tilde{v}_i - \tilde{v}_j)$ in (5.26) imply that the agents need to use the virtual outputs of the consensus protocols from their neighbors via the communication topology $\mathcal{G}$.

The adaptive dynamic protocols (5.25) and (5.26) are extensions of the observer-type consensus protocol (2.21) in Chapter 2 by dynamically tuning the coupling weights of the communication graph. Alternative distributed adaptive consensus protocols based on the observer-type protocol (2.33) can be similarly designed. The details are omitted here for conciseness.

5.2.1 Consensus Using Edge-Based Adaptive Protocols

In this subsection, we study the consensus problem of the agents in (5.1) under the edge-based adaptive protocol (5.25). Let $z_i = [x_i^T, v_i^T]^T$, $\zeta_i = z_i - \frac{1}{N} \sum_{j=1}^{N} z_j$, $z = [z_1^T, \cdots, z_N^T]^T$, and $\zeta = [\zeta_1^T, \cdots, \zeta_N^T]^T$. Then, we get $\zeta = [(I_N - \frac{1}{N}\mathbf{11}^T) \otimes I_{2n}]z$. Because 0 is a simple eigenvalue of $I_N - \frac{1}{N}\mathbf{11}^T$ with $\mathbf{1}$ as a corresponding eigenvector and 1 is the other eigenvalue with multiplicity $N - 1$, it follows that $\zeta = 0$ if and only if $z_1 = \cdots = z_N$. Therefore, the consensus problem of the agents in (5.1) under the protocol (5.25) is solved if ζ converges to zero. Because $\varepsilon_{ij} = \varepsilon_{ji}$ and $c_{ij}(0) = c_{ji}(0)$, it follows from (5.25) that $c_{ij}(t) = c_{ji}(t)$, $\forall t \geq 0$. It is not difficult to obtain that ζ_i and c_{ij} satisfy

$$\dot{\zeta}_i = \mathcal{M}\zeta_i + \sum_{j=1}^{N} c_{ij} a_{ij} \mathcal{H}(\zeta_i - \zeta_j),$$

$$\dot{c}_{ij} = \varepsilon_{ij} a_{ij} (\zeta_i - \zeta_j)^T \mathcal{R}(\zeta_i - \zeta_j), \quad i = 1, \cdots, N,$$

(5.27)

where

$$\mathcal{M} = \begin{bmatrix} A & BF \\ 0 & A + BF \end{bmatrix}, \quad \mathcal{H} = \begin{bmatrix} 0 & 0 \\ -LC & LC \end{bmatrix}, \quad \mathcal{R} = (I_2 \otimes C^T)\Gamma(I_2 \otimes C).$$

The following theorem designs the adaptive protocol (5.25) to achieve consensus.

Theorem 33 *Suppose that the communication graph $\mathcal{G}$ is undirected and connected. Then, the N agents in (5.1) reach consensus under the edge-based adaptive protocol (5.25) with F satisfying that $A + BF$ is Hurwitz, $\Gamma = \begin{bmatrix} I_q & -I_q \\ -I_q & I_q \end{bmatrix}$, and $L = -Q^{-1}C^T$, where $Q > 0$ is a solution to the following LMI:*

$$A^T Q + QA - 2C^T C < 0.$$

(5.28)

Moreover, the protocol states v_i, $i = 1, \cdots, N$, converge to zero and each coupling weight c_{ij} converges to some finite steady-state value.

Proof 33 *Consider the Lyapunov function candidate*

$$V_4 = \frac{1}{2} \sum_{i=1}^{N} \zeta_i^T \mathcal{Q}\zeta_i + \sum_{i=1}^{N} \sum_{j=1, j \neq i}^{N} \frac{(c_{ij} - \alpha)^2}{4\varepsilon_{ij}},$$

(5.29)

where $\mathcal{Q} \triangleq \begin{bmatrix} \varsigma\tilde{Q} + Q & -Q \\ -Q & Q \end{bmatrix}$, $\tilde{Q} > 0$ satisfies that $\tilde{Q}(A + BF) + (A + BF)^T\tilde{Q} < 0$, and α and ς are positive constants to be determined later. In light of Lemma 19 (Schur Complement Lemma), it is easy to know that $\mathcal{Q} > 0$.

The time derivative of V_4 along the trajectory of (5.27) can be obtained as

$$\dot{V}_4 = \sum_{i=1}^{N} \zeta_i^T \mathcal{Q} \dot{\zeta}_i + \sum_{i=1}^{N} \sum_{j=1, j \neq i}^{N} \frac{c_{ij} - \alpha}{2\varepsilon_{ij}} \dot{c}_{ij}$$

$$= \sum_{i=1}^{N} \zeta_i^T \mathcal{Q}[\mathcal{M}\zeta_i + \sum_{j=1}^{N} c_{ij} a_{ij} \mathcal{H}(\zeta_i - \zeta_j)] \qquad (5.30)$$

$$+ \frac{1}{2} \sum_{i=1}^{N} \sum_{j=1}^{N} (c_{ij} - \alpha) a_{ij} (\zeta_i - \zeta_j)^T \mathcal{R}(\zeta_i - \zeta_j).$$

Let $\tilde{\zeta}_i = T\zeta_i$, $i = 1, \cdots, N$, with $T = \begin{bmatrix} I & 0 \\ -I & I \end{bmatrix}$. Then, (5.30) can be rewritten as

$$\dot{V}_4 = \sum_{i=1}^{N} \tilde{\zeta}_i^T \tilde{\mathcal{Q}}[\widetilde{\mathcal{M}}\tilde{\zeta}_i + \sum_{i=1}^{N} \sum_{j=1}^{N} c_{ij} a_{ij} \widetilde{\mathcal{H}}(\tilde{\zeta}_i - \tilde{\zeta}_j)]$$

$$+ \frac{1}{2} \sum_{i=1}^{N} \sum_{j=1}^{N} (c_{ij} - \alpha) a_{ij} (\tilde{\zeta}_i - \tilde{\zeta}_j)^T \widetilde{\mathcal{R}}(\tilde{\zeta}_i - \tilde{\zeta}_j), \qquad (5.31)$$

where

$$\tilde{\mathcal{Q}} = T^{-T} \mathcal{Q} T^{-1} = \begin{bmatrix} \varsigma\tilde{Q} & 0 \\ 0 & Q \end{bmatrix}, \quad \widetilde{\mathcal{M}} = T\mathcal{M}T^{-1} = \begin{bmatrix} A + BF & BF \\ 0 & A \end{bmatrix},$$

$$\widetilde{\mathcal{H}} = T\mathcal{H}T^{-1} = \begin{bmatrix} 0 & 0 \\ 0 & LC \end{bmatrix}, \quad \widetilde{\mathcal{R}} = T^{-T}\mathcal{R}T^{-1} = \begin{bmatrix} 0 & 0 \\ 0 & C^T C \end{bmatrix}.$$

Because $c_{ij}(t) = c_{ji}(t)$, $\forall t \geq 0$, we have

$$\sum_{i=1}^{N} \sum_{j=1}^{N} (c_{ij} - \alpha) a_{ij} (\tilde{\zeta}_i - \tilde{\zeta}_j)^T \widetilde{\mathcal{R}}(\tilde{\zeta}_i - \tilde{\zeta}_j)$$

$$= 2 \sum_{i=1}^{N} \sum_{j=1}^{N} (c_{ij} - \alpha) a_{ij} \tilde{\zeta}_i^T \widetilde{\mathcal{R}}(\tilde{\zeta}_i - \tilde{\zeta}_j). \qquad (5.32)$$

It is easy to see that $\tilde{\mathcal{Q}}\widetilde{\mathcal{H}} = -\widetilde{\mathcal{R}}$. Then, by letting $\tilde{\zeta} = [\tilde{\zeta}_1^T, \cdots, \tilde{\zeta}_N^T]^T$, it follows from (5.31) and (5.32) that

$$\dot{V}_4 = \sum_{i=1}^{N} \tilde{\zeta}_i^T \tilde{\mathcal{Q}}\widetilde{\mathcal{M}}\tilde{\zeta}_i - \alpha \sum_{i=1}^{N} \sum_{j=1}^{N} a_{ij} \tilde{\zeta}_i^T \widetilde{\mathcal{R}}(\tilde{\zeta}_i - \tilde{\zeta}_j)$$

$$= \frac{1}{2} \tilde{\zeta}^T [I_N \otimes (\tilde{\mathcal{Q}}\widetilde{\mathcal{M}} + \widetilde{\mathcal{M}}^T \tilde{\mathcal{Q}}) - 2\alpha \mathcal{L} \otimes \widetilde{\mathcal{R}}]\tilde{\zeta}, \qquad (5.33)$$

where $\mathcal{L}$ is the Laplacian matrix associated with $\mathcal{G}$.

By the definitions of ζ and $\tilde{\zeta}$, it is easy to see that $(\mathbf{1}^T \otimes I)\tilde{\zeta} =$

$(\mathbf{1}^T \otimes T)\zeta = 0$. *Because* $\mathcal{G}$ *is connected, it then follows from Lemma 2 that* $\tilde{\zeta}^T(\mathcal{L} \otimes I)\tilde{\zeta} \geq \lambda_2(\mathcal{L})\tilde{\zeta}^T\tilde{\zeta}$, *where* $\lambda_2(\mathcal{L})$ *is the smallest nonzero eigenvalue of* $\mathcal{L}$. *Therefore, we can get from (5.33) that*

$$\dot{V}_4 \leq \frac{1}{2}\tilde{\zeta}^T[I_N \otimes (\tilde{\mathcal{Q}}\widetilde{\mathcal{M}} + \widetilde{\mathcal{M}}^T\tilde{\mathcal{Q}} - 2\alpha\lambda_2(\mathcal{L})\tilde{\mathcal{R}})]\tilde{\zeta}. \tag{5.34}$$

Note that

$$\tilde{\mathcal{Q}}\widetilde{\mathcal{M}} + \widetilde{\mathcal{M}}^T\tilde{\mathcal{Q}} - 2\alpha\lambda_2(\mathcal{L})\tilde{\mathcal{R}}$$

$$= \begin{bmatrix} \varsigma[\tilde{Q}(A+BF) + (A+BF)^T\tilde{Q}] & \varsigma\tilde{Q}BF \\ \varsigma F^TB^T\tilde{Q} & \Pi \end{bmatrix}, \tag{5.35}$$

where $\Pi \triangleq QA + A^TQ - 2\alpha\lambda_2(\mathcal{L})C^TC$. *By choosing* α *sufficiently large such that* $\alpha\lambda_2(\mathcal{L}) \geq 1$, *it follows from (5.28) that that* $\Pi < 0$. *Then, choosing* $\varsigma > 0$ *sufficiently small and in virtue of Lemma 19, we can obtain from (5.35) and (5.34) that* $\tilde{\mathcal{Q}}\widetilde{\mathcal{M}} + \widetilde{\mathcal{M}}^T\tilde{\mathcal{Q}} - 2\alpha\lambda_2(\mathcal{L})\tilde{\mathcal{R}} < 0$ *and* $\dot{V}_4 \leq 0$.

Since $\dot{V}_4 \leq 0$, $V_1(t)$ *is bounded, implying that each* c_{ij} *is also bounded. By noting that* $\mathcal{R} \geq 0$, *we can see from (5.27) that* c_{ij} *is monotonically increasing. Then, it follows that each coupling weight* c_{ij} *converges to some finite value. Note that* $\dot{V}_4 \equiv 0$ *implies that* $\tilde{\zeta} = 0$ *and* $\zeta = 0$. *Hence, by Lemma 11 (LaSalle's Invariance principle), it follows that* $\zeta(t) \to 0$ *as* $t \to \infty$. *That is, the consensus problem is solved. By (5.25) and noting the fact that* $A + BF$ *is Hurwitz, it is easy to see that the protocol states* v_i, $i = 1, \cdots, N$, *converge to zero.*

Remark 35 *As shown in Proposition 3 of Chapter 2, a necessary and sufficient condition for the existence of a* $Q > 0$ *to the LMI (5.28) is that* (A, C) *is detectable. Therefore, a sufficient condition for the existence of an adaptive protocol (5.25) satisfying Theorem 33 is that* (A, B, C) *is stabilizable and detectable. Because the distributed adaptive protocol (5.25) is essentially nonlinear, the separation principle which holds for linear multi-agent systems as in Chapter 2 generally does not hold anymore. A favorable feature of the adaptive protocol (5.25) is that its feedback gain matrices* F, L, *and* Γ *can be independently designed. Similarly as shown in Theorem 32, it can be seen that the edge-based adaptive protocol (5.25) designed in Theorem 33 is applicable to arbitrary switching communication graphs which are connected at each time instant.*

5.2.2 Consensus Using Node-Based Adaptive Protocols

This subsection considers the consensus problem of the agents in (5.1) under the node-based adaptive protocol (5.26). Let $\tilde{z}_i = [x_i^T, \tilde{v}_i^T]^T$, $\bar{\zeta}_i = \tilde{z}_i - \frac{1}{N}\sum_{j=1}^{N}\tilde{z}_j$, and $\bar{\zeta} = [\bar{\zeta}_1^T, \cdots, \bar{\zeta}_N^T]^T$. As shown in the last subsection, the consensus problem of agents (5.1) under the protocol (5.26) is solved if $\bar{\zeta}$

converges to zero. It is not difficult to obtain that $\bar{\zeta}$ and d_i satisfy the following dynamics:

$$\dot{\bar{\zeta}} = [I_N \otimes \mathcal{M} + ((I_N - \frac{1}{N}\mathbf{1}\mathbf{1}^T)D\mathcal{L}) \otimes \mathcal{H}]\bar{\zeta},$$

$$\dot{d}_i = \tau_i[\sum_{j=1}^{N} \mathcal{L}_{ij}\bar{\zeta}_j^T]\mathcal{R}[\sum_{j=1}^{N} \mathcal{L}_{ij}\bar{\zeta}_j], \quad i = 1, \cdots, N, \tag{5.36}$$

where $D(t) = \text{diag}(d_1(t), \cdots, d_N(t))$, $\mathcal{M}$, $\mathcal{H}$, and $\mathcal{R}$ are defined in (5.27).

Theorem 34 *Assume that the communication graph $\mathcal{G}$ is undirected and connected. Then, the N agents in (5.1) reach consensus under the node-based adaptive protocol (5.26) with L, F, and Γ given as in Theorem 33. Moreover, the protocol states $\tilde{v}_i$, $i = 1, \cdots, N$, converge to zero and each coupling weight d_i converges to some finite steady-state value.*

Proof 34 *Consider the Lyapunov function candidate*

$$V_5 = \frac{1}{2}\bar{\zeta}^T(\mathcal{L} \otimes \mathcal{Q})\bar{\zeta} + \sum_{i=1}^{N} \frac{(d_i - \beta)^2}{2\tau_i}, \tag{5.37}$$

where $\mathcal{Q}$ is defined in (5.29) and β is a positive constant to be determined later. For a connected graph $\mathcal{G}$, it follows from Lemma 2 and the definition of $\bar{\zeta}$ that $\bar{\zeta}^T(\mathcal{L} \otimes \mathcal{Q})\bar{\zeta} \geq \lambda_2(\mathcal{L})\bar{\zeta}^T(I_N \otimes \mathcal{Q})\bar{\zeta}$. Therefore, it is easy to see that $\Omega_c = \{\bar{\zeta}, d_i | V_5 \leq c\}$ is compact for any positive c.

Following similar steps to those in the proof of Theorem 33, we can obtain the time derivative of V_5 along the trajectory of (5.36) as

$$\dot{V}_5 = \hat{\zeta}^T[\mathcal{L} \otimes \tilde{\mathcal{Q}}\tilde{\mathcal{M}} - \mathcal{L}D\mathcal{L} \otimes \tilde{\mathcal{R}}]\hat{\zeta}$$

$$+ \sum_{i=1}^{N}(d_i - \beta)(\sum_{j=1}^{N} \mathcal{L}_{ij}\hat{\zeta}_j^T)\tilde{\mathcal{R}}(\sum_{j=1}^{N} \mathcal{L}_{ij}\hat{\zeta}_j), \tag{5.38}$$

where $\hat{\zeta} \triangleq [\hat{\zeta}_1^T, \cdots, \hat{\zeta}_N^T]^T = (I_N \otimes T)\bar{\zeta}$, T, $\tilde{\mathcal{Q}}$, $\tilde{\mathcal{M}}$, and $\tilde{\mathcal{R}}$ are the same as in (5.31). Observe that

$$\hat{\zeta}^T(\mathcal{L}D\mathcal{L} \otimes \tilde{\mathcal{R}})\hat{\zeta} = \sum_{i=1}^{N} d_i(\sum_{j=1}^{N} \mathcal{L}_{ij}\hat{\zeta}_j^T)\tilde{\mathcal{R}}(\sum_{j=1}^{N} \mathcal{L}_{ij}\hat{\zeta}_j). \tag{5.39}$$

Substituting (5.39) into (5.38) yields

$$\dot{V}_5 = \frac{1}{2}\hat{\zeta}^T[\mathcal{L} \otimes (\tilde{\mathcal{Q}}\tilde{\mathcal{M}} + \tilde{\mathcal{M}}^T\tilde{\mathcal{Q}}^T) - 2\beta\mathcal{L}^2 \otimes \tilde{\mathcal{R}}]\hat{\zeta}. \tag{5.40}$$

Because $\mathcal{G}$ is connected, let unitary matrices U and U^T be defined as in the proof of Theorem 31 such that $U^T\mathcal{L}U = \Lambda \triangleq \text{diag}(0, \lambda_2, \cdots, \lambda_N)$, where $\lambda_2 \leq \cdots \leq \lambda_N$ are the nonzero eigenvalues of $\mathcal{L}$. Let $\check{\zeta} \triangleq [\check{\zeta}_1^T, \cdots, \check{\zeta}_N^T]^T = (U^T \otimes I)\hat{\zeta}$.

By the definitions of $\check{\zeta}$ and $\hat{\zeta}$, it is easy to see that $\check{\zeta}_1 = (\frac{1^T}{\sqrt{N}} \otimes T)\zeta = 0$. Then, it follows from (5.40) that

$$
\begin{aligned}
\dot{V}_5 &= \frac{1}{2}\check{\zeta}^T [\Lambda \otimes (\widetilde{\mathcal{Q}}\widetilde{\mathcal{M}} + \widetilde{\mathcal{M}}^T \widetilde{\mathcal{Q}}) - 2\beta\Lambda^2 \otimes \widetilde{\mathcal{R}}]\check{\zeta} \\
&= \frac{1}{2}\sum_{i=2}^{N} \lambda_i \check{\zeta}_i^T (\widetilde{\mathcal{Q}}\widetilde{\mathcal{M}} + \widetilde{\mathcal{M}}^T \widetilde{\mathcal{Q}} - 2\beta\lambda_i\widetilde{\mathcal{R}})\check{\zeta}_i.
\end{aligned}
\tag{5.41}
$$

As shown in the proof of Theorem 33, by choosing $\varsigma > 0$ sufficiently small and β sufficiently large such that $\beta\lambda_2(\mathcal{L}) \geq 1$, we can obtain from (5.41) that $\dot{V}_5 \leq 0$. Note that $\dot{V}_5 \equiv 0$ implies that $\check{\zeta}_i = 0$, $i = 2, \cdots, N$, which, together with $\check{\zeta}_1 = 0$, further implies that $\zeta = 0$. Therefore, it follows from Lemma 11 (Lasalle's Invariance principle) that $\zeta \to 0$ as $t \to \infty$. The convergence of d_i and $\tilde{v}_i$, $i = 1, \cdots, N$, can be shown by following similar steps in the proof of Theorem 33, which is omitted here for brevity.

Remark 36 *The adaptive consensus protocols (5.2), (5.3), (5.25), and (5.26) are extensions of their counterparts with constant coupling weights in Chapter 2 by using adaptive coupling weights. Compared to the consensus protocols in Chapter 2 whose design generally requires the knowledge of the smallest nonzero eigenvalue of the Laplacian matrix, one advantage of these adaptive consensus protocols in the current chapter is that they depend on only local information, which thereby are fully distributed and scalable in the sense that new agents can be added or existing agents can be removed without redesigning the protocol. As mentioned earlier, the reason why this can be done lies in the decoupling property of the consensus region approach.*

5.2.3   Simulation Examples

In the following, a numerical example is presented for illustration.

Example 17 *Consider a network of third-order integrators, described by (5.1), with*

$$
A = \begin{bmatrix} 0 & 1 & 0 \\ 0 & 0 & 1 \\ 0 & 0 & 0 \end{bmatrix}, B = \begin{bmatrix} 0 \\ 0 \\ 1 \end{bmatrix}, C = \begin{bmatrix} 1 & 0 & 0 \end{bmatrix}.
$$

Choose $F = -[\,3\ 6.5\ 4.5\,]$ such that $A + BF$ is Hurwitz. Solving the LMI (5.28) by using the SeDuMi toolbox [158] gives the feedback gain matrix L in (5.25) and (5.26) as $L = -[\,2.1115\ 1.3528\ 0.6286\,]^T$.

To illustrate Theorem 37, let the communication graph be $\mathcal{G}_1$ in FIGURE 5.2(a) and $\tau_i = 1$, $i = 1, \cdots, 6$, in (5.3). The consensus errors $x_i - x_1$, $i = 2, \cdots, 6$, of the third-order integrators under the protocol (5.3) with F and L as above and $\Gamma = \begin{bmatrix} 1 & -1 \\ -1 & 1 \end{bmatrix}$ are depicted in in FIGURE 5.3(a). The coupling weights d_i associated with the nodes are drawn in FIGURE 5.3(b).

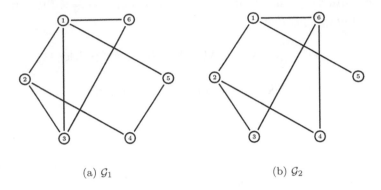

(a) $\mathcal{G}_1$ (b) $\mathcal{G}_2$

FIGURE 5.2: Undirected communication graphs $\mathcal{G}_1$ and $\mathcal{G}_2$.

To illustrate that Theorem 34 is applicable to switching graphs, let $\mathcal{G}_{\sigma(t)}$ switch randomly every 0.1 second between $\mathcal{G}_1$ and $\mathcal{G}_2$ as shown in FIGURE 5.2. Note that both $\mathcal{G}_1$ and $\mathcal{G}_2$ are connected. Let $\varepsilon_{ij} = 1$, $i, j = 1, \cdots, 6$, in (5.25), and $c_{ij}(0) = c_{ji}(0)$ be randomly chosen. The consensus errors $x_i - x_1$, $i = 2, \cdots, 6$, under the protocol (5.25) designed as above are depicted in FIGURE 5.4(a). The coupling weights c_{ij} associated with the edges in this case are shown in FIGURES 5.4(b). FIGURES 5.3(a) and 5.4(a) state that consensus is indeed achieved in both cases. From FIGURES 5.3(b) and 5.4(b), it can be observed that the coupling weights converge to finite steady-state values.

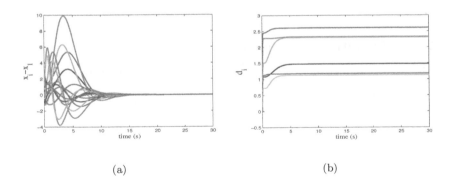

(a)   (b)

FIGURE 5.3: (a) The consensus errors $x_i - x_1$ of third-order integrators under (5.26); (b) the coupling weights d_i in (5.26).

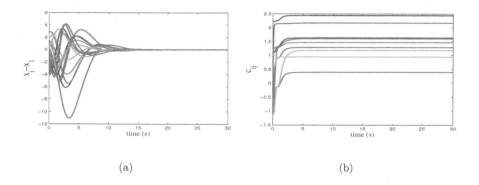

(a)   (b)

FIGURE 5.4: (a) The consensus errors $x_i - x_1$ of third-order integrators under (5.25); (b) the coupling weights c_{ij} in (5.25).

5.3 Extensions to Leader-Follower Graphs

For undirected communication graphs in the previous sections, the final consensus value reached by the agents under the adaptive protocols (5.2) and (5.3), which depends on the initial values and the agent dynamics, might be unknown a priori. In some cases, it might be desirable for the agents' states to converge onto a reference trajectory. This is actually the leader-follower consensus problem.

In this section, we intend to consider the case where the N agents in (5.1) maintain a leader-follower communication graph $\mathcal{G}$. Without loss of generality, assume that the agent indexed by 1 is the leader and the agents indexed by $2, \cdots, N$, are followers. The leader does not receive any information from the followers, i.e., it has no neighbor. For simplicity, the control input of the leader is assumed to be zero, i.e., $u_1 = 0$. The case with nonzero u_1 will be investigated in the next chapter.

In the sequel, the following assumption is needed.

Assumption 5.1 *The subgraph associated with the followers is undirected and the graph $\mathcal{G}$ contains a directed spanning tree with the leader as the root.*

Denote by $\mathcal{L}$ the Laplacian matrix associated with $\mathcal{G}$. Because the leader has no neighbors, $\mathcal{L}$ can be partitioned as $\mathcal{L} = \begin{bmatrix} 0 & 0_{1 \times (N-1)} \\ \mathcal{L}_2 & \mathcal{L}_1 \end{bmatrix}$, where $\mathcal{L}_2 \in \mathbf{R}^{(N-1) \times 1}$ and $\mathcal{L}_1 \in \mathbf{R}^{(N-1) \times (N-1)}$ is symmetric.

It is said that the leader-follower consensus problem is solved if the states of the followers converge to the state of the leader, i.e., $\lim_{t \to \infty} \|x_i(t) - x_1(t)\| = 0, \forall i = 2, \cdots, N$.

Similarly as in the previous sections, both the edge-based and node-based adaptive consensus protocols can be brought out to solve the leader-follower consensus problem. For conciseness, only the case with relative state information is discussed in this section. The case with relative output information can be similarly done, which will be also briefly mentioned in the latter part of the next chapter.

The edge-based adaptive consensus protocol is proposed for the followers as

$$u_i = K \sum_{j=1}^{N} c_{ij} a_{ij} (x_i - x_j),$$

$$\dot{c}_{ij} = \kappa_{ij} a_{ij} (x_i - x_j)^T \Gamma (x_i - x_j), \quad i = 2, \cdots, N,$$

(5.42)

where $K \in \mathbf{R}^{p \times n}$ and $\Gamma \in \mathbf{R}^{n \times n}$ are the feedback gain matrices, c_{ij} is the coupling weight associated with the edge (i, j) with $c_{ij}(0) = c_{ji}(0)$ for $i = 2, \cdots, N$, $j = 1, \cdots, N$, $\kappa_{ij} = \kappa_{ji} > 0$, and the rest of the variables are defined in (5.3).

Theorem 35 *For any graph $\mathcal{G}$ satisfying Assumption 5.1, the N agents in (5.1) reach leader-follower consensus under the edge-based protocol (5.42) with $K = -B^T P^{-1}$ and $\Gamma = P^{-1} B B^T P^{-1}$, where $P > 0$ is a solution to (5.6) and the coupling weights c_{ij} converge to finite steady-state values.*

Proof 35 *Let the consensus error $v_i = x_i - x_1$, $i = 2, \cdots, N$. Then, we can get from (5.1) and (5.42) that*

$$\dot{v}_i = A v_i + \sum_{j=2}^{N} c_{ij} a_{ij} K (v_i - v_j) + c_{i1} a_{i1} K v_i,$$

$$\dot{c}_{i1} = \epsilon_{i1} a_{i1} v_i^T \Gamma v_i,$$

$$\dot{c}_{ij} = \epsilon_{ij} a_{ij} (v_i - v_j)^T \Gamma (v_i - v_j), \ i = 2, \cdots, N.$$

(5.43)

Clearly, the leader-follower consensus problem of (5.1) is solved by (5.42) if the states v_i of (5.43) converge to zero.

Consider the Lyapunov function candidate

$$V_6 = \frac{1}{2} \sum_{i=2}^{N} v_i^T P^{-1} v_i + \sum_{i=2}^{N} \sum_{j=2, j \neq i}^{N} \frac{(c_{ij} - \beta)^2}{4 \kappa_{ij}} + \sum_{i=2}^{N} \frac{(c_{i1} - \beta)^2}{2 \kappa_{i1}},$$

where β is a positive constant. Following similar steps as in the proof of The-

orem 29, the time derivative of V_6 along (5.43) can be obtained as

$$\dot{V}_6 = \sum_{i=2}^{N} v_i^T P^{-1} \dot{v}_i + \sum_{i=2}^{N} \sum_{j=2, j \neq i}^{N} \frac{c_{ij} - \beta}{2\kappa_{ij}} \dot{c}_{ij} + \sum_{i=2}^{N} \frac{(c_{i1} - \beta)}{\kappa_{i1}} \dot{c}_{i1}$$

$$= \sum_{i=2}^{N} v_i^T P^{-1} [A v_i + \sum_{j=2}^{N} c_{ij} a_{ij} K (v_i - v_j) + c_{i1} a_{i1} K v_i]$$

$$+ \frac{1}{2} \sum_{i=2}^{N} \sum_{j=2, j \neq i}^{N} (c_{ij} - \beta) a_{ij} (v_i - v_j)^T \Gamma (v_i - v_j)$$

$$+ \sum_{i=2}^{N} (c_{i1} - \beta) a_{i1} v_i^T \Gamma v_i.$$

(5.44)

Similarly to (5.9), the following holds:

$$\sum_{i=2}^{N} \sum_{j=2, j \neq i}^{N} (c_{ij} - \beta) a_{ij} (v_i - v_j)^T \Gamma (v_i - v_j)$$

$$= 2 \sum_{i=2}^{N} \sum_{j=2}^{N} (c_{ij} - \beta) a_{ij} v_i^T \Gamma (v_i - v_j).$$

Then, we can get from (5.44) that

$$\dot{V}_6 = \frac{1}{2} v^T [I_{N-1} \otimes (P^{-1} A + A^T P^{-1}) - 2\beta \mathcal{L}_1 \otimes \Gamma] v,$$

where $v = [v_2^T, \cdots, v_N^T]^T$. For any graph $\mathcal{G}$ satisfying Assumption 5.2, it follows from Lemma 2 that $\mathcal{L}_1 > 0$. Thus $v^T (\mathcal{L}_1 \otimes \Gamma) v \geq \lambda_2(\mathcal{L}) v^T (I_{N-1} \otimes \Gamma) v$, with $\lambda_2(\mathcal{L})$ being the smallest eigenvalue of $\mathcal{L}_1$. Then, we have

$$\dot{V}_6 \leq \frac{1}{2} v^T [I_{N-1} \otimes (P^{-1} A + A^T P^{-1} - 2\beta \lambda_2(\mathcal{L}) \Gamma] v.$$

The rest of the proof is similar to that of Theorem 29, which is omitted for brevity.

Besides, the node-based adaptive consensus protocol is proposed for each follower as

$$u_i = d_i K \sum_{j=1}^{N} a_{ij} (x_i - x_j),$$

$$\dot{d}_i = \epsilon_i [\sum_{j=1}^{N} a_{ij} (x_i - x_j)^T] \Gamma [\sum_{j=1}^{N} a_{ij} (x_i - x_j)], \quad i = 2, \cdots, N,$$

(5.45)

where $\bar{d}_i$ denotes the coupling weight associated with follower i and $\epsilon_i > 0$.

Theorem 36 *For any graph $\mathcal{G}$ satisfying Assumption 5.1, the N agents in (5.1) reach leader-follower consensus under the node-based protocol (5.45) with K and Γ given as in Theorem 35. Besides, each coupling weight d_i converges to some finite steady-state value.*

Proof 36 *Clearly, the leader-follower consensus problem of (5.1) under (5.45) is solved if the consensus error v_i, $i = 2, \cdots, N$, defined as in the proof of Theorem 35, converge to zero. It is easy to see that v_i and $\bar{d}_i$ satisfy*

$$\dot{v}_i = Av_i + d_i K \sum_{j=2}^{N} \mathcal{L}_{ij} v_j,$$

$$\dot{d}_i = \epsilon_i [\sum_{j=2}^{N} \mathcal{L}_{ij} v_j]^T \Gamma [\sum_{j=2}^{N} \mathcal{L}_{ij} v_j], \ i = 2, \cdots, N,$$

where $\mathcal{L}_{ij}$ denotes the (i,j)-th entry of the Laplacian matrix $\mathcal{L}$.
Consider the Lyapunov function candidate

$$V_7 = \frac{1}{2} \sum_{i=2}^{N} v_i^T P^{-1} v_i + \sum_{i=2}^{N} \frac{(d_i - \sigma)^2}{2\epsilon_i},$$

where $\sigma > 0$ is to be determined. The rest of the proof can be completed by following similar steps in proving Theorem 36.

5.4   Robust Redesign of Distributed Adaptive Protocols

In this section, we investigate the robustness of the proposed adaptive consensus protocols (5.2) and (5.3) with respect to external disturbances. The dynamics of the agents subject to external disturbances are described by

$$\dot{x}_i = Ax_i + Bu_i + \omega_i, \quad i = 1, \cdots, N, \tag{5.46}$$

where $x_i \in \mathbf{R}^n$ is the state, $u_i \in \mathbf{R}^p$ is the control input, and $\omega_i \in \mathbf{R}^n$ denotes external disturbances associated with the i-th agent, which satisfies the following assumption.

Assumption 5.2 *There exist positive constants θ_i such that $\|\omega_i\| \le \theta_i$, $i = 1, \cdots, N$.*

The communication graph among the N agents is represented by an undirected graph $\mathcal{G}$, which is assumed to be connected throughout this section.

It is worth noting that due to the existence of nonzero ω_i, the relative states under the adaptive protocols (5.2) and (5.3) generally will not converge to zero any more but rather can only be expected to converge into some small neighborhood of the origin. Since the coupling weights c_{ij} and d_i are integrals of the nonnegative quadratic functions of the relative states, it is easy to see that in this case c_{ij} and d_i will grow unbounded, which is called the parameter drift phenomenon in the classic adaptive control literature [63]. Therefore, the adaptive protocols (5.2) and (5.3) are fragile in the presence of the bounded disturbances ω_i.

The main objective of this section is to make modifications on (5.2) and (5.3) in order to present some distributed robust adaptive protocols which can guarantee the ultimate boundedness of the consensus error and the coupling weights for the agents in (5.46) in the presence of bounded disturbances ω_i.

Motivated by the σ-modification technique in the robust adaptive control literature [63], we propose a modified edge-based adaptive protocol as follows:

$$
\begin{aligned}
u_i &= \sum_{j=1}^{N} c_{ij} a_{ij} K(x_i - x_j), \\
\dot{c}_{ij} &= \kappa_{ij} a_{ij} [-\phi_{ij} c_{ij} + (x_i - x_j)^T \Gamma(x_i - x_j)], \quad i = 1, \cdots, N,
\end{aligned}
\tag{5.47}
$$

where ϕ_{ij}, $i, j = 1, \cdots, N$, are small positive constants and the rest of the variables are defined as in (5.2).

Similarly, a modified node-based adaptive protocol can be described by

$$
\begin{aligned}
u_i &= d_i \sum_{j=1}^{N} a_{ij} K(x_i - x_j), \\
\dot{d}_i &= \tau_i [-\varphi_i d_i + (\sum_{j=1}^{N} a_{ij}(x_i - x_j)^T) \Gamma(\sum_{j=1}^{N} a_{ij}(x_i - x_j))], \quad i = 1, \cdots, N,
\end{aligned}
\tag{5.48}
$$

where φ_i, $i = 1, \cdots, N$, are small positive constants and the rest of the variables are defined as in (5.3).

5.4.1 Robust Edge-Based Adaptive Protocols

In this subsection, we study the consensus problem of the agents in (5.46) under the modified edge-based adaptive protocol (5.47). Let the consensus error ξ be defined as in (5.4). From (5.46) and (5.47), it is not difficult to obtain that ξ_i and c_{ij} satisfy

$$
\begin{aligned}
\dot{\xi}_i &= A\xi_i + \sum_{j=1}^{N} c_{ij} a_{ij} BK(\xi_i - \xi_j) + \omega_i - \frac{1}{N} \sum_{j=1}^{N} \omega_j, \\
\dot{c}_{ij} &= \kappa_{ij} a_{ij} [-\phi_{ij} c_{ij} + (\xi_i - \xi_j)^T \Gamma(\xi_i - \xi_j)], \quad i = 1, \cdots, N.
\end{aligned}
\tag{5.49}
$$

The following theorem presents a sufficient condition to achieve consensus for (5.49).

Theorem 37 *Suppose that $\mathcal{G}$ is undirected and connected and Assumption 5.2 holds. The feedback gain matrices of the modified edge-based adaptive protocol (5.47) are designed as $K = -B^T Q^{-1}$ and $\Gamma = Q^{-1} B B^T Q^{-1}$, where $Q > 0$ is a solution to the following LMI:*

$$AQ + QA^T + \varepsilon Q - 2BB^T < 0, \tag{5.50}$$

where $\varepsilon > 1$. Then, both the consensus error ξ and the adaptive gains c_{ij} $i, j = 1, \cdots, N$, in (5.49) are uniformly ultimately bounded and the following statements hold.

i) *For any ϕ_{ij}, the parameters ξ and c_{ij} exponentially converge to the residual set*

$$\mathcal{D}_1 \triangleq \{\xi, c_{ij} : V_8 < \frac{1}{2\delta\lambda_{\max}(Q)} \sum_{i=1}^N \theta_i^2 + \frac{\alpha^2}{4\delta} \sum_{i=1}^N \sum_{j=1}^N \phi_{ij} a_{ij}\}, \tag{5.51}$$

where $\delta \triangleq \min_{i,j=1,\cdots,N}\{\varepsilon - 1, \kappa_{ij}\phi_{ij}\}$ and

$$V_8 = \frac{1}{2} \sum_{i=1}^N \xi_i^T Q^{-1} \xi_i + \sum_{i=1}^N \sum_{j=1,j\neq i}^N \frac{\tilde{c}_{ij}^2}{4\kappa_{ij}}, \tag{5.52}$$

where $\tilde{c}_{ij} = c_{ij} - \alpha$, α is a positive scalar satisfying $\alpha \geq \frac{1}{\lambda_2}$, and λ_2 is the smallest nonzero eigenvalue of the Laplacian matrix $\mathcal{L}$ associated with $\mathcal{G}$.

ii) *If ϕ_{ij} is chosen such that $\varrho \triangleq \min_{i,j=1,\cdots,N}\{\kappa_{ij}\phi_{ij}\} < \varepsilon - 1$, then in addition to i), ξ exponentially converges to the residual set*

$$\mathcal{D}_2 \triangleq \{\xi : \|\xi\|^2 \leq \frac{\lambda_{\max}(Q)}{\varepsilon - 1 - \varrho}[\frac{1}{\lambda_{\min}(Q)} \sum_{i=1}^N \theta_i^2 + \frac{\alpha^2}{2} \sum_{i=1}^N \sum_{j=1}^N \phi_{ij} a_{ij}]\}. \tag{5.53}$$

Proof 37 *Choose the Lyapunov function candidate as in (5.52), which clearly is positive definite. The time derivative of V_8 along the trajectory of (5.49) is given by*

$$\dot{V}_8 = \sum_{i=1}^N \xi_i^T Q^{-1} \dot{\xi}_i + \sum_{i=1}^N \sum_{j=1,j\neq i}^N \frac{\tilde{c}_{ij}}{2\kappa_{ij}} \dot{\tilde{c}}_{ij}$$

$$= \sum_{i=1}^N \xi_i^T Q^{-1}[A\xi_i + \sum_{j=1}^N (\tilde{c}_{ij} + \alpha) a_{ij} BF(\xi_i - \xi_j) + \omega_i - \frac{1}{N} \sum_{j=1}^N \omega_j] \tag{5.54}$$

$$+ \frac{1}{2} \sum_{i=1}^N \sum_{j=1,j\neq i}^N \tilde{c}_{ij} a_{ij}[-\phi_{ij}(\tilde{c}_{ij} + \alpha) + (\xi_i - \xi_j)^T \Gamma(\xi_i - \xi_j)].$$

In light of (5.9), substituting $K = -B^T Q^{-1}$ and $\Gamma = Q^{-1} B B^T Q^{-1}$ into (5.54) yields

$$\dot{V}_8 \leq \frac{1}{2} \xi^T \mathcal{X} \xi + \xi^T (M \otimes Q^{-1}) \omega + \frac{1}{4} \sum_{i=1}^{N} \sum_{j=1}^{N} \phi_{ij} a_{ij} (-\tilde{c}_{ij}^2 + \alpha^2), \tag{5.55}$$

where $\omega = [\omega_1^T, \cdots, \omega_N^T]^T$, $M = I_N - \frac{1}{N} \mathbf{1} \mathbf{1}^T$, $\mathcal{X} \triangleq I_N \otimes (Q^{-1} A + A^T Q^{-1}) - 2\alpha \mathcal{L} \otimes Q^{-1} B B^T Q^{-1}$, and we have used the fact that $-\tilde{c}_{ij}^2 - \tilde{c}_{ij} \alpha \leq -\frac{1}{2} \tilde{c}_{ij}^2 + \frac{1}{2} \alpha^2$. Using the following fact:

$$-\frac{1}{2} \xi^T (M \otimes Q^{-1}) \xi + \xi^T (M \otimes Q^{-1}) \omega - \frac{1}{2} \omega^T (M \otimes Q^{-1}) \omega$$
$$= -\frac{1}{2} (\xi - \omega)^T (M \otimes Q^{-1}) (\xi - \omega) \leq 0, \tag{5.56}$$

we can get from (5.55) that

$$\dot{V}_8 \leq \frac{1}{2} \xi^T (\mathcal{X} + M \otimes Q^{-1}) \xi + \frac{1}{2 \lambda_{\min}(Q)} \sum_{i=1}^{N} \theta_i^2$$
$$+ \frac{1}{4} \sum_{i=1}^{N} \sum_{j=1}^{N} \phi_{ij} a_{ij} (-\tilde{c}_{ij}^2 + \alpha^2). \tag{5.57}$$

where we have used the assertion that $\omega^T (M \otimes Q^{-1}) \omega \leq \frac{1}{\lambda_{\min}(Q)} \sum_{i=1}^{N} \theta_i^2$. Note that (5.57) can be rewritten into

$$\dot{V}_8 \leq -\delta V_8 + \delta V_8 + \frac{1}{2} \xi^T (\mathcal{X} + M \otimes Q^{-1}) \xi$$
$$+ \frac{1}{2 \lambda_{\min}(Q)} \sum_{i=1}^{N} \theta_i^2 + \frac{1}{4} \sum_{i=1}^{N} \sum_{j=1}^{N} \phi_{ij} a_{ij} (-\tilde{c}_{ij}^2 + \alpha^2)$$
$$= -\delta V_8 + \frac{1}{2} \xi^T [\mathcal{X} + (\delta I_N + M) \otimes Q^{-1}] \xi + \frac{1}{2 \lambda_{\min}(Q)} \sum_{i=1}^{N} \theta_i^2$$
$$- \frac{1}{4} \sum_{i=1}^{N} \sum_{j=1}^{N} (\phi_{ij} - \frac{\delta}{\kappa_{ij}}) a_{ij} \tilde{c}_{ij}^2 + \frac{\alpha^2}{4} \sum_{i=1}^{N} \sum_{j=1}^{N} \phi_{ij} a_{ij}. \tag{5.58}$$

By noting that $\delta \leq \min_{i,j=1,\cdots,N} \{\kappa_{ij} \phi_{ij}\}$, it follows from (5.58) that

$$\dot{V}_8 \leq -\delta V_8 + \frac{1}{2} \xi^T [\mathcal{X} + (\delta I_N + M) \otimes Q^{-1}] \xi$$
$$+ \frac{1}{2 \lambda_{\min}(Q)} \sum_{i=1}^{N} \theta_i^2 + \frac{\alpha^2}{4} \sum_{i=1}^{N} \sum_{j=1}^{N} \phi_{ij} a_{ij}. \tag{5.59}$$

Because $\mathcal{G}$ is connected, it follows from Lemma 2 that zero is a simple *eigenvalue of $\mathcal{L}$ and all the other eigenvalues are positive. By the definitions of M and $\mathcal{L}$, we have $M\mathcal{L} = \mathcal{L} = \mathcal{L}M$. Thus there exists a unitary matrix $U \in \mathbf{R}^{N \times N}$ such that $U^T M U$ and $U^T \mathcal{L} U$ are both diagonal [58]. Since $\mathcal{L}$ and M have the same right and left eigenvectors corresponding to the zero eigenvalue, namely, $\mathbf{1}$, we can choose $U = [\frac{1}{\sqrt{N}} \ Y]$, $U^T = \begin{bmatrix} \frac{\mathbf{1}^T}{\sqrt{N}} \\ W \end{bmatrix}$, with $Y \in \mathbf{R}^{N \times (N-1)}$, $W \in \mathbf{R}^{(N-1) \times N}$, satisfying*

$$U^T M U = \Pi = \begin{bmatrix} 0 & 0 \\ 0 & I_{N-1} \end{bmatrix},$$

$$U^T \mathcal{L} U = \Lambda = \mathrm{diag}(0, \lambda_2, \cdots, \lambda_N),$$

(5.60)

where λ_i, $i = 2, \cdots, N$, are the nonzero eigenvalues of $\mathcal{L}$. Let $\bar{\xi} \triangleq [\bar{\xi}_1^T, \cdots, \bar{\xi}_N^T]^T = (U^T \otimes Q^{-1})\xi$. By the definitions of ξ and $\bar{\xi}$, it is easy to see that $\bar{\xi}_1 = (\frac{\mathbf{1}^T}{\sqrt{N}} \otimes Q^{-1})\xi = 0$. Then, by using (5.60), it follows that

$$\xi^T[\mathcal{X} + (\delta I_N + M) \otimes Q^{-1}]\xi = \sum_{i=2}^{N} \bar{\xi}_i^T[AQ + QA^T + (\delta + 1)Q - 2\alpha\lambda_i BB^T]\bar{\xi}_i$$

$$\leq \sum_{i=2}^{N} \bar{\xi}_i^T[AQ + QA^T + \varepsilon Q - 2BB^T]\bar{\xi}_i \leq 0,$$

(5.61)

where we have used the fact that $\alpha\lambda_i \geq 1$, $i = 2, \cdots, N$, and $\delta \leq \varepsilon - 1$. Then, it follows from (5.59) and (5.61) that

$$\dot{V}_8 \leq -\delta V_8 + \frac{1}{2\lambda_{\min}(Q)} \sum_{i=1}^{N} \theta_i^2 + \frac{\alpha^2}{4} \sum_{i=1}^{N} \sum_{j=1}^{N} \phi_{ij} a_{ij}.$$

(5.62)

By using Lemma 16 (the Comparison lemma), we can obtain from (5.62) that

$$V_8 \leq [V_8(0) - \frac{1}{2\delta\lambda_{\min}(Q)} \sum_{i=1}^{N} \theta_i^2 - \frac{\alpha^2}{4\delta} \sum_{i=1}^{N} \sum_{j=1}^{N} \phi_{ij} a_{ij}]e^{-\delta t}$$

$$+ \frac{1}{2\delta\lambda_{\min}(Q)} \sum_{i=1}^{N} \theta_i^2 + \frac{\alpha^2}{4\delta} \sum_{i=1}^{N} \sum_{j=1}^{N} \phi_{ij} a_{ij}.$$

(5.63)

Therefore, V_8 exponentially converges to the residual set $\mathcal{D}_1$ in (5.51) with a convergence rate faster than $e^{-\delta t}$, implying that ξ and c_{ij} are uniformly ultimately bounded.

For the case where $\varrho \triangleq \min_{i,j=1,\cdots,N}\{\kappa_{ij}\phi_{ij}\} < \varepsilon - 1$, we can obtain a *smaller residual set for ξ by rewriting (5.57) into*

$$\dot{V}_8 \leq -\varrho V_8 + \frac{1}{2}\xi^T[\mathcal{X} + ((\varepsilon - 1)I_N + M) \otimes Q^{-1}]\xi + \frac{\alpha^2}{4}\sum_{i=1}^{N}\sum_{j=1}^{N}\phi_{ij}a_{ij}$$

$$- \frac{\varepsilon - 1 - \varrho}{2}\xi^T(I_N \otimes Q^{-1})\xi + \frac{1}{2\lambda_{\min}(Q)}\sum_{i=1}^{N}\theta_i^2$$

$$- \frac{1}{4}\sum_{i=1}^{N}\sum_{j=1}^{N}(\phi_{ij} - \frac{\varrho}{\kappa_{ij}})a_{ij}\tilde{c}_{ij}^2$$

$$\leq -\varrho V_8 - \frac{\varepsilon - 1 - \varrho}{2\lambda_{\max}(Q)}\|\xi\|^2 + \frac{1}{2\lambda_{\min}(Q)}\sum_{i=1}^{N}\theta_i^2 + \frac{\alpha^2}{4}\sum_{i=1}^{N}\sum_{j=1}^{N}\phi_{ij}a_{ij}.$$

(5.64)

Obviously, if $\|\xi\|^2 > \frac{\lambda_{\max}(Q)}{\varepsilon - 1 - \varrho}[\frac{1}{\lambda_{\min}(Q)}\sum_{i=1}^{N}\theta_i^2 + \frac{\alpha^2}{2}\sum_{i=1}^{N}\sum_{j=1}^{N}\phi_{ij}a_{ij}]$, it follows from (5.64) that $\dot{V}_8 < 0$, which implies that ξ exponentially converges to the residual set $\mathcal{D}_2$ in (5.53) with a convergence rate faster than $e^{-\varrho t}$.

Remark 37 *As shown in Proposition 2 in Chapter 2, there exists a $Q > 0$ satisfying (5.28) if and only if (A, B) is controllable. Thus, a sufficient condition for the existence of (5.47) satisfying Theorem 37 is that (A, B) is controllable, which, compared to the existence condition of (5.2) satisfying Theorem 29, is more stringent. Theorem 37 shows that the modified edge-based adaptive protocol (5.47) can ensure the boundedness of the consensus error ξ and the adaptive gains c_{ij} for the agents described by (5.46). That is, (5.47) is indeed robust in the presence of bounded external disturbances.*

Remark 38 *Note that the residual sets $\mathcal{D}_1$ and $\mathcal{D}_2$ in Theorem 37 depends on the communication graph $\mathcal{G}$, the upper bound of the external disturbances ω_i, the parameters Q, ϕ_{ij}, and κ_{ij} of the adaptive protocol (5.47). For small ϕ_{ij} satisfying $\min \kappa_{ij}\phi_{ij} \leq \varepsilon - 1$, we can observed from (5.53) that smaller ϕ_{ij} implies a smaller bound for the consensus error ξ and from (5.51) that smaller ϕ_{ij} meanwhile yields a larger bound for coupling gains c_{ij}. For the case where $\phi_{ij} = 0$, c_{ij} will tend to infinity. In real implementations, we should choose relatively small ϕ_{ij} in order to guarantee a small ξ and have tolerable c_{ij}.*

5.4.2 Robust Node-Based Adaptive Protocols

In this subsection, we study the consensus problem of the agents in (5.46) under the modified node-based adaptive protocol (5.48). Let the consensus error ξ be defined as in (5.4). From (5.46) and (5.47), we can obtain that ξ

and d_i satisfy

$$\dot{\xi} = (I_N \otimes A + MD\mathcal{L} \otimes BK)\xi + (M \otimes I_n)\omega,$$

$$\dot{d}_i = \tau_i[-\varphi_i d_i + (\sum_{j=1}^{N} \mathcal{L}_{ij}\xi_j^T)\Gamma(\sum_{j=1}^{N} \mathcal{L}_{ij}\xi_j)], \quad i = 1, \cdots, N, \tag{5.65}$$

where $D = \text{diag}(d_1, \cdots, d_N)$ and $M = I_N - \frac{1}{N}\mathbf{1}\mathbf{1}^T$.

Theorem 38 *Suppose that $\mathcal{G}$ is connected and Assumption 5.2 holds. Both the consensus error ξ and the adaptive gains d_i $i = 1, \cdots, N$, in (5.65) are uniformly ultimately bounded under the modified node-based adaptive protocol (5.48) with K and Γ given as in Theorem 37. Moreover, the following two statements hold.*

i) *For any φ_i, the parameters ξ and d_i exponentially converge to the residual set*

$$\mathcal{D}_3 \triangleq \{\xi, d_i : V_9 < \frac{\lambda_{\max}(\mathcal{L})}{2\delta\lambda_{\min}(Q)} \sum_{i=1}^{N} \theta_i^2 + \frac{\alpha^2}{2\delta} \sum_{i=1}^{N} \varphi_i\}, \tag{5.66}$$

where $\delta \triangleq \min_{i=1,\cdots,N}\{\varepsilon - 1, \tau_i\varphi_i\}$ and

$$V_9 = \frac{1}{2}\xi^T(\mathcal{L} \otimes Q^{-1})\xi + \sum_{i=1}^{N} \frac{\tilde{d}_i^2}{2\tau_i}, \tag{5.67}$$

where $\tilde{d}_i = d_i - \alpha$, and Q and α are the same as in Theorem 37.

ii) *If small φ_i satisfies $\vartheta \triangleq \min_{i=1,\cdots,N}\{\tau_i\varphi_i\} < \varepsilon - 1$, then in addition to i), ξ exponentially converges to the residual set*

$$\mathcal{D}_4 \triangleq \{\xi : \|\xi\|^2 \le \frac{\lambda_{\max}(Q)}{(\varepsilon - 1 - \vartheta)\lambda_2}[\frac{\lambda_{\max}(\mathcal{L})}{\lambda_{\min}(Q)} \sum_{i=1}^{N} \theta_i^2 + \alpha^2 \sum_{i=1}^{N} \varphi_i]\}. \tag{5.68}$$

Proof 38 *Choose the Lyapunov function candidate as in (5.67). The time derivative of V_9 along the trajectory of (5.65) can be obtained as*

$$\dot{V}_9 = \xi^T(\mathcal{L} \otimes Q^{-1})\dot{\xi} + \sum_{i=1}^{N} \frac{\tilde{d}_i}{\tau_i}\dot{d}_i$$

$$= \xi^T(\mathcal{L} \otimes Q^{-1}A + \mathcal{L}D\mathcal{L} \otimes Q^{-1}BK)\xi + \xi^T(\mathcal{L} \otimes Q^{-1})\omega \tag{5.69}$$

$$+ \sum_{i=1}^{N} \tilde{d}_i[-\varphi_i(\tilde{d}_i + \alpha) + (\sum_{j=1}^{N} \mathcal{L}_{ij}\xi_j^T)\Gamma(\sum_{j=1}^{N} \mathcal{L}_{ij}\xi_j)].$$

By using (5.18) and the assertion that $-\tilde{d}_i^2 - \tilde{d}_i\alpha \le -\frac{1}{2}\tilde{d}_i^2 + \frac{1}{2}\alpha^2$, it follows from (5.69) that

$$\dot{V}_9 \le \frac{1}{2}\xi^T \mathcal{Y}\xi + \xi^T(\mathcal{L} \otimes Q^{-1})\omega - \frac{1}{2}\sum_{i=1}^{N} \varphi_i(\tilde{d}_i^2 - \alpha^2), \tag{5.70}$$

where $\mathcal{Y} \triangleq \mathcal{L} \otimes (Q^{-1}A + A^T Q^{-1}) - 2\alpha \mathcal{L}^2 \otimes Q^{-1} BB^T Q^{-1}$. *By using the technique of completing the square as in (5.56), we can obtain from (5.70) that*

$$\dot{V}_9 \leq \frac{1}{2}\xi^T (\mathcal{Y} + \mathcal{L} \otimes Q^{-1})\xi + \frac{1}{2}\omega^T (\mathcal{L} \otimes Q^{-1})\omega - \frac{1}{2}\sum_{i=1}^{N} \varphi_i (\tilde{d}_i^2 - \alpha^2)$$

$$\leq \frac{1}{2}\xi^T (\mathcal{Y} + \mathcal{L} \otimes Q^{-1})\xi + \frac{\lambda_{\max}(\mathcal{L})}{2\lambda_{\min}(Q)} \sum_{i=1}^{N} \theta_i^2 - \frac{1}{2}\sum_{i=1}^{N} \varphi_i (\tilde{d}_i^2 - \alpha^2),$$

(5.71)

Note that (5.71) can be rewritten into

$$\dot{V}_9 \leq -\delta V_9 + \frac{1}{2}\xi^T [\mathcal{Y} + (\delta + 1)\mathcal{L} \otimes Q^{-1}]\xi$$

$$+ \frac{\lambda_{\max}(\mathcal{L})}{2\lambda_{\min}(Q)} \sum_{i=1}^{N} \theta_i^2 - \frac{1}{2}\sum_{i=1}^{N}(\varphi_i - \frac{\delta}{\tau_i})\tilde{d}_i^2 + \frac{\alpha^2}{2}\sum_{i=1}^{N}\varphi_i$$

$$\leq -\delta V_9 + \frac{1}{2}\xi^T [\mathcal{Y} + (\delta + 1)\mathcal{L} \otimes Q^{-1}]\xi$$

$$+ \frac{\lambda_{\max}(\mathcal{L})}{2\lambda_{\min}(Q)} \sum_{i=1}^{N} \theta_i^2 + \frac{\alpha^2}{2}\sum_{i=1}^{N}\varphi_i,$$

(5.72)

where we have used $\tau_i \varphi_i \leq \delta$, $i = 1, \cdots, N$, to get the last inequality. Letting $\bar{\xi}$ be defined as in the proof of Theorem 37, we have

$$\xi^T [\mathcal{Y} + (\delta + 1)\mathcal{L} \otimes Q^{-1}]\xi = \sum_{i=2}^{N} \lambda_i \bar{\xi}_i^T [AQ + QA^T + (\delta + 1)Q - 2\alpha\lambda_i BB^T]\bar{\xi}_i$$

$$\leq \sum_{i=2}^{N} \lambda_i \bar{\xi}_i^T [AQ + QA^T + \varepsilon Q - 2BB^T]\bar{\xi}_i \leq 0.$$

(5.73)

Then, it follows from (5.72) and (5.73) that

$$\dot{V}_9 \leq -\delta V_9 + \frac{\lambda_{\max}(\mathcal{L})}{2\lambda_{\min}(Q)} \sum_{i=1}^{N} \theta_i^2 + \frac{\alpha^2}{2}\sum_{i=1}^{N}\varphi_i.$$

(5.74)

Therefore, by using Lemma 16 (the Comparison lemma), we get from (5.74) that V_9 exponentially converges to the residual set $\mathcal{D}_3$ in (5.66) with a rate faster than $e^{-\delta t}$, which, in light of $V_9 \geq \frac{\lambda_2}{2\lambda_{\max}(Q)}\|\xi\|^2 + \sum_{i=1}^{N} \frac{\tilde{d}_i^2}{2\tau_i}$, further implies that ξ and d_i are uniformly ultimately bounded.

Furthermore, for the case where $\vartheta \triangleq \min_{i=1,\cdots,N}\{\tau_i \varphi_i\} < \varepsilon - 1$, we rewrite

(5.71) into

$$\dot{V}_9 \leq -\vartheta V_9 + \xi^T[\mathcal{Y} + \varepsilon\mathcal{L} \otimes Q^{-1}]\xi$$

$$- \frac{\varepsilon - 1 - \vartheta}{2}\xi^T(\mathcal{L} \otimes Q^{-1})\xi + \frac{\lambda_{\max}(\mathcal{L})}{2\lambda_{\min}(Q)}\sum_{i=1}^{N}\theta_i^2 + \frac{\alpha^2}{2}\sum_{i=1}^{N}\varphi_i \qquad (5.75)$$

$$\leq -\vartheta V_2 - \frac{(\varepsilon - 1 - \vartheta)\lambda_2}{2\lambda_{\max}(Q)}\|\xi\|^2 + \frac{\lambda_{\max}(\mathcal{L})}{2\lambda_{\min}(Q)}\sum_{i=1}^{N}\theta_i^2 + \frac{\alpha^2}{2}\sum_{i=1}^{N}\varphi_i.$$

Obviously, if $\|\xi\|^2 > \frac{\lambda_{\max}(Q)}{(\varepsilon-1-\vartheta)\lambda_2}[\frac{\lambda_{\max}(\mathcal{L})}{\lambda_{\min}(Q)}\sum_{i=1}^{N}\theta_i^2 + \alpha^2\sum_{i=1}^{N}\varphi_i]$, *it follows from (5.75) that* $\dot{V}_9 \leq -\vartheta V_9$. *Then, by noting* $V_9 \geq \frac{\lambda_2}{2\lambda_{\max}(Q)}\|\xi\|^2$, *we can get that* ξ *exponentially converges to the residual set* $\mathcal{D}_4$ *in (5.68) with a rate faster than* $e^{-\vartheta t}$.

Remark 39 *Similar to the modified edge-based adaptive protocol (5.47), the modified node-based adaptive protocol (5.48) can be also implemented in a fully distributed fashion and, as shown in Theorem 37, is also robust with respect to bounded external disturbances.*

Remark 40 *One weakness of the robust adaptive protocols (5.47) and (5.48) is that in the absence of disturbances we can no longer ensure the asymptotical convergence of the consensus error to the origin, which is actually an inherent drawback of the σ-modification technique [63]. Other robust adaptive control techniques, e.g., the classic dead-zone modification and the projection operator techniques [62], or the new low-frequency learning method [191], might also be able to be used to develop alternative robust adaptive consensus protocols.*

5.4.3 Simulation Examples

Example 18 *Consider a network of six agents whose the communication graph is given as in FIGURE 5.2(b). The dynamics of the agents are given by (5.46), with*

$$x_i = \begin{bmatrix} x_{i1} \\ x_{i2} \\ x_{i3} \end{bmatrix}, \quad A = \begin{bmatrix} 0 & 1 & 0 \\ 0 & 0 & 1 \\ 0 & 0 & 0 \end{bmatrix}, \quad B = \begin{bmatrix} 0 \\ 0 \\ 1 \end{bmatrix}.$$

For illustration, the disturbances associated with the agents are assumed to be

$$\omega_1 = \begin{bmatrix} 0 \\ 0 \\ 0.1\sin(t) \end{bmatrix}, \quad \omega_2 = \begin{bmatrix} 0 \\ -0.5\cos(5t) \\ 0.3\sin(t) \end{bmatrix}, \quad \omega_3 = \begin{bmatrix} -0.1\sin(3t) \\ 0 \\ 0.3\sin(3t) \end{bmatrix}, \quad \omega_4 = \begin{bmatrix} 0.25\cos(2t) \\ 0 \\ 0.2\cos(2t) \end{bmatrix},$$

$$\omega_5 = \begin{bmatrix} 0 \\ -0.2\sin(x_{51}) \\ 0 \end{bmatrix}, \quad \text{and } \omega_6 = \begin{bmatrix} e^{-2t} \\ 0 \\ 0 \end{bmatrix}.$$

First, we use the node-based adaptive protocol (5.3). Solving the LMI (5.6)

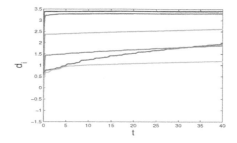

FIGURE 5.5: The coupling weights d_i in (5.2).

gives the gain matrices K and Γ *in (5.3) as*

$$K = - \begin{bmatrix} 0.6285 & 1.3525 & 2.1113 \end{bmatrix}, \quad \Gamma = \begin{bmatrix} 0.3950 & 0.8500 & 1.3269 \\ 0.8500 & 1.8292 & 2.8554 \\ 1.3269 & 2.8554 & 4.4574 \end{bmatrix}.$$

Let $\tau_i = 5$, $i = 1, \cdots, 6$, *in (5.3). The coupling weights* d_i *associated with the nodes using (5.3) designed as above are drawn in FIGURE 5.5, from which it can be observed that* d_i, $i = 1, \cdots, 6$, *do keep increasing to infinity. Thus, (5.3) is not robust in the presence of external disturbances. Next, we use the modified node-based adaptive protocol (5.3) to achieve consensus. Solving the LMI (5.50) with* $\varepsilon = 1.5$ *gives the gain matrices K and* Γ *in (5.48) as*

$$K = - \begin{bmatrix} 5.6227 & 9.0648 & 4.4676 \end{bmatrix}, \quad \Gamma = \begin{bmatrix} 31.6151 & 50.9690 & 25.1200 \\ 50.9690 & 82.1707 & 40.4978 \\ 25.1200 & 40.4978 & 19.9593 \end{bmatrix}.$$

To illustrate Theorem 38, select $\varphi_i = 0.005$ *and* $\tau_i = 5$, $i = 1, \cdots, 6$, *in (5.48). The relative states* $x_i - x_1$, $i = 2, \cdots, 6$, *of the agents and the coupling weights* d_i *under (5.48) designed as above are shown in FIGURE 5.6, both of which are clearly bounded.*

5.5 Distributed Adaptive Protocols for Graphs Containing Directed Spanning Trees

The distributed adaptive protocols proposed in the previous sections are applicable to only undirected communication graphs or leader-follower graphs where the subgraph among the followers are undirected. How to design fully distributed adaptive consensus protocols for the general case with directed

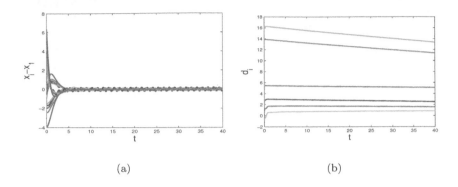

(a)   (b)

FIGURE 5.6: (a) $x_i - x_1$, $i = 2, \cdots, 6$, under (5.48); (b) the coupling weights d_i in (5.48).

graphs is quite challenging. The main difficulty lies in that the Laplacian matrices of directed graphs are generally asymmetric, which renders the construction of adaptive consensus protocol and the selection of appropriate Lyapunov function far from being easy.

5.5.1   Distributed Adaptive Consensus Protocols

In this section, we extend the node-based adaptive protocol (5.3) in Section 5.3 to the case with general directed leader-follower graphs. The dynamics of the agents are described by (5.1). As we did in Section 5.3, let the agent in (5.1) indexed by 1 be the leader whose control input is assumed to be zero and the agents indexed by $2, \cdots, N$, be the followers. The communication graph $\mathcal{G}$ is supposed to satisfy the following assumption.

Assumption 5.3 *The graph $\mathcal{G}$ contains a directed spanning tree with the leader as the root node.*

Denote by $\mathcal{L}$ the Laplacian matrix associated with $\mathcal{G}$. Because the node indexed by 1 is the leader which has no neighbors, $\mathcal{L}$ can be partitioned as

$$\mathcal{L} = \begin{bmatrix} 0 & 0_{1 \times (N-1)} \\ \mathcal{L}_2 & \mathcal{L}_1 \end{bmatrix}, \tag{5.76}$$

where $\mathcal{L}_2 \in \mathbf{R}^{(N-1) \times 1}$ and $\mathcal{L}_1 \in \mathbf{R}^{(N-1) \times (N-1)}$. Since $\mathcal{G}$ satisfies Assumption 5.3, it follows from Lemma 1 that all eigenvalues of $\mathcal{L}_1$ have positive real parts. It then can be verified that $\mathcal{L}_1$ is a nonsingular M-matrix (see Definition 6 in Chapter 1) and is diagonally dominant.

Based on the relative states of neighboring agents, we propose the following distributed adaptive consensus protocol with time-varying coupling weights:

$$u_i = d_i \rho_i(\nu_i^T P^{-1} \nu_i) K \nu_i,$$
$$\dot{d}_i = \nu_i^T \Gamma \nu_i, \quad i = 2, \cdots, N, \tag{5.77}$$

where $\nu_i \triangleq \sum_{j=1}^N a_{ij}(x_i - x_j)$, $d_i(t)$ denotes the time-varying coupling weight associated with the i-th follower with $d_i(0) \geq 1$, $P > 0$ is a solution to the LMI (5.6), $K \in \mathbf{R}^{p \times n}$ and $\Gamma \in \mathbf{R}^{n \times n}$ are the feedback gain matrices, $\rho_i(\cdot)$ are smooth and monotonically nondecreasing functions to be determined later which satisfy that $\rho_i(s) \geq 1$ for $s > 0$.

Let $\nu = [\nu_2^T, \cdots, \nu_N^T]^T$. Then,

$$\nu = (\mathcal{L}_1 \otimes I_n) \begin{bmatrix} x_2 - x_1 \\ \vdots \\ x_N - x_1 \end{bmatrix}, \tag{5.78}$$

where $\mathcal{L}_1$ is defined in (5.76). Because $\mathcal{L}_1$ is nonsingular for $\mathcal{G}$ satisfying Assumption 5.3, it is easy to see that the leader-follower consensus problem is solved if and only if ν converges to zero. Hereafter, we refer to ν as the consensus error. In light of (5.1) and (5.77), it is not difficult to obtain that ν and d_i satisfy the following dynamics:

$$\dot{\xi} = [I_{N-1} \otimes A + \mathcal{L}_1 \widehat{D} \hat{\rho}(\nu) \otimes BK] \nu,$$
$$\dot{d}_i = \nu_i^T \Gamma \nu_i, \tag{5.79}$$

where $\hat{\rho}(\nu) \triangleq \text{diag}(\rho_2(\nu_2^T P^{-1} \nu_2), \cdots, \rho_N(\nu_N^T P^{-1} \nu_N))$, $\widehat{D} \triangleq \text{diag}(d_2, \cdots, d_N)$.

Before moving on to present the main result of this subsection, we first introduce a property of the nonsingular M-matrix $\mathcal{L}_1$.

Lemma 29 *There exists a positive diagonal matrix G such that $G\mathcal{L}_1 + \mathcal{L}_1^T G > 0$. One such G is given by $\text{diag}(q_2, \cdots, q_N)$, where $q = [q_2, \cdots, q_N]^T = (\mathcal{L}_1^T)^{-1} \mathbf{1}$.*

Proof 39 *The first assertion is well known; see Theorem 4.25 in [127] or Theorem 2.3 in [10]. The second assertion is shown in the following. Note that the specific form of G given here is different from that in [30, 127, 196].*

Since $\mathcal{L}_1$ is a nonsingular M-matrix, it follows from Theorem 4.25 in [127] that $(\mathcal{L}_1^T)^{-1}$ exists, is nonnegative, and thereby cannot have a zero row. Then, it is easy to verify that $q > 0$ and hence $G\mathcal{L}_{11} \geq 0$ [1]. By noting that $\mathcal{L}_1^T G \mathbf{1} = \mathcal{L}_1^T q = \mathbf{1}$, we can conclude that $(G\mathcal{L}_1 + \mathcal{L}_1^T G)\mathbf{1} > 0$, implying that $G\mathcal{L}_1 + \mathcal{L}_1^T G$ is strictly diagonally dominant. Since the diagonal entries of $G\mathcal{L}_1 + \mathcal{L}_1^T G$ are positive, it then follows from Gershgorin's disc theorem [58] that every eigenvalue of $G\mathcal{L}_1 + \mathcal{L}_1^T G$ is positive, implying that $G\mathcal{L}_1 + \mathcal{L}_1^T G > 0$.

---

[1] For a vector x, $x > (\geq)0$ means that every entry of x is positive (nonnegative).

The following result provides a sufficient condition to design the adaptive consensus protocol (5.77).

Theorem 39 *Suppose that the communication graph $\mathcal{G}$ satisfies Assumption 5.3. Then, the leader-follower consensus problem of the agents in (5.1) can be solved by the adaptive protocol (5.77) with $K = -B^T P^{-1}$, $\Gamma = P^{-1} B B^T P^{-1}$, and $\rho_i(\nu_i^T P^{-1} \nu_i) = (1 + \nu_i^T P^{-1} \nu_i)^3$, where $P > 0$ is a solution to the LMI (5.6). Moreover, each coupling weight d_i converges to some finite steady-state value.*

Proof 40 *Consider the following Lyapunov function candidate:*

$$V_{10} = \sum_{i=2}^{N} \frac{d_i}{2q_i} \int_0^{\nu_i^T P^{-1} \nu_i} \rho_i(s) ds + \frac{\hat{\lambda}_0}{24} \sum_{i=2}^{N} \tilde{d}_i^2, \qquad (5.80)$$

where $\tilde{d}_i = d_i - \alpha$, α is a positive scalar to be determined later, $\hat{\lambda}_0$ denotes the smallest eigenvalue of $G\mathcal{L}_1 + \mathcal{L}_1^T G$, and G and q_i are given as in Lemma 29. Since $\mathcal{G}$ satisfies Assumption 5.3, it follows from Lemma 29 that $G > 0$ and $\hat{\lambda}_0 > 0$. Because $d_i(0) \geq 1$, it follows from the second equation in (5.79) that $d_i(t) \geq 1$ for $t > 0$. Furthermore, by noting that $\rho_i(\cdot)$ are smooth and monotonically increasing functions satisfying $\rho_i(s) \geq 1$ for $s > 0$, it is not difficult to see that V_{10} is positive definite with respect to ν_i and $\tilde{d}_i$, $i = 2, \cdots, N$.

The time derivative of V_{10} along the trajectory of (5.79) is given by

$$
\begin{aligned}
\dot{V}_{10} &= \sum_{i=2}^{N} \frac{d_i}{q_i} \rho_i(\nu_i^T P^{-1} \nu_i) \nu_i^T P^{-1} \dot{\nu}_i \\
&\quad + \sum_{i=2}^{N} \frac{\dot{d}_i}{2q_i} \int_0^{\nu_i^T P^{-1} \nu_i} \rho_i(s) ds \\
&\quad + \frac{\hat{\lambda}_0}{12} \sum_{i=2}^{N} (d_i - \alpha) \nu_i^T P^{-1} B B^T P^{-1} \nu_i.
\end{aligned}
\qquad (5.81)
$$

In the sequel, for conciseness we shall use $\hat{\rho}$ and ρ_i instead of $\hat{\rho}(\nu)$ and $\rho_i(\nu_i^T P^{-1} \nu_i)$, respectively, whenever without causing any confusion.

By using (5.79) and after some mathematical manipulations, we can get that

$$
\begin{aligned}
\sum_{i=2}^{N} \frac{d_i}{q_i} \rho_i \nu_i^T P^{-1} \dot{\nu}_i &= \nu^T (\hat{D} \hat{\rho} G \otimes P^{-1}) \dot{\nu} \\
&= \frac{1}{2} \nu^T [\hat{D} \hat{\rho} G \otimes (P^{-1} A + A^T P^{-1}) \\
&\quad - \hat{D} \hat{\rho} (G\mathcal{L}_1 + \mathcal{L}_1^T G) \hat{D} \hat{\rho} \otimes P^{-1} B B^T P^{-1}] \nu \\
&\leq \frac{1}{2} \nu^T [\hat{D} \hat{\rho} Q \otimes (P^{-1} A + A^T P^{-1}) \\
&\quad - \hat{\lambda}_0 \hat{D}^2 \hat{\rho}^2 \otimes P^{-1} B B^T P^{-1}] \nu,
\end{aligned}
\qquad (5.82)
$$

where we have used the fact that $Q\mathcal{L}_1 + \mathcal{L}_1^T Q > 0$ to get the first inequality.

Because ρ_i are monotonically increasing and satisfy $\rho(s) \geq 1$ for $s > 0$, it follows that

$$
\begin{aligned}
\sum_{i=2}^{N} \frac{\dot{d}_i}{q_i} \int_0^{\nu_i^T P^{-1} \nu_i} \rho_i(s)ds &\leq \sum_{i=2}^{N} \frac{\dot{d}_i}{q_i} \rho_i \nu_i^T P^{-1} \nu_i \\
&\leq \sum_{i=2}^{N} \frac{\dot{d}_i}{3 q_i^3 \hat{\lambda}_0^2} + \sum_{i=2}^{N} \frac{2}{3} \hat{\lambda}_0 \dot{d}_i \rho_i^{\frac{3}{2}} (\nu_i^T P^{-1} \nu_i)^{\frac{3}{2}} \\
&\leq \sum_{i=2}^{N} \frac{\dot{d}_i}{3 q_i^3 \hat{\lambda}_0^2} + \sum_{i=1}^{N} \frac{2}{3} \hat{\lambda}_0 \dot{d}_i \rho_i^{\frac{3}{2}} (1 + \nu_i^T P^{-1} \nu_i)^{\frac{3}{2}} \\
&= \sum_{i=2}^{N} (\frac{1}{3 q_i^3 \hat{\lambda}_0^2} + \frac{2}{3} \hat{\lambda}_0 \rho_i^2) \nu_i^T P^{-1} B B^T P^{-1} \nu_i,
\end{aligned}
\tag{5.83}
$$

where we have used the mean value theorem for integrals to get the first inequality and used Lemma 17 (Young's inequality) to get the second inequality.

Substituting (5.82) and (5.83) into (5.81) yields

$$
\begin{aligned}
\dot{V}_{10} \leq &\frac{1}{2} \nu^T [\widehat{D}\rho G \otimes (P^{-1}A + A^T P^{-1})] \nu \\
&- \sum_{i=2}^{N} [\hat{\lambda}_0 (\frac{1}{2} d_i^2 \rho_i^2 - \frac{1}{12} d_i - \frac{1}{3} \rho_i^2) \\
&+ \frac{1}{12} (\hat{\lambda}_0 \alpha - \frac{2}{q_i^3 \hat{\lambda}_0^2})] \nu_i^T P^{-1} B B^T P^{-1} \nu_i.
\end{aligned}
\tag{5.84}
$$

Choose $\alpha \geq \hat{\alpha} + \max_{i=2,\cdots,N} \frac{2}{q_i^3 \hat{\lambda}_0^3}$, where $\hat{\alpha} > 0$ will be determined later. Then, by noting that $\rho_i \geq 1$ and $d_i \geq 1$, $i = 2, \cdots, N$, it follows from (5.84) that

$$
\begin{aligned}
\dot{V}_{10} \leq &\frac{1}{2} \nu^T [\widehat{D}\rho G \otimes (P^{-1}A + A^T P^{-1})] \nu \\
&- \frac{\hat{\lambda}_0}{12} \sum_{i=2}^{N} (d_i^2 \rho_i^2 + \hat{\alpha}) \nu_i^T P^{-1} B B^T P^{-1} \nu_i \\
\leq &\frac{1}{2} \nu^T [\widehat{D}\rho G \otimes (P^{-1}A + A^T P^{-1}) \\
&- \frac{1}{3} \sqrt{\hat{\alpha}} \hat{\lambda}_0 \widehat{D}\hat{\rho} \otimes P^{-1} B B^T P^{-1}] \nu.
\end{aligned}
\tag{5.85}
$$

Let $\tilde{\nu} = (\sqrt{\widehat{D}\rho G} \otimes I)\nu$ and choose $\hat{\alpha}$ to be sufficiently large such that $\sqrt{\hat{\alpha}} \hat{\lambda}_0 G^{-1} \geq 6I$. Then, we can get from (5.85) that

$$
\dot{V}_{10} \leq \frac{1}{2} \tilde{\nu}^T [I_N \otimes (P^{-1}A + A^T P^{-1} - 2P^{-1} B B^T P^{-1})] \tilde{\nu} \leq 0,
\tag{5.86}
$$

where to get the last inequality we have used the assertion that $P^{-1}A + A^T P^{-1} - 2P^{-1}BB^T P^{-1} < 0$, which follows readily form (5.6).

Since $\dot{V}_{10} \leq 0$, $V_{10}(t)$ is bounded and so is each d_i. By noting $\dot{d}_i \geq 0$, it can be seen from (5.79) that d_i are monotonically increasing. Then, it follows that each coupling weight d_i converges to some finite value. Note that $\dot{V}_{10} \equiv 0$ implies that $\tilde{\nu} = 0$ and thereby $\nu = 0$. Hence, by Lemma 11 (LaSalle's Invariance principle), it follows that the consensus error ν asymptotically converges to zero. That is, the leader-follower consensus problem is solved.

Remark 41 *In comparison to the adaptive protocols in the previous sections, a distinct feature of (5.77) is that inspired by the changing supply functions in [157], monotonically increasing functions ρ_i are introduced into (5.77) to provide extra freedom for design. As the consensus error ν converges to zero, the functions ρ_i will converge to 1, in which case the adaptive protocol (5.77) will reduce to the adaptive protocol (5.45) for undirected graphs in Section 5.3. It should be mentioned that the Lyapunov function used in the proof of Theorem 39 is partly inspired by [175].*

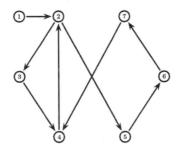

FIGURE 5.7: The directed communication graph.

Example 19 *Consider a network of double integrators, described by (5.1), with*

$$A = \begin{bmatrix} 0 & 1 \\ 0 & 0 \end{bmatrix}, \quad B = \begin{bmatrix} 0 \\ 1 \end{bmatrix}.$$

The communication graph is given as in FIGURE 5.7, where the node indexed by 1 is the leader which is only accessible to the node indexed by 2. It is easy to verify that the graph in FIGURE 5.7 satisfies Assumption 5.3.

Solving the LMI (5.6) by using the SeDuMi toolbox [158] gives a solution

$$P = \begin{bmatrix} 2.7881 & -0.8250 \\ -0.8250 & 0.9886 \end{bmatrix}.$$

Thus, the feedback gain matrices in (5.77) are obtained as

$$K = -\begin{bmatrix} 0.3974 & 1.3432 \end{bmatrix}, \quad \Gamma = \begin{bmatrix} 0.1580 & 0.5339 \\ 0.5339 & 1.8042 \end{bmatrix}.$$

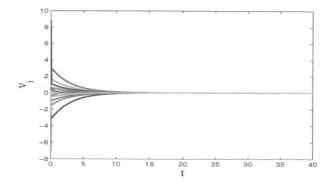

FIGURE 5.8: ν_i, $i = 2, \cdots, 7$ of double integrators under the protocol (5.77).

To illustrate Theorem 39, let the initial states $d_i(0)$ and $\rho_i(0)$ be randomly chosen within the interval $[1, 3]$. The consensus errors ν_i, $i = 2, \cdots, 7$, of the double integrators, defined as in (5.78), under the adaptive protocol (5.77) with K, Γ, and ρ_i chosen as above, are depicted in in FIGURE 5.8, which states that leader-follower consensus is indeed achieved. The coupling weights d_i associated with the nodes are drawn in FIGURE 5.9, from which it can be observed that the coupling weights converge to finite steady-state values.

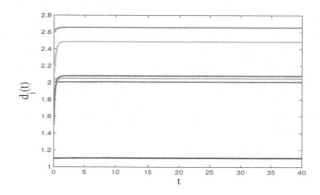

FIGURE 5.9: The coupling weights d_i in (5.77).

5.5.2 Robust Redesign in the Presence of External Disturbances

In this subsection, we revisit the distributed adaptive protocol (5.77) for the case where the agents are subject to external disturbances. The dynamics of the i-th agent are described by

$$\dot{x}_i = Ax_i + Bu_i + \tilde{\omega}_i, \quad i = 1, \cdots, N, \qquad (5.87)$$

where x_i and u_i are defined as in (5.1), and $\omega_i \in \mathbf{R}^n$ denotes external disturbances associated with the i-th agent, which is assumed to satisfy the following matching condition [43, 183].

Assumption 5.4 *There exist functions ω_i such that $\tilde{\omega}_i = B\omega_i$, $i = 1, \cdots, N$.*

The disturbances satisfying Assumption 5.4 are called matching disturbances. Under Assumption 5.4, the dynamics of the i-th agent in (5.87) can be rewritten into

$$\dot{x}_i = Ax_i + B[u_i + \omega_i], \quad i = 1, \cdots, N. \qquad (5.88)$$

In the following, we consider the case where the matching disturbances ω_i are bounded. That is, the following assumption holds.

Assumption 5.5 *There exist positive constants θ_i such that $\|\omega_i\| \leq \theta_i$, $i = 1, \cdots, N$.*

Similarly as discussed in Section 5.4, due to existences of ω_i, the relative states of the agents under the protocol (5.77) generally will not converge to zero any more but rather can only be expected to converge into some small neighborhood of the origin. Since the derivatives of the adaptive gains d_i are of nonnegative quadratic forms in terms of the relative states, in this case it is easy to see from (5.77) that the adaptive gains d_i will keep growing to infinity, i.e., the parameter drift phenomenon arises. Therefore, the adaptive protocol (5.77) is not robust in the presence of external disturbances.

The objective of this subsection is to make modifications on (5.77) in order to present novel distributed robust adaptive protocols which can guarantee the ultimate boundedness of the consensus error and the adaptive gains for the agents in (5.88). Similarly as in Section 5.4, by utilizing the σ-modification technique [62], we propose a new distributed adaptive consensus protocol as follows:

$$\begin{aligned} u_i &= d_i \rho_i(\nu_i^T Q \nu_i) K \nu_i, \\ \dot{d}_i &= -\varphi_i(d_i - 1) + \nu_i^T \Gamma \nu_i, \quad i = 2, \cdots, N, \end{aligned} \qquad (5.89)$$

where φ_i, $i = 2, \cdots, N$, are small positive constants, $Q > 0$ is a solution to the following algebraic Riccati equation (ARE):

$$A^T Q + QA + I - QBB^T Q = 0. \qquad (5.90)$$

and the rest of the variables are defined as in (5.77). Different from the previous subsection where the LMI (5.6) is used to design the adaptive protocol (5.77), here we use the algebraic Ricatti equation (5.90) to design the new adaptive protocol (5.89). These two approaches are nearly equivalent.

The consensus error ν is defined as in (5.78). In light of (5.88) and (5.89), it is not difficult to obtain that ν and d_i satisfy the following dynamics:

$$
\begin{aligned}
\dot{\nu} &= [I_{N-1} \otimes A + \mathcal{L}_1 \widehat{D}\hat{\rho}(\nu) \otimes BK]\nu + (\mathcal{L}_1 \otimes B)\omega, \\
\dot{d}_i &= -\varphi_i(d_i - 1) + \nu_i^T \Gamma \nu_i,
\end{aligned}
\tag{5.91}
$$

where $\omega \triangleq [\omega_2^T - \omega_1^T, \cdots, \omega_N^T - \omega_1^T]^T$, $\hat{\rho}(\nu)$ and $\widehat{D}$ are defined as in (5.79).

In light of Assumption 5.5, it is easy to see that

$$
\|\omega\| \leq \sqrt{\sum_{i=2}^{N}(\theta_i + \theta_1)^2}.
\tag{5.92}
$$

By the second equation in (5.91), we have

$$
\begin{aligned}
d_i(t) &= d_i(0)e^{-\varphi_i t} + \int_0^t e^{-\varphi_i(t-s)}(\varphi_i + \nu_i^T \Gamma \nu_i)ds \\
&= (d_i(0) - 1)e^{-\varphi_i t} + 1 + \int_0^t e^{-\varphi_i(t-s)}\nu_i^T \Gamma \nu_i ds \\
&\geq 1,
\end{aligned}
\tag{5.93}
$$

where we have used the fact that $d_i(0) \geq 1$ to get the last inequality.

Theorem 40 *Suppose that the communication graph $\mathcal{G}$ satisfies Assumption 5.3. The parameters of the adaptive protocol (5.89) are designed as $K = -B^T Q$, $\Gamma = QBB^T Q$, and $\rho_i(\nu_i^T Q \nu_i) = (1 + \nu_i^T Q \nu_i)^3$. Then, both the consensus error ν and the coupling gains d_i, $i = 2, \cdots, N$, in (5.91), under the adaptive protocol (5.89), are uniformly ultimately bounded. Furthermore, if φ_i is chosen to be small enough such that $\delta \triangleq \min_{i=2,\cdots,N} \varphi_i < \tau \triangleq \frac{1}{\lambda_{\max}(Q)}$, then ν exponentially converges to the residual set*

$$
\mathcal{D}_5 \triangleq \{\nu : \|\nu\|^2 \leq \frac{2\Xi}{(\tau - \delta)\lambda_{\min}(Q) \min_{i=2,\cdots,N} \frac{1}{q_i}}\},
\tag{5.94}
$$

where

$$
\Xi \triangleq \frac{\hat{\lambda}_0}{24} \sum_{i=1}^{N} \varphi_i(\alpha - 1)^2 + \frac{12}{\hat{\lambda}_0}\bar{\sigma}^2(G\mathcal{L}_1) \sum_{i=2}^{N}(\theta_i + \theta_1)^2,
\tag{5.95}
$$

$\alpha = \frac{72}{\hat{\lambda}_0^2} \max_{i=2,\cdots,N} q_i^2 + \max_{i=2,\cdots,N} \frac{2}{q_i^3 \hat{\lambda}_0^3}$, $\bar{\sigma}(G\mathcal{L}_1)$ *denotes the largest singular value of $G\mathcal{L}_1$, and q_i, G, and $\hat{\lambda}_0$ are chosen as in the proof of Theorem 39.*

Proof 41 *Let the Lyapapunov function candidate be V_{10} as in the proof of Theorem 39. Because $d_i(t) \geq 1$ for $t > 0$ in (5.93), by following the steps in the proof of Theorem 39, it is not difficult to see that V_{10} is still positive*

definite with respect to ν_i *and* $\tilde{d}_i$, $i = 2, \cdots, N$. *The time derivative of* V_{10}
along the trajectory of (5.91) is given by

$$\dot{V}_{10} = \sum_{i=2}^{N} \frac{d_i}{q_i} \rho_i (\nu_i^T Q \nu_i) \nu_i^T Q \dot{\nu}_i$$

$$+ \sum_{i=2}^{N} \frac{d_i}{2q_i} \int_0^{\nu_i^T Q \nu_i} \rho_i(s) ds \qquad (5.96)$$

$$+ \frac{\hat{\lambda}_0}{12} \sum_{i=2}^{N} (d_i - \alpha)[-\varphi_i(d_i - 1) + \nu_i^T \Gamma \nu_i].$$

*By following similar steps in the proof of Theorem 39, we can get the
following results:*

$$\sum_{i=2}^{N} \frac{d_i}{q_i} \rho_i \nu_i^T Q \dot{\nu}_i \le \frac{1}{2} \nu^T [\hat{D}\hat{\rho}G \otimes (QA + A^T Q) - \hat{\lambda}_0 \hat{D}^2 \hat{\rho}^2 \otimes QBB^T Q] \nu \qquad (5.97)$$

$$+ \nu^T (\hat{D}\hat{\rho}G\mathcal{L}_1 \otimes QB)\omega,$$

and

$$\sum_{i=2}^{N} \frac{d_i}{q_i} \int_0^{\nu_i^T Q \nu_i} \rho_i(s) ds \le \sum_{i=2}^{N} \frac{\dot{d}_i}{q_i} \int_0^{\nu_i^T Q \nu_i} \rho_i(s) ds$$

$$\le \sum_{i=2}^{N} \left(\frac{1}{3q_i^3 \hat{\lambda}_0^2} + \frac{2}{3} \hat{\lambda}_0 \rho_i^2 \right) \dot{\tilde{d}}_i \qquad (5.98)$$

$$\le \sum_{i=2}^{N} \left(\frac{1}{3q_i^3 \hat{\lambda}_0^2} + \frac{2}{3} \hat{\lambda}_0 \rho_i^2 \right) \nu_i^T QBB^T Q \nu_i$$

we have used the fact that $\dot{d}_i \le \dot{\tilde{d}}_i \triangleq \nu_i^T \Gamma \nu_i$ *to get the first inequality. Substituting (5.97) and (5.98) into (5.96) yields*

$$\dot{V}_{10} \le \frac{1}{2} \nu^T [\hat{D}\hat{\rho}G \otimes (QA + A^T Q)] \nu$$

$$- \frac{\hat{\lambda}_0}{12} \sum_{i=2}^{N} (d_i^2 \rho_i^2 + 2\hat{\alpha}) \nu_i^T QBB^T Q \nu_i \qquad (5.99)$$

$$+ \frac{\hat{\lambda}_0}{24} \sum_{i=2}^{N} \varphi_i[-\tilde{d}_i^2 + (\alpha - 1)^2] + \nu^T (\hat{D}\hat{\rho}G\mathcal{L}_1 \otimes QB)\omega,$$

where we have used the facts that $\rho_i \ge 1$ *and* $d_i \ge 1$, $i = 2, \cdots, N$, $\alpha = 2\hat{\alpha} + \max_{i=2,\cdots,N} \frac{2}{q_i^3 \hat{\lambda}_0^3}$, *where* $\hat{\alpha} = \frac{36}{\hat{\lambda}_0^2} \max_{i=2,\cdots,N} q_i^2$ *and used the following assertion:*

$$-(d_i - \alpha)(d_i - 1) = -\tilde{d}_i(\tilde{d}_i + \alpha - 1) \le -\frac{1}{2}\tilde{d}_i^2 + \frac{1}{2}(\alpha - 1)^2.$$

Note that

$$2\nu^T(\hat{D}\hat{\rho}G\mathcal{L}_1 \otimes QB)\omega = 2\nu^T(\sqrt{\frac{\hat{\lambda}_0}{12}}\hat{D}\hat{\rho} \otimes QB)(\sqrt{\frac{12}{\hat{\lambda}_0}}G\mathcal{L}_1 \otimes I)\omega$$

$$\leq \frac{\hat{\lambda}_0}{12}\nu^T(\hat{D}^2\hat{\rho}^2 \otimes QBB^TQ)\nu + \frac{12}{\hat{\lambda}_0}\|(G\mathcal{L}_1 \otimes I)\omega\|^2 \qquad (5.100)$$

$$\leq \frac{\hat{\lambda}_0}{12}\nu^T(\hat{D}^2\hat{\rho}^2 \otimes QBB^TQ)\nu + \frac{12}{\hat{\lambda}_0}\bar{\sigma}^2(G\mathcal{L}_1)\sum_{i=2}^{N}(\theta_i + \theta_1)^2,$$

where we have used (5.92) to get the last inequality. Then, substituting (5.100) into (5.99) gives

$$\dot{V}_{10} \leq \frac{1}{2}\nu^T[\hat{D}\hat{\rho}G \otimes (QA + A^TQ) - \frac{\hat{\lambda}_0}{24}(\hat{D}^2\hat{\rho}^2 + 4\hat{\alpha}I) \otimes QBB^TQ)]\nu$$

$$- \frac{\hat{\lambda}_0}{24}\sum_{i=2}^{N}\varphi_i\tilde{d}_i^2 + \Xi \qquad (5.101)$$

$$\leq \frac{1}{2}W(\nu) - \frac{\hat{\lambda}_0}{24}\sum_{i=2}^{N}\varphi_i\tilde{d}_i^2 + \Xi,$$

where we have used the assertion that $\frac{\hat{\lambda}_0}{12}(\hat{D}^2\hat{\rho}^2 + 4\hat{\alpha}I) \geq \frac{\hat{\lambda}_0}{6}\sqrt{\hat{\alpha}}\hat{D}\hat{\rho} \geq \hat{D}\hat{\rho}G$ if $\sqrt{\hat{\alpha}}I \geq \frac{6}{\hat{\lambda}_0}G$ to get the last inequality, Ξ is defined as in (5.95), and

$$W(\nu) \triangleq \nu^T[\hat{D}\hat{\rho}G \otimes (QA + A^TQ - QBB^TQ)]\nu$$
$$= -\nu^T(\hat{D}\hat{\rho}G \otimes I)\nu \leq 0.$$

Therefore, we can verify that $\frac{1}{2}W(\nu) - \frac{\hat{\lambda}_0}{24}\sum_{i=2}^{N}\varphi_i\tilde{d}_i^2$ is negative definite. In virtue of Lemma 15, we get that both the consensus error ν and the adaptive gains d_i are uniformly ultimately bounded.

Note that (5.99) can be rewritten into

$$\dot{V}_{10} \leq -\delta V_{10} + \delta V_{10} + \frac{1}{2}W(\nu) - \frac{\hat{\lambda}_0}{24}\sum_{i=2}^{N}\varphi_i\tilde{d}_i^2 + \Xi. \qquad (5.102)$$

Because ρ_i are monotonically increasing and satisfy $\rho_i(s) \geq 1$ for $s > 0$, we have

$$\sum_{i=2}^{N}\frac{d_i}{q_i}\int_0^{\nu_i^TQ\nu_i}\rho_i(s)ds \leq \sum_{i=2}^{N}\frac{d_i}{q_i}\rho_i\nu_i^TQ\nu_i \qquad (5.103)$$

$$= \nu^T(\hat{D}\hat{\rho}G \otimes Q)\nu.$$

Substituting (5.103) into (5.102) yields

$$\dot{V}_{10} \leq -\delta V_{10} + \frac{1}{2}\tilde{W}(\nu) + \frac{\hat{\lambda}_0}{24}\sum_{i=2}^{N}(\varphi_i - \delta)\tilde{d}_i^2$$

$$-\frac{\tau - \delta}{2}\nu^T(\widehat{D}\hat{\rho}G \otimes Q)\nu + \Xi, \tag{5.104}$$

where $\tilde{W}(\nu) \triangleq \nu^T[\widehat{D}\hat{\rho}G \otimes (-I + \tau Q)]\nu$. *Because* $\tau = \frac{1}{\lambda_{\max}(Q)}$, *we can obtain that* $\tilde{W}(\nu) \leq 0$. *Then, it follows from (5.104) that*

$$\dot{V}_{10} \leq -\delta V_{10} - \frac{\tau - \delta}{2}\lambda_{\min}(Q)(\min_{i=2,\cdots,N}\frac{1}{q_i})\|\nu\|^2 + \Xi, \tag{5.105}$$

where we have used the facts that $\varphi_i \geq \delta$, $\delta < \tau$, $\widehat{C} \geq I$, $\hat{\rho} \geq I$, *and* $G > 0$. *Obviously, it follows from (5.104) that* $\dot{V}_{10} \leq -\delta V_{10}$ *if* $\|\nu\|^2 > \frac{2\Xi}{(\tau - \delta)\lambda_{\min}(Q)\min_{i=2,\cdots,N}\frac{1}{q_i}}$. *Then, we can get that if* $\delta \leq \tau$ *then* ν *exponentially converges to the residual set* $\mathcal{D}_5$ *in (5.94) with a convergence rate faster than* $e^{-\delta t}$.

Remark 42 *Theorem 40 shows that the modified adaptive protocol (5.89) can ensure the ultimate boundedness of the consensus error* ν *and the adaptive gains* d_i *for the agents in (5.88), implying that (5.89) is indeed robust in the presence of bounded external disturbances. As discussed in Remark 38,* φ_i *should be chosen to be relatively small in order to ensure a smaller bound for the consensus error* ν.

Remark 43 *In contrast to Section 5.4 with undirected graphs and general bounded disturbances, the results in this subsection for directed graphs are applicable to matching disturbances. Extending the results in this subsection to the case with disturbances which does not necessarily satisfy the matching condition is not an easy task. Note that the idea used in the proof of Theorem 40 can be derive similar results for the adaptive protocols (5.47) and (5.48) in Section 5.4 for the special case with matching disturbances.*

Example 20 *Consider a network of double integrators, described by (5.88), where the disturbances associated with the agents are assumed to be* $\omega_1 = 0$, $\omega_2 = 0.5\sin(t)$, $\omega_3 = 0.1\sin(t)$, $\omega_4 = 0.2\cos(2t)$, $\omega_5 = -0.3e^{-2t}$, $\omega_6 = -0.2\sin(x_{51})$, *and* $\omega_7 = 0$. *The communication graph is given as in FIGURE 5.7.*

Solving the ARE (5.90) by using MATLAB gives a solution $Q = \begin{bmatrix} 1.7321 & 1 \\ -1 & 1.7321 \end{bmatrix}$. *Thus, the feedback gain matrices in (5.89) are obtained as*

$$K = -\begin{bmatrix} 1 & 1.7321 \end{bmatrix}, \quad \Gamma = \begin{bmatrix} 1 & 1.7321 \\ 1.7321 & 3 \end{bmatrix}.$$

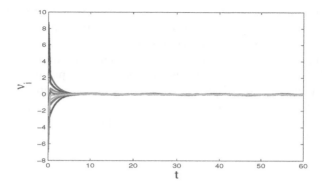

FIGURE 5.10: The consensus error ν_i, $i = 2, \cdots, 7$, of double integrators under the protocol (5.89).

To illustrate Theorem 40, let let $\varphi_i = 0.02$ in (5.89) and the initial states $d_i(0)$ be randomly chosen within the interval $[1, 3]$. The consensus errors ν_i, $i = 2, \cdots, 7$, of the double integrators, defined as in (5.78), and the coupling weights d_i associated with the followers are depicted in in FIGURE 5.10 and FIGURE 5.11, respectively, both of which are clearly bounded.

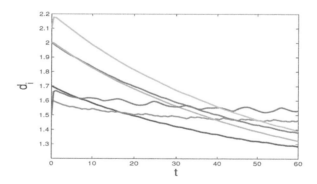

FIGURE 5.11: The coupling weights d_i in (5.89).

5.6 Notes

The results in this section are mainly based on [78, 80, 93, 95]. There exist some related works which have proposed similar adaptive protocols to achieve consensus or synchronization. For instance, some edge-based adaptive strategies were proposed for the synchronization of complex networks in [32, 33]. However, the inner coupling matrix is required to be given a priori and the input matrix of each node is essentially assumed to be an identity matrix in [32, 33], which is quite restrictive. Distributed adaptive protocols based on the relative state information of neighboring agents were presented in [159, 190] to achieve second-order consensus with nonlinear dynamics. In [175], distributed adaptive consensus protocols were proposed for first-order multi-agent systems with uncertainties and directed graphs.

6

Distributed Tracking of Linear Multi-Agent Systems with a Leader of Possibly Nonzero Input

CONTENTS

6.1	Problem Statement	138
6.2	Distributed Discontinuous Tracking Controllers	139
	6.2.1 Discontinuous Static Controllers	139
	6.2.2 Discontinuous Adaptive Controllers	142
6.3	Distributed Continuous Tracking Controllers	144
	6.3.1 Continuous Static Controllers	144
	6.3.2 Adaptive Continuous Controllers	147
6.4	Distributed Output-Feedback Controllers	152
6.5	Simulation Examples	156
6.6	Notes	158

This chapter considers the distributed tracking (i.e., leader-follower consensus) problem for multi-agent systems with general continuous-time linear agent dynamics. Actually, the distributed tracking problem has been discussed in Chapters 2 and 4. The distributed tracking problem is a consensus problem for the special case where the communication graph contains a directed spanning tree and a root node. It is worth noting that one common assumption for the results on tracking in Chapters 2 and 4 is that the leader's control input is either equal to zero or available to all the followers. In many circumstances, nonzero control actions might be implemented on the leader in order to achieve certain objectives, e.g., to reach a desirable consensus value or to avoid hazardous obstacles. However, it is restrictive and impractical to assume that all the followers know the leader's control input, especially when the scale of the network is large. Actually, the leader's control input might not be available to any follower, e.g., for the case where the leader is an uncooperative target.

In this chapter, we consider the distributed tracking control problem for multi-agent systems with general linear dynamics with a leader whose control input might be nonzero and time varying and available to at most a subset of followers. Due to the nonzero control input, the dynamics of the leader are different from those of the followers. Thus, contrary to the homogeneous multi-agent systems in previous chapters, the resulting multi-agent system

137

in this chapter is in essence heterogeneous, which makes the problem much more challenging. The distributed controllers in the previous chapters are not applicable any more.

The distributed tracking control problem is formulated in Section 6.1. Without loss of generalization, it is assumed that the control input of the leader is bounded. Such an assumption is not strong, since unbounded control inputs are not implementable. In Section 6.2, based on the relative states of neighboring agents, a distributed static tracking controller, including a discontinuous nonlinear term to deal with the effect of the nonzero control input of the leader, is proposed and designed, under which the states of the followers converge to the state of the leader, if the subgraph among the followers is undirected, the leader has directed paths to all followers. It is worthwhile mentioning that the design of the static controller requires the upper bound of the leader's control input and the nonzero eigenvalue information of the communication graph, which actually are global information. In order to remove this limitation, a distributed discontinuous adaptive controller with time-varying coupling gains is further designed, which is fully distributed and also solves the distributed tracking problem.

Note that the proposed discontinuous controllers might cause the undesirable chattering phenomenon in real implementations. To eliminate the chattering effect, by using the boundary layer concept and the σ-modification technique, distributed continuous static and adaptive protocol are constructed in Section 6.3, under which the tracking error and the coupling gains are shown to be uniformly ultimately bounded. The upper bound of the tracking error is also explicitly given, which can be made to be reasonably small by properly selecting the design parameters of the proposed controllers. As an extension, distributed tracking controllers based on the local output information are further discussed in Section 6.4. A simulation example is presented for illustration in Section 6.5.

6.1   Problem Statement

Consider a group of N identical agents with general linear dynamics, consisting of $N-1$ followers and a leader. The dynamics of the i-th agent are described by

$$\dot{x}_i = Ax_i + Bu_i,$$
$$y_i = Cx_i \quad i = 1, \cdots, N, \tag{6.1}$$

where $x_i \in \mathbf{R}^n$ is the state, $u_i \in \mathbf{R}^p$ is the control input, and $y_i \in \mathbf{R}^q$ is the output of the i-th agent, respectively.

Without loss of generality, let the agent in (6.1) indexed by 1 be the leader and the agents indexed by $2, \cdots, N$, be the followers. It is assumed that the leader receives no information from any follower and the state (but not the control input) of the leader is available to only a subset of the followers. The interaction graph among the N agents is represented by a directed graph $\mathcal{G}$, which satisfies the following assumption.

Assumption 6.1 *The subgraph $\mathcal{G}_s$ associated with the followers is undirected and in the graph $\mathcal{G}$ the leader has directed paths to all followers. (Equivalently, $\mathcal{G}$ contains a directed spanning tree with the leader as the root)*

Denote by $\mathcal{L}$ the Laplacian matrix associated with $\mathcal{G}$. Because the leader has no neighbors, $\mathcal{L}$ can be partitioned as

$$\mathcal{L} = \begin{bmatrix} 0 & 0_{1\times(N-1)} \\ \mathcal{L}_2 & \mathcal{L}_1 \end{bmatrix}, \tag{6.2}$$

where $\mathcal{L}_2 \in \mathbf{R}^{(N-1)\times 1}$ and $\mathcal{L}_1 \in \mathbf{R}^{(N-1)\times(N-1)}$. Since $\mathcal{G}_s$ is undirected, $\mathcal{L}_1$ is symmetric.

The objective of this chapter is to solve the distributed tracking problem for the agents in (6.1), i.e., to design some distributed controllers under which the states of the N followers converge to the state of the leader in the sense of $\lim_{t\to\infty} \|x_i(t) - x_1(t)\| = 0, \forall i = 2, \cdots, N$.

Here we consider the general case where u_1 is possibly nonzero and time varying and accessible to at most a subset of followers, under the following mild assumption:

Assumption 6.2 *The leader's control input u_1 is bounded, i.e., there exists a positive constant γ such that $\|u_1\| \leq \gamma$.*

In the following sections, several distributed tracking controllers will be presented.

6.2 Distributed Discontinuous Tracking Controllers

6.2.1 Discontinuous Static Controllers

In this section, it is assumed that each follower can have access to the local state information of its own and its neighbors. Based on the relative states of neighboring agents, the following distributed tracking controller is proposed for each follower:

$$u_i = c_1 K \sum_{j=1}^{N} a_{ij}(x_i - x_j) + c_2 g(K \sum_{j=1}^{N} a_{ij}(x_i - x_j)), \ i = 2, \cdots, N, \tag{6.3}$$

where $c_1 > 0$ and $c_2 > 0 \in \mathbf{R}$ are constant coupling gains, $K \in \mathbf{R}^{p\times n}$ is the feedback gain matrix, a_{ij} is the (i,j)-th entry of the adjacency matrix associated with $\mathcal{G}$, and $g(\cdot)$ is a nonlinear function defined such that for $w \in \mathbf{R}^n$,

$$g(w) = \begin{cases} \frac{w}{\|w\|} & \text{if } \|w\| \neq 0 \\ 0 & \text{if } \|w\| = 0. \end{cases} \tag{6.4}$$

Let $\xi_i = x_i - x_1$, $i = 2, \cdots, N$, and $\xi = [\xi_2^T, \cdots, \xi_N^T]^T$. Using (6.3) for (6.1), we obtain the closed-loop network dynamics as

$$\dot{\xi} = (I_{N-1} \otimes A + c_1 \mathcal{L}_1 \otimes BK)\xi \\ + c_2(I_{N-1} \otimes B)G(\xi) - (1 \otimes B)u_1, \tag{6.5}$$

where $\mathcal{L}_1$ is defined as in (6.2) and $G(\xi) \triangleq \begin{bmatrix} g(K \sum_{j=2}^N \mathcal{L}_{2j}\xi_j) \\ \vdots \\ g(K \sum_{j=2}^N \mathcal{L}_{Nj}\xi_j) \end{bmatrix}$. Since the

function $g(\cdot)$ in the static controller (6.3) is nonsmooth, the well-posedness and the existence of the solution to (6.5) can be understood in the Filippov sense [150]. Clearly, the distributed tracking problem is solved by (6.3) if the closed-loop system (6.5) is asymptotically stable. Hereafter, we refer to ξ as the tracking error.

Theorem 41 *Suppose that Assumptions 6.1 and 6.2 hold. The distributed tracking control problem of the agents described by (6.1) is solved under the controller (6.3) with $c_1 \geq \frac{1}{\lambda_{\min}(\mathcal{L}_1)}$, $c_2 \geq \gamma$, and $K = -B^T P^{-1}$, where $\lambda_{\min}(\mathcal{L}_1)$ denotes the smallest eigenvalue of $\mathcal{L}_1$ and $P > 0$ is a solution to the following linear matrix inequality (LMI):*

$$AP + PA^T - 2BB^T < 0. \tag{6.6}$$

Proof 42 *Consider the following Lyapunov function candidate:*

$$V_1 = \frac{1}{2}\xi^T (\mathcal{L}_1 \otimes P^{-1})\xi. \tag{6.7}$$

For a communication graph $\mathcal{G}$ satisfying Assumption 6.1, it follows from Lemma 1 and (6.2) that $\mathcal{L}_1 > 0$. Clearly, V_1 is positive definite and continuously differentiable. The time derivative of V_1 along the trajectory of (6.5) can be obtained as

$$\dot{V}_1 = \xi^T (\mathcal{L}_1 \otimes P^{-1})\dot{\xi} \\ = \frac{1}{2}\xi^T \mathcal{X}\xi + c_2\xi^T (\mathcal{L}_1 \otimes P^{-1}B)G(\xi) - \xi^T (\mathcal{L}_1 1 \otimes P^{-1}B)u_1, \tag{6.8}$$

where

$$\mathcal{X} = \mathcal{L}_1 \otimes (P^{-1}A + A^T P^{-1}) - 2c_1 \mathcal{L}_1^2 \otimes P^{-1}BB^T P^{-1}. \tag{6.9}$$

In virtue of Assumption 6.2, we have

$$-\xi^T(\mathcal{L}_1 \mathbf{1} \otimes P^{-1}B)u_1 = -\sum_{i=2}^{N}\sum_{j=2}^{N}\mathcal{L}_{ij}\xi_j^T P^{-1}Bu_1$$

$$\leq \sum_{i=2}^{N}\|B^T P^{-1}\sum_{j=2}^{N}\mathcal{L}_{ij}\xi_j\|\|u_1\| \qquad (6.10)$$

$$\leq \gamma\sum_{i=2}^{N}\|B^T P^{-1}\sum_{j=2}^{N}\mathcal{L}_{ij}\xi_j\|.$$

In this case, it follows from (6.4) that

$$\xi^T(\mathcal{L}_1 \otimes P^{-1}B)G(\xi) = \left[\sum_{j=2}^{N}\mathcal{L}_{2j}\xi_j^T P^{-1}B \quad \cdots \quad \sum_{j=2}^{N}\mathcal{L}_{Nj}\xi_j^T P^{-1}B\right]$$

$$\times \begin{bmatrix} \dfrac{-B^T P^{-1}\sum_{j=2}^{N}\mathcal{L}_{2j}\xi_j}{\|B^T P^{-1}\sum_{j=1}^{N}\mathcal{L}_{2j}\xi_j\|} \\ \vdots \\ \dfrac{-B^T P^{-1}\sum_{j=2}^{N}\mathcal{L}_{Nj}\xi_j}{\|B^T P^{-1}\sum_{j=2}^{N}\mathcal{L}_{Nj}\xi_j\|} \end{bmatrix}$$

$$= -\sum_{i=2}^{N}\|B^T P^{-1}\sum_{j=2}^{N}\mathcal{L}_{ij}\xi_j\|.$$

$$(6.11)$$

Then, we can get from (6.8), (6.10), and (6.11) that

$$\dot{V}_1 \leq \frac{1}{2}\xi^T \mathcal{X}\xi - (c_2 - \gamma)\sum_{i=2}^{N}\|B^T P^{-1}\sum_{j=2}^{N}\mathcal{L}_{ij}\xi_j\|$$

$$\leq \frac{1}{2}\xi^T \mathcal{X}\xi. \qquad (6.12)$$

Because the graph $\mathcal{L}$ satisfies Assumption 6.1, by (6.6), we obtain that

$$(\mathcal{L}_1^{-\frac{1}{2}} \otimes P)\mathcal{X}(\mathcal{L}_1^{-\frac{1}{2}} \otimes P) = I_N \otimes (AP + PA^T) - 2c_1\mathcal{L}_1 \otimes BB^T$$

$$\leq I_N \otimes [AP + PA^T - 2c_1\lambda_{\min}(\mathcal{L}_1)BB^T] \qquad (6.13)$$

$$< 0,$$

which implies $\mathcal{X} < 0$ and $\dot{V}_1 < 0$. Therefore, the state ξ of (6.5) is asymptotically stable, i.e., the tracking control problem is solved.

Remark 44 *As shown in Proposition 1 in Chapter 2, a necessary and sufficient condition for the existence of a $P > 0$ to the LMI (6.6) is that (A, B) is stabilizable. Therefore, a sufficient condition for the existence of (6.3) satisfying Theorem 41 is that (A, B) is stabilizable. The design of the parameters c_1, c_2, and K of (6.3) in Theorem 41 is essentially also based on the consensus* *region approach. These three parameters can be independently designed. Specifically, K relies on only the agent dynamics, c_1 deals with the communication graph $\mathcal{G}$, and c_2 is related to only the upper bound of u_1.*

Remark 45 *The distributed controller (6.3) consists of a linear term $c_1 K \sum_{j=1}^{N} a_{ij}(x_i - x_j)$ and a nonlinear term $c_2 g(K \sum_{j=1}^{N} a_{ij}(x_i - x_j))$, where the nonlinear term is used to suppress the effect of the leader's nonzero input. Without the nonlinear term in (6.3), it can be seen from (6.5) that even though K is designed such that $I_{N-1} \otimes A + c_1 \mathcal{L}_1 \otimes BK$ is Hurwitz, the tracking problem cannot be solved due to the nonzero u_1. Furthermore, by letting $c_2 = 0$, (6.3) is reduced to the distributed controller in Chapter 2 for the special case where $u_1 = 0$.*

Remark 46 *To present the nonlinear controller (6.3), we borrow ideas from the sliding mode control literature (e.g., [187, 43]) and treat the nonzero control input of the leader as an external disturbances satisfying the matching condition [43]. The nonlinear function $g(\cdot)$ in (6.3) is a unit vector nonlinearity, i.e., $\|g(\cdot)\| = 1$. Actually, $g(\cdot)$ in (6.3) can be replaced by the signum function $\mathrm{sgn}(\cdot)$, defined as $\mathrm{sgn}(s) = \begin{cases} 1 & \text{if } s > 0 \\ 0 & \text{if } s = 0 \\ -1 & \text{if } s < 0 \end{cases}$. In this case, the design of the new controller and the derivation of the theorem can be accordingly modified. Please refer to [92] for details.*

6.2.2   Discontinuous Adaptive Controllers

In the last subsection, the design of the distributed controller (6.3) depends on the minimal eigenvalue $\lambda_{\min}(\mathcal{L}_1)$ of $\mathcal{L}_1$ and the upper bound γ of the leader's control input u_1. However, $\lambda_{\min}(\mathcal{L}_1)$ is global information of the communication graph, which is not easy to compute especially when the multi-agent network is of a large scale. Furthermore, the value of γ might not be explicitly known for each follower in some applications. The objective of this subsection is to solve the distributed tracking problem without requiring that $\lambda_{\min}(\mathcal{L}_1)$ and γ be explicitly known. To this end, we propose the following distributed tracking controller with an adaptive law for updating the coupling gain for each follower:

$$u_i = d_i K \sum_{j=1}^{N} a_{ij}(x_i - x_j) + d_i g(K \sum_{j=1}^{N} a_{ij}(x_i - x_j)),$$

$$\dot{d}_i = \tau_i [(\sum_{j=1}^{N} a_{ij}(x_i - x_j)^T)\Gamma(\sum_{j=1}^{N} a_{ij}(x_i - x_j)) \tag{6.14}$$

$$+ \|K \sum_{j=1}^{N} a_{ij}(x_i - x_j)\|], \quad i = 2, \cdots, N,$$

where $K \in \mathbf{R}^{p \times n}$ and a_{ij} are defined in (6.3), $d_i(t)$ denotes the time-varying coupling gain (weight) associated with the i-th follower, $\Gamma \in \mathbf{R}^{n \times n}$ is the feedback gain matrix, and τ_i are positive scalars.

Let ξ be defined as in (6.5) and $D = \mathrm{diag}(d_2, \cdots, d_N)$. From (6.1) and (6.14), we can get the closed-loop network dynamics as

$$\dot{\xi} = (I_{N-1} \otimes A + D\mathcal{L}_1 \otimes BK)\xi + (D \otimes B)G(\xi) - (1 \otimes B)u_1,$$

$$\dot{d}_i = \tau_i [(\sum_{j=2}^{N} \mathcal{L}_{ij}\xi_j^T)\Gamma(\sum_{j=2}^{N} \mathcal{L}_{ij}\xi_j) + \|K \sum_{j=2}^{N} \mathcal{L}_{ij}\xi_j\|], \quad i = 2, \cdots, N. \tag{6.15}$$

Theorem 42 *Suppose that Assumptions 6.1 and 6.2 hold. Then, the distributed tracking control problem of the agents described by (6.1) is solved under the adaptive controller (6.14) with $K = -B^T P^{-1}$ and $\Gamma = P^{-1}BB^T P^{-1}$, where $P > 0$ is a solution to the LMI (6.6). Moreover, each coupling gain d_i converges to some finite steady-state value.*

Proof 43 *Consider the following Lyapunov function candidate*

$$V_2 = \frac{1}{2}\xi^T(\mathcal{L}_1 \otimes P^{-1})\xi + \sum_{i=2}^{N} \frac{1}{2\tau_i}(d_i - \beta)^2, \tag{6.16}$$

where β is a positive constant. As shown in the proof of Theorem 41, V_2 is positive semi-definite. The time derivative of V_2 along (6.15) can be obtained as

$$\begin{aligned}\dot{V}_2 &= \xi^T(\mathcal{L}_1 \otimes P^{-1})\dot{\xi} + \sum_{i=2}^{N} \frac{d_i - \beta}{\tau_i}\dot{d}_i \\ &= \xi^T(\mathcal{L}_1 \otimes P^{-1}A + \mathcal{L}_1 D\mathcal{L}_1 \otimes P^{-1}BK)\xi \\ &\quad + \xi^T(\mathcal{L}_1 D \otimes P^{-1}B)G(\xi) - \xi^T(\mathcal{L}_1 1 \otimes P^{-1}B)u_1 \\ &\quad + \sum_{i=2}^{N}(d_i - \beta)[(\sum_{j=2}^{N} \mathcal{L}_{ij}\xi_j^T)\Gamma(\sum_{j=2}^{N} \mathcal{L}_{ij}\xi_j) + \|K \sum_{j=2}^{N} \mathcal{L}_{ij}\xi_j\|]. \end{aligned} \tag{6.17}$$

By substituting $K = -BP^{-1}$, it is easy to get that

$$\begin{aligned}&\xi^T(\mathcal{L}_1 D\mathcal{L}_1 \otimes P^{-1}BK)\xi \\ &= -\sum_{i=2}^{N} d_i[\sum_{j=2}^{N} \mathcal{L}_{ij}\xi_j]^T P^{-1}BB^T P^{-1}[\sum_{j=2}^{N} \mathcal{L}_{ij}\xi_j]. \end{aligned} \tag{6.18}$$

By the definitions of $g(\cdot)$ and $G(\cdot)$, we can get that

$$\xi^T(\mathcal{L}_1 D \otimes P^{-1}B)G(\xi) = -\sum_{i=2}^{N} d_i\|B^T P^{-1} \sum_{j=2}^{N} \mathcal{L}_{ij}\xi_j\|. \tag{6.19}$$

Substituting (6.18), (6.19), and (6.10) into (6.17) yields

$$\dot{V}_2 \le \frac{1}{2}\xi^T \mathcal{Y}\xi - (\beta - \gamma) \sum_{i=2}^{N} \|B^T P^{-1} \sum_{j=2}^{N} \mathcal{L}_{ij}\xi_j\|, \tag{6.20}$$

where $\mathcal{Y} \triangleq \mathcal{L}_1 \otimes (P^{-1}A + A^T P^{-1}) - 2\beta \mathcal{L}_1^2 \otimes P^{-1}BB^T P^{-1}$.

As shown in the proof of Theorem 41, by selecting β sufficiently large such that $\beta \ge \gamma$ and $\beta\lambda_i \ge 1$, $i = 2, \cdots, N$, we get from (6.20) that

$$\dot{V}_2 \le \frac{1}{2}\xi^T \mathcal{Y}\xi, \tag{6.21}$$

and from (6.13) that $\mathcal{Y} > 0$. Then, it follows that $\dot{V}_2 \le 0$, implying that $V_2(t)$ is nonincreasing. Therefore, in view of (6.16), we know that d_i and ξ are bounded. Since by Assumption 6.2, u_1 is bounded, this implies from the first equation in (6.15) that $\dot{\xi}$ is bounded. As $V_2(t)$ is nonincreasing and bounded from below by zero, it has a finite limit V_2^∞ as $t \to \infty$. Integrating (6.21), we have $\int_0^\infty \frac{1}{2}\xi^T \mathcal{Y}\xi d\tau \le V_2(\zeta(t_1)) - V_2^\infty$. Thus, $\int_0^\infty \frac{1}{2}\xi^T \mathcal{Y}\xi d\tau$ exists and is finite. Because ξ and $\dot{\xi}$ are bounded, it is easy to see from (6.21) that $\xi^T \mathcal{Y}\dot{\xi}$ is also bounded, which in turn guarantees the uniform continuity of $\xi^T \mathcal{Y}\xi$. Therefore, by Lemma 13 (Barbalat's Lemma), we get that $\xi^T \mathcal{Y}\xi \to 0$ as $t \to \infty$, i.e., $\xi \to 0$ as $t \to \infty$. By noting that $\Gamma \ge 0$ and $\tau_i > 0$, it follows from (6.15) that d_i is monotonically increasing. Thus, the boundedness of d_i implies that each d_i converges to some finite value.

Remark 47 *Compared to the static controller (6.3), the adaptive controller (6.14) requires neither the minimal eigenvalue* $\lambda_{\min}(\mathcal{L}_1)$ *of* $\mathcal{L}_1$ *nor the upper bound* γ *of* u_1*, as long as* u_1 *is bounded. On the other hand, the coupling gains need to be dynamically updated in (6.14), implying that the adaptive controller (6.14) is more complex than the static controller (6.3). The term* $d_i g(\sum_{j=1}^{N} a_{ij}K(x_i - x_j))$ *in (6.14) is used to tackle the effect of the leader's nonzero control input* u_1 *on consensus. For the special case where* $u_1 = 0$*, we can accordingly remove the terms* $d_i g(\sum_{j=1}^{N} a_{ij}K(x_i - x_j))$ *and* $\tau_i \|\sum_{j=1}^{N} a_{ij}K(x_i - x_j)\|$ *from (6.14), which will reduce to the continuous protocol (5.45) in the previous chapter.*

6.3 Distributed Continuous Tracking Controllers

6.3.1 Continuous Static Controllers

An inherent drawback of the discontinuous controller (6.3) is that it will result in the undesirable chattering effect in real implementation, due to imperfections in switching devices [187, 43]. To avoid the chattering effect, one feasible approach is to use the boundary layer technique [187, 43] to give a continuous approximation of the discontinuous function $g(\cdot)$.

Using the boundary layer technique, we propose a distributed continuous static controller as

$$u_i = c_1 K \sum_{j=1}^{N} a_{ij}(x_i - x_j) + c_2 \hat{g}_i(K \sum_{j=1}^{N} a_{ij}(x_i - x_j)), \quad i = 2, \cdots, N,$$

(6.22)

where the nonlinear functions $\hat{g}(\cdot)$ are defined such that for $w \in \mathbf{R}^n$,

$$\hat{g}_i(w) = \begin{cases} \frac{w}{\|w\|} & \text{if } \|w\| > \kappa_i \\ \frac{w}{\kappa_i} & \text{if } \|w\| \le \kappa_i \end{cases}$$

(6.23)

with κ_i being small positive scalars, denoting the widths of the boundary layers, and the rest of the variables are the same as in (6.3).

Let the tracking error ξ be defined as in (6.5). From (6.1) and (6.22), we can obtain that ξ in this case satisfies

$$\dot{\xi} = (I_{N-1} \otimes A + c_1 \mathcal{L}_1 \otimes BK)\xi + c_2(I_{N-1} \otimes B)\widehat{G}(\xi) - (\mathbf{1} \otimes B)u_1, \quad (6.24)$$

where

$$\widehat{G}(\xi) \triangleq \begin{bmatrix} \hat{g}_2(K \sum_{j=2}^{N} \mathcal{L}_{2j}\xi_j) \\ \vdots \\ \hat{g}_N(K \sum_{j=2}^{N} \mathcal{L}_{Nj}\xi_j) \end{bmatrix}, \quad (6.25)$$

The following theorem states the ultimate boundedness of the tracking error ξ.

Theorem 43 *Assume that Assumptions 6.1 and 6.2 hold. Then, the tracking error ξ of (6.24) under the continuous controller (6.22) with c_1, c_2, and K chosen as in Theorem 41 is uniformly ultimately bounded and exponentially converges to the residual set*

$$\mathcal{D}_1 \triangleq \{\xi : \|\xi\|^2 \le \frac{2\lambda_{\max}(P)\gamma}{\alpha\lambda_{\min}(\mathcal{L}_1)} \sum_{i=2}^{N} \kappa_i\}, \quad (6.26)$$

where

$$\alpha = \frac{-\lambda_{\max}(AP + PA^T - 2BB^T)}{\lambda_{\max}(P)}. \quad (6.27)$$

Proof 44 *Consider the Lyapunov function V_1 as in the proof of Theorem 41. The time derivative of V_1 along the trajectory of (6.24) is given by*

$$\dot{V}_1 = \xi^T(\mathcal{L}_1 \otimes P^{-1})\dot{\xi}$$
$$= \frac{1}{2}\xi^T \mathcal{X}\xi + c_2\xi^T(\mathcal{L}_1 \otimes P^{-1}B)\widehat{G}(\xi) - \xi^T(\mathcal{L}_1\mathbf{1} \otimes P^{-1}B)u_1,$$

(6.28)

where $\mathcal{X}$ is defined in (6.9).

Next, consider the following three cases:

i) $\|K\sum_{j=2}^{M}\mathcal{L}_{ij}\xi_j\| > \kappa_i$, i.e., $\|B^T P^{-1}\sum_{j=2}^{M}\mathcal{L}_{ij}\xi_j\| > \kappa_i$, $i = 2, \cdots, N$.
In this case, it follows from (6.23) and (6.25) that

$$\xi^T(\mathcal{L}_1 \otimes P^{-1}B)\widehat{G}(\xi) = -\sum_{i=2}^{N}\|B^T P^{-1}\sum_{j=2}^{N}\mathcal{L}_{ij}\xi_j\|. \tag{6.29}$$

Then, we can get from (6.28), (6.10), and (6.29) that

$$\dot{V}_1 \leq \frac{1}{2}\xi^T \mathcal{X}\xi.$$

ii) $\|B^T P^{-1}\sum_{j=2}^{M}\mathcal{L}_{ij}\xi_j\| \leq \kappa_i$, $i = 2, \cdots, N$.
From (6.10), we can obtain that

$$-\xi^T(\mathcal{L}_1\mathbf{1} \otimes P^{-1}B)u_1 \leq \gamma\sum_{i=2}^{N}\|B^T P^{-1}\sum_{j=2}^{N}\mathcal{L}_{ij}\xi_j\|$$
$$\leq \gamma\sum_{i=2}^{N}\kappa_i. \tag{6.30}$$

Further, it follows from (6.23), (6.25), and (6.29) that

$$\xi^T(\mathcal{L}_1 \otimes P^{-1}B)\widehat{G}(\xi) = -\sum_{i=2}^{N}\frac{1}{\kappa_i}\|B^T P^{-1}\sum_{j=2}^{N}\mathcal{L}_{ij}\xi_j\|^2 \leq 0. \tag{6.31}$$

Thus, we get from (6.28), (6.30), and (6.31) that

$$\dot{V}_1 \leq \frac{1}{2}\xi^T \mathcal{X}\xi + \gamma\sum_{i=2}^{N}\kappa_i. \tag{6.32}$$

iii) ξ satisfies neither Case i) nor Case ii).

Without loss of generality, assume that $\|B^T P^{-1}\sum_{j=2}^{M}\mathcal{L}_{ij}\xi_j\| > \kappa_i$, $i = 2, \cdots, l$, and $\|B^T P^{-1}\sum_{j=2}^{M}\mathcal{L}_{ij}\xi_j\| \leq \kappa_i$, $i = l+1, \cdots, N$, where $3 \leq l \leq N-1$. It is easy to see from (6.10), (6.23), and (6.29) that

$$-\xi^T(\mathcal{L}_1\mathbf{1} \otimes P^{-1}B)u_1 \leq \gamma\sum_{i=2}^{l}\|B^T P^{-1}\sum_{j=2}^{N}\mathcal{L}_{ij}\xi_j\| + \gamma\sum_{i=2}^{N-l}\kappa_i,$$

$$\xi^T(\mathcal{L}_1 \otimes P^{-1}B)\widehat{G}(\xi) \leq -\sum_{i=2}^{l}\|B^T P^{-1}\sum_{j=2}^{N}\mathcal{L}_{ij}\xi_j\|.$$

Clearly, in this case we have

$$\dot{V}_1 \leq \frac{1}{2}\xi^T \mathcal{X}\xi + \gamma\sum_{i=2}^{N-l}\kappa_i.$$

Therefore, by analyzing the above three cases, we get that $\dot{V}_1$ satisfies (6.32) for all $\xi \in \mathbf{R}^{Nn}$. Note that (6.32) can be rewritten as

$$\dot{V}_1 \leq -\alpha V_1 + \alpha V_1 + \frac{1}{2}\xi^T \mathcal{X}\xi + \gamma \sum_{i=2}^{N} \kappa_i$$

$$= -\alpha V_1 + \frac{1}{2}\xi^T(\mathcal{X} + \alpha \mathcal{L}_1 \otimes P^{-1})\xi + \gamma \sum_{i=2}^{N} \kappa_i. \tag{6.33}$$

Because $\alpha = \frac{-\lambda_{\max}(AP+PA^T-2BB^T)}{\lambda_{\max}(P)}$, in light of (6.6), we can obtain that

$$(\mathcal{L}_1^{-\frac{1}{2}} \otimes P)(\mathcal{X} + \alpha \mathcal{L}_1 \otimes P^{-1})(\mathcal{L}_1^{-\frac{1}{2}} \otimes P)$$
$$\leq I_N \otimes [AP + PA^T + \alpha P - 2BB^T] < 0.$$

Then, it follows from (6.33) that

$$\dot{V}_1 \leq -\alpha V_1 + \gamma \sum_{i=2}^{N} \kappa_i. \tag{6.34}$$

By using Lemma 16 (the Comparison lemma), we can obtain from (6.34) that

$$V_1(\xi) \leq [V_1(\xi(0)) - \frac{\gamma \sum_{i=2}^{N} \kappa_i}{\alpha}]e^{-\alpha t} + \frac{\gamma \sum_{i=2}^{N} \kappa_i}{\alpha}, \tag{6.35}$$

which implies that ξ exponentially converges to the residual set $\mathcal{D}_1$ in (6.26) with a convergence rate not less than $e^{-\alpha t}$.

Remark 48 *Contrary to the discontinuous controller (6.3), the chattering effect can be avoided by using the continuous controller (6.22). The tradeoff is that the continuous controller (6.22) does not guarantee asymptotic stability of the tracking error ξ. Note that the upper bound of ξ depends on the communication graph $\mathcal{G}$, the number of followers, the upper bounds of the leader's control input and the widths κ_i of the boundary layers. By choosing sufficiently small κ_i, the tracking error ξ under the continuous controller (6.22) can be arbitrarily small, which is acceptable in most circumstances.*

6.3.2 Adaptive Continuous Controllers

In this subsection, we will present a continuous approximation of the discontinuous adaptive controller (6.14). It is worth noting that simply replacing the discontinuous function $g(\cdot)$ by the continuous functions $\hat{g}_i(\cdot)$ in (6.23) as did in the last subsection does not work. Using the adaptive controller with $g(\cdot)$ replaced by $\hat{g}_i(\cdot)$, the tracking error ξ will not asymptotically converge to zero but rather can only be expected to converge into some small neighborhood of zero. In this case, it can be observed from the last equation in (6.14) that the adaptive weights $d_i(t)$ will slowly grow unbounded. This argument can also been easily verified by simulation results. To tackle this problem, we propose to use the so-called σ-modification technique [63, 183] to modify the discontinuous adaptive controller (6.14).

Using the boundary layer concept and the σ-modification technique, a distributed continuous adaptive controller is proposed as

$$u_i = d_i K \sum_{j=1}^{N} a_{ij}(x_i - x_j) + d_i r_i(K \sum_{j=1}^{N} a_{ij}(x_i - x_j)),$$

$$\dot{d}_i = \tau_i[-\varphi_i d_i + (\sum_{j=1}^{N} a_{ij}(x_i - x_j)^T)\Gamma(\sum_{j=1}^{N} a_{ij}(x_i - x_j)) \qquad (6.36)$$

$$+ \|K \sum_{j=1}^{N} a_{ij}(x_i - x_j)\|], \quad i = 2, \cdots, N,$$

where φ_i are small positive constants, the nonlinear functions $r_i(\cdot)$ are defined as follows: for $w \in \mathbf{R}^n$,

$$r_i(w) = \begin{cases} \frac{w}{\|w\|} & \text{if } d_i\|w\| > \kappa_i \\ \frac{w}{\kappa_i}d_i & \text{if } d_i\|w\| \leq \kappa_i \end{cases} \qquad (6.37)$$

and the rest of the variables are defined as in (6.14).

Let the tracking error ξ and the coupling gains $D(t)$ be as defined in (6.15). Then, it follows from (6.1) and (6.36) that ξ and $D(t)$ satisfy the following dynamics:

$$\dot{\xi} = (I_{N-1} \otimes A + D\mathcal{L}_1 \otimes BK)\xi + (D \otimes B)R(\xi) - (\mathbf{1} \otimes B)u_1,$$

$$\dot{d}_i = \tau_i[-\varphi_i d_i + (\sum_{j=2}^{N} \mathcal{L}_{ij}\xi_j^T)\Gamma(\sum_{j=2}^{N} \mathcal{L}_{ij}\xi_j) + \|K \sum_{j=2}^{N} \mathcal{L}_{ij}\xi_j\|], \quad i = 2, \cdots, N,$$

$$(6.38)$$

where $R(x) \triangleq \begin{bmatrix} r_2(K \sum_{j=2}^{N} \mathcal{L}_{2j}\xi_j) \\ \vdots \\ r_N(K \sum_{j=2}^{N} \mathcal{L}_{Nj}\xi_j) \end{bmatrix}.$

The following theorem presents the ultimate boundedness of the states ξ and d_i of (6.38).

Theorem 44 *Suppose that Assumptions 6.1 and 6.2 hold. Then, both the tracking error ξ and the coupling gains d_i, $i = 2, \cdots, N$, in (6.38), under the continuous adaptive controller (6.36) with K and Γ designed as in Theorem 42, are uniformly ultimately bounded . Moreover, the following two assertions hold.*

i) For any φ_i, both ξ and d_i exponentially converge to the residual set

$$\mathcal{D}_2 \triangleq \{\xi, d_i : V_3 < \frac{1}{2\delta} \sum_{i=2}^{N} (\beta^2 \varphi_i + \frac{\kappa_i}{2})\}, \qquad (6.39)$$

where $\delta \triangleq \min_{i=2,\cdots,N}\{\alpha, \tau_i\varphi_i\}$ *and*

$$V_3 = \frac{1}{2}\xi^T(\mathcal{L} \otimes P^{-1})\xi + \sum_{i=2}^{N} \frac{\tilde{d}_i^2}{2\tau_i}, \tag{6.40}$$

where α *is defined as in (6.27),* $\tilde{d}_i = d_i - \beta$, *and* $\beta = \max\{\frac{1}{\lambda_{\min}(\mathcal{L}_1)}, \gamma\}$.

ii) *If* φ_i *satisfies* $\varrho \triangleq \min_{i=2,\cdots,N}\{\tau_i\varphi_i\} < \alpha$, *then in addition to i),* ξ *exponentially converges to the residual set*

$$\mathcal{D}_3 \triangleq \{\xi \;:\; \|\xi\|^2 \leq \frac{\lambda_{\max}(P)}{\lambda_{\min}(\mathcal{L}_1)(\alpha - \varrho)} \sum_{i=2}^{N}(\beta^2\varphi_i + \frac{1}{2}\kappa_i)\}. \tag{6.41}$$

Proof 45 *Consider the Lyapunov function candidate* V_3 *as in (6.40). The time derivative of* V_3 *along (6.38) can be obtained as*

$$\dot{V}_3 = \xi^T(\mathcal{L}_1 \otimes P^{-1})\dot{\xi} + \sum_{i=2}^{N} \frac{\tilde{d}_i}{\tau_i}\dot{\tilde{d}}_i$$

$$= \xi^T(\mathcal{L}_1 \otimes P^{-1}A + \mathcal{L}_1 D\mathcal{L}_1 \otimes P^{-1}BK)\xi$$

$$+ \xi^T(\mathcal{L}_1 D \otimes P^{-1}B)R(\xi) - \xi^T(\mathcal{L}_1\mathbf{1} \otimes P^{-1}B)u_1$$

$$+ \sum_{i=2}^{N} \tilde{d}_i[-\varphi_i(\tilde{d}_i + \beta) + (\sum_{j=2}^{N}\mathcal{L}_{ij}\xi_j^T)\Gamma(\sum_{j=2}^{N}\mathcal{L}_{ij}\xi_j) + \|K\sum_{j=2}^{N}\mathcal{L}_{ij}\xi_j\|],$$

$$\tag{6.42}$$

where $D = \text{diag}(\tilde{d}_1 + \beta, \cdots, \tilde{d}_N + \beta)$.
  Next, consider the following three cases:
  i) $d_i\|K\sum_{j=2}^{M}\mathcal{L}_{ij}\xi_j\| > \kappa_i$, $i = 2, \cdots, N$.
In this case, we can get from (6.37) that

$$\xi^T(\mathcal{L}_1 D \otimes P^{-1}B)R(\xi) = -\sum_{i=2}^{N}(\tilde{d}_i + \beta)\|B^T P^{-1}\sum_{j=2}^{N}\mathcal{L}_{ij}\xi_j\|. \tag{6.43}$$

In light of (6.10) and (6.18), substituting (6.43) into (6.42) yields

$$\dot{V}_3 \leq \frac{1}{2}\xi^T\mathcal{Z}\xi + \frac{1}{2}\sum_{i=2}^{N}\varphi_i(\beta^2 - \tilde{d}_i^2) - (\beta - \gamma)\sum_{i=2}^{N}\|B^T P^{-1}\sum_{j=2}^{N}\mathcal{L}_{ij}\xi_j\|$$

$$\leq \frac{1}{2}\xi^T\mathcal{Z}\xi + \frac{1}{2}\sum_{i=2}^{N}\varphi_i(\beta^2 - \tilde{d}_i^2),$$

where we have used the fact that $-\tilde{d}_i^2 - \tilde{d}_i\beta \leq -\frac{1}{2}\tilde{d}_i^2 + \frac{1}{2}\beta^2$ *to get the first inequality and* $\mathcal{Z} \triangleq \mathcal{L}_1 \otimes (P^{-1}A + A^T P^{-1}) - 2\beta\mathcal{L}_1^2 \otimes P^{-1}BB^T P^{-1}$.

ii) $d_i \| K \sum_{j=2}^{N} \mathcal{L}_{ij} \xi_j \| \le \kappa_i, \ i = 2, \cdots, N.$

In this case, we can get from (6.37) that

$$\xi^T (\mathcal{L}_1 D \otimes P^{-1} B) R(\xi) = -\sum_{i=2}^{N} \frac{(\tilde{d}_i + \beta)^2}{\kappa_i} \| B^T P^{-1} \sum_{j=2}^{N} \mathcal{L}_{ij} \xi_j \|. \qquad (6.44)$$

Then, it follows from (6.43), (6.10), (6.18), and (6.42) that

$$
\begin{aligned}
\dot{V}_3 &\le \frac{1}{2} \xi^T \mathcal{Z} \xi + \frac{1}{2} \sum_{i=2}^{N} \varphi_i (\beta^2 - \tilde{d}_i^2) \\
&\quad - \sum_{i=2}^{N} \frac{(\tilde{d}_i + \beta)^2}{\kappa_i} \| B^T P^{-1} \sum_{j=2}^{N} \mathcal{L}_{ij} \xi_j \|^2 \\
&\quad + \sum_{i=2}^{N} (\tilde{d}_i + \beta) \| B^T P^{-1} \sum_{j=2}^{N} \mathcal{L}_{ij} \xi_j \| \\
&\le \frac{1}{2} \xi^T \mathcal{Z} \xi + \frac{1}{2} \sum_{i=2}^{N} \varphi_i (\beta^2 - \tilde{d}_i^2) + \frac{1}{4} \sum_{i=2}^{N} \kappa_i.
\end{aligned}
\qquad (6.45)
$$

Note that to get the last inequality in (6.45), we have used the following fact:

$$-\frac{d_i^2}{\kappa_i} \| B^T P^{-1} \sum_{j=2}^{N} \mathcal{L}_{ij} \xi_j \|^2 + d_i \| B^T P^{-1} \sum_{j=2}^{N} \mathcal{L}_{ij} \xi_j \| \le \frac{1}{4} \kappa_i,$$

for $d_i \| K \sum_{j=2}^{N} \mathcal{L}_{ij} \xi_j \| \le \kappa_i, \ i = 2, \cdots, N.$

iii) ξ *satisfies neither Case i) nor Case ii).*

Without loss of generality, assume that $d_i \| K \sum_{j=2}^{N} \mathcal{L}_{ij} \xi_j \| \le \kappa_i, \ i = 2, \cdots, l,$ *and* $d_i \| K \sum_{j=2}^{N} \mathcal{L}_{ij} \xi_j \| > \kappa_i, \ i = l+1, \cdots, N,$ *where* $3 \le l \le N - 1.$ *By following similar steps in the two cases above, it is easy to get that*

$$\dot{V}_3 \le \frac{1}{2} \xi^T \mathcal{Z} \xi + \frac{1}{2} \sum_{i=2}^{N} \varphi_i (\beta^2 - \tilde{d}_i^2) + \frac{1}{4} \sum_{i=2}^{N-l} \kappa_i.$$

Therefore, based on the above three cases, we can get that $\dot{V}_3$ *satisfies (6.45) for all* $\xi \in \mathbf{R}^{Nn}.$ *Note that (6.45) can be rewritten into*

$$
\begin{aligned}
\dot{V}_3 &\le -\delta V_3 + \delta V_3 + \frac{1}{2} \xi^T \mathcal{Z} \xi + \frac{1}{2} \sum_{i=2}^{N} \varphi_i (\beta^2 - \tilde{d}_i^2) + \frac{1}{4} \sum_{i=2}^{N} \kappa_i \\
&= -\delta V_3 + \frac{1}{2} \xi^T (\mathcal{Z} + \delta \mathcal{L} \otimes P^{-1}) \xi - \frac{1}{2} \sum_{i=2}^{N} (\varphi_i - \frac{\delta}{\tau_i}) \tilde{d}_i^2 \\
&\quad + \frac{1}{2} \sum_{i=2}^{N} (\beta^2 \varphi_i + \frac{1}{2} \kappa_i).
\end{aligned}
\qquad (6.46)
$$

Because $\beta\lambda_{\min}(\mathcal{L}_1) \geq 1$ and $0 < \delta \leq \alpha$, by following similar steps in the proof of Theorem 43, we can show that $\xi^T(\mathcal{Z} + \delta\mathcal{L} \otimes P^{-1})\xi \leq 0$. Further, by noting that $\delta \leq \min_{i=2,\cdots,N} \varphi_i\tau_i$, it follows from (6.46) that

$$\dot{V}_3 \leq -\delta V_3 + \frac{1}{2}\sum_{i=2}^{N}(\beta^2\varphi_i + \frac{1}{2}\kappa_i), \qquad (6.47)$$

which implies that

$$V_3 \leq [V_3(0) - \frac{1}{2\delta}(\sum_{i=2}^{N} \beta^2\varphi_i + \frac{1}{2}\kappa_i)]e^{-\delta t}$$

$$+ \frac{1}{2\delta}\sum_{i=2}^{N}(\beta^2\varphi_i + \frac{1}{2}\kappa_i). \qquad (6.48)$$

Therefore, V_3 exponentially converges to the residual set $\mathcal{D}_2$ in (6.39) with a convergence rate faster than $e^{-\delta t}$, which, in light of $V_3 \geq \frac{\lambda_{\min}(\mathcal{L}_1)}{2\lambda_{\max}(P)}\|\xi\|^2 + \sum_{i=2}^{N}\frac{\tilde{d}_i^2}{2\tau_i}$, implies that ξ and $\tilde{d}_i$ are uniformly ultimately bounded.

Next, if $\varrho \triangleq \min_{i=2,\cdots,N} \varphi_i\tau_i < \alpha$, we can obtain a smaller residual set for ξ by rewriting (6.46) into

$$\dot{V}_3 \leq -\varrho V_3 + \frac{1}{2}\xi^T(\mathcal{Z} + \alpha\mathcal{L} \otimes P^{-1})\xi - \frac{1}{2}\sum_{i=2}^{N}(\varphi_i - \frac{\varrho}{\tau_i})\tilde{d}_i^2$$

$$- \frac{\alpha - \varrho}{2}\xi^T(\mathcal{L} \otimes P^{-1})\xi + \frac{1}{2}\sum_{i=2}^{N}(\beta^2\varphi_i + \frac{1}{2}\kappa_i) \qquad (6.49)$$

$$\leq -\varrho V_3 - \frac{\lambda_{\min}(\mathcal{L}_1)(\alpha - \varrho)}{2\lambda_{\max}(P)}\|\xi\|^2 + \frac{1}{2}\sum_{i=2}^{N}(\beta^2\varphi_i + \frac{1}{2}\kappa_i).$$

Obviously, it follows from (6.49) that $\dot{V}_3 \leq -\varrho V_3$ if $\|\xi\|^2 > \frac{\lambda_{\max}(P)}{\lambda_{\min}(\mathcal{L}_1)(\alpha-\varrho)}\sum_{i=2}^{N}(\beta^2\varphi_i + \frac{1}{2}\kappa_i)$. Then, by noting $V_3 \geq \frac{\lambda_{\min}(\mathcal{L}_1)}{2\lambda_{\max}(P)}\|\xi\|^2$, we can get that if $\varrho \leq \alpha$ then ξ exponentially converges to the residual set $\mathcal{D}_3$ in (6.41) with a convergence rate faster than $e^{-\varrho t}$.

Remark 49 *It is worth mentioning that adding $-\varphi_i d_i$ into (6.36) is essentially motivated by the so-called σ-modification technique in [63, 183], which plays a vital role to guarantee the ultimate boundedness of the tracking error ξ and the adaptive gains d_i. From (6.39) and (6.41), we can observe that the residual sets $\mathcal{D}_2$ and $\mathcal{D}_3$ decrease as κ_i decrease. Given κ_i, smaller φ_i gives a smaller bound for ξ and at the same time yields a larger bound for d_i. For the case where $\varphi_i = 0$, d_i will tend to infinity. In real implementations, if large d_i is acceptable, we can choose φ_i and κ_i to be relatively small in order to guarantee a small tracking error ξ.*

Remark 50 *Similarly as in Section 5.4 in the previous chapter, it is not difficult to show that the proposed adaptive controller (6.36) can be redesigned to be robust with respect to bounded external disturbances, i.e., for the agents described by (5.46) in Section 5.4 of the previous chapter. The details are omitted here for conciseness.*

6.4   Distributed Output-Feedback Controllers

In this section, we intend to consider the case where each agent has only access to the local output information, rather than the local state information, of itself and its neighbors. In this case, based on the relative estimates of the states of neighboring agents, the following distributed continuous adaptive controller is proposed for each follower:

$$
\dot{\hat{v}}_i = A\hat{v}_i + Bu_i + \hat{L}(C\hat{v}_i - y_i),
$$

$$
u_i = d_i \sum_{j=1}^{N} a_{ij}\hat{F}(\hat{v}_i - \hat{v}_j) + d_i r_i(\sum_{j=1}^{N} a_{ij}\hat{F}(\hat{v}_i - \hat{v}_j)),
$$

$$
\dot{d}_i = \tau_i[-\varphi_i d_i + (\sum_{j=1}^{N} a_{ij}(\hat{v}_i - \hat{v}_j)^T)\hat{\Gamma}(\sum_{j=1}^{N} a_{ij}(\hat{v}_i - \hat{v}_j)) \tag{6.50}
$$

$$
+ \|\sum_{j=1}^{N} a_{ij}\hat{F}(\hat{v}_i - \hat{v}_j)\|], \quad i = 2, \cdots, N,
$$

where $\hat{v}_i \in \mathbf{R}^n$ is the estimate of the state of the i-th follower, $\hat{v}_1 \in \mathbf{R}^n$ denotes the estimate of the state of the leader, given by $\dot{\hat{v}}_1 = A\hat{v}_1 + Bu_1 + L(C\hat{v}_1 - y_1)$, φ_i are small positive constants, τ_i are positive scalars, d_i denotes the time-varying coupling gain associated with the i-th follower, $\hat{L} \in \mathbf{R}^{n \times q}$, $\hat{F} \in \mathbf{R}^{p \times n}$, and $\hat{\Gamma} \in \mathbf{R}^{n \times n}$ are the feedback gain matrices, and $r_i(\cdot)$ are defined as in (6.37). Note that the term $\sum_{j=1}^{N} a_{ij}(\hat{v}_i - \hat{v}_j)$ in (6.50) imply that the agents need to get the estimates of the states of their neighbors, which can be transmitted via the communication graph $\mathcal{G}$.

Let $\zeta_i = \begin{bmatrix} x_i - x_1 \\ \hat{v}_i - \hat{v}_1 \end{bmatrix}$, $i = 2, \cdots, N$, and $\zeta = [\zeta_2^T, \cdots, \zeta_N^T]^T$. Using (6.50) for (6.1), we obtain the closed-loop network dynamics as

$$
\dot{\zeta} = (I \otimes \mathcal{S} + D\mathcal{L}_1 \otimes \mathcal{F})\zeta + (D \otimes \mathcal{B})R(\zeta) - (\mathbf{1} \otimes \mathcal{B})u_1,
$$

$$
\dot{d}_i = \tau_i[-\varphi_i d_i + (\sum_{j=2}^{N} \mathcal{L}_{ij}\zeta_j^T)\mathcal{I}(\sum_{j=2}^{N} \mathcal{L}_{ij}\zeta_j) + \|\mathcal{O}\sum_{j=2}^{N} \mathcal{L}_{ij}\zeta_j\|], i = 2, \cdots, N,
$$

$$
\tag{6.51}
$$

where $D = \text{diag}(d_2, \cdots, d_N)$, and

$$R(\zeta) \triangleq \begin{bmatrix} r_2(\mathcal{O} \sum_{j=2}^{N} \mathcal{L}_{2j}\zeta_j) \\ \vdots \\ r_N(\mathcal{O} \sum_{j=2}^{N} \mathcal{L}_{Nj}\zeta_j) \end{bmatrix}, \quad \mathcal{S} = \begin{bmatrix} A & 0 \\ -\hat{L}C & A + \hat{L}C \end{bmatrix},$$

$$\mathcal{F} = \begin{bmatrix} 0 & B\hat{F} \\ 0 & B\hat{F} \end{bmatrix}, \quad \mathcal{I} = \begin{bmatrix} 0 & 0 \\ 0 & \hat{\Gamma} \end{bmatrix}, \quad \mathcal{B} = \begin{bmatrix} B \\ B \end{bmatrix}, \quad \mathcal{O} = \begin{bmatrix} 0 & \hat{F} \end{bmatrix}.$$

Clearly, the distributed tracking problem is solved if the tracking error ζ of (6.51) converges to zero.

Theorem 45 *Suppose that Assumptions 6.1 and 6.2 hold. Then, both the tracking error ζ and the coupling gains d_i, $i = 2, \cdots, N$, in (6.51) are u-niformly ultimately bounded under the distributed continuous adaptive protocol (6.50) with $\hat{L}$ satisfying that $A + \hat{L}C$ is Hurwitz, $\hat{F} = -B^T P^{-1}$, and $\hat{\Gamma} = P^{-1} B B^T P^{-1}$, where $P > 0$ is a solution to the LMI (6.6).*

Proof 46 *Consider the following Lyapunov function candidate*

$$V_4 = \frac{1}{2}\zeta^T (\mathcal{L}_1 \otimes \mathcal{P})\zeta + \sum_{i=2}^{N} \frac{\tilde{d}_i^2}{2\tau_i},$$

where $\mathcal{P} \triangleq \begin{bmatrix} \vartheta\hat{P} & -\vartheta\hat{P} \\ -\vartheta\hat{P} & \vartheta\hat{P}+P^{-1} \end{bmatrix}$, $\hat{P} > 0$ satisfies that $(A+\hat{L}C)\hat{P}+(A+\hat{L}C)^T\hat{P} < 0$, $\tilde{d}_i = d_i - \beta$, $i = 2, \cdots, N$, β is sufficiently large such that $\beta\lambda_{\min}(\mathcal{L}_1) \geq 1$ and $\beta \geq \gamma$, and ϑ is a positive constant to be determined later. In light of Lemma 19 (Schur Complement Lemma), it is easy to know that $\mathcal{P} > 0$. Since $\mathcal{P} > 0$, it is easy to see that V_4 is positive definite.

The time derivative of V_4 along (6.51) can be obtained as

$$\begin{aligned} \dot{V}_4 = &\hat{\zeta}^T[(\mathcal{L}_1 \otimes \widehat{\mathcal{P}}\widehat{\mathcal{S}} + \mathcal{L}_1 D \mathcal{L}_1 \otimes \widehat{\mathcal{P}}\widehat{\mathcal{F}})\hat{\zeta} - (\mathcal{L}_1 \mathbf{1} \otimes \widehat{\mathcal{P}}\widehat{\mathcal{B}})u_1] \\ &+ \hat{\zeta}^T(\mathcal{L}_1 D \otimes \widehat{\mathcal{P}}\widehat{\mathcal{B}})R(\hat{\zeta}) + \sum_{i=2}^{N} \tilde{d}_i[-\varphi_i(\tilde{d}_i + \beta) \\ &+ (\sum_{j=2}^{N} \mathcal{L}_{ij}\hat{\zeta}_j^T)\mathcal{I}(\sum_{j=2}^{N} \mathcal{L}_{ij}\hat{\zeta}_j) + \|\mathcal{O} \sum_{j=2}^{N} \mathcal{L}_{ij}\hat{\zeta}_j\|], \end{aligned} \qquad (6.52)$$

where $D = \text{diag}(\tilde{d}_2 + \beta, \cdots, \tilde{d}_N + \beta)$, $\hat{\zeta} \triangleq [\hat{\zeta}_1^T, \cdots, \hat{\zeta}_N^T]^T = (I \otimes \hat{T})\zeta$ with $\hat{T} = \begin{bmatrix} I & -I \\ 0 & I \end{bmatrix}$, $\widehat{\mathcal{P}} = \begin{bmatrix} \vartheta\hat{P} & 0 \\ 0 & P^{-1} \end{bmatrix}$, $\widehat{\mathcal{S}} = \begin{bmatrix} A+\hat{L}C & 0 \\ -\hat{L}C & A \end{bmatrix}$, $\widehat{\mathcal{F}} = \begin{bmatrix} 0 & 0 \\ 0 & B\hat{F} \end{bmatrix}$, and $\widehat{\mathcal{B}} = \begin{bmatrix} 0 \\ B \end{bmatrix}$.
Note that

$$\hat{\zeta}^T(\mathcal{L}_1 D \mathcal{L}_1 \otimes \widehat{\mathcal{P}}\widehat{\mathcal{F}})\hat{\zeta} = \sum_{i=2}^{N}(\tilde{d}_i + \beta)[\sum_{j=2}^{N} \mathcal{L}_{ij}\hat{\zeta}_j^T]\widehat{\mathcal{P}}\widehat{\mathcal{F}}[\sum_{j=2}^{N} \mathcal{L}_{ij}\hat{\zeta}_j]. \qquad (6.53)$$

In virtue of Assumption 6.1, we have

$$-\hat{\zeta}^T(\mathcal{L}_1 \mathbf{1} \otimes \widehat{\mathcal{P}}\widehat{\mathcal{B}})u_1 \leq \gamma \sum_{i=2}^{N} \|\widehat{\mathcal{B}}^T\widehat{\mathcal{P}}\sum_{j=2}^{N}\mathcal{L}_{ij}\hat{\zeta}_j\|. \tag{6.54}$$

Next, consider the following three cases.
i) $d_i\|\mathcal{O}\sum_{j=2}^{N}\mathcal{L}_{ij}\hat{\zeta}_j\| > \kappa_i$, $i = 2, \cdots, N$.
By observing that $\widehat{\mathcal{P}}\widehat{\mathcal{B}} = -\mathcal{O}^T$, it is not difficult to get from the definitions of $r_i(\cdot)$ and $R(\cdot)$ that

$$\hat{\zeta}^T(\mathcal{L}_1 D \otimes \widehat{\mathcal{P}}\widehat{\mathcal{B}})R(\hat{\zeta}) = -\sum_{i=2}^{N}(\tilde{d}_i + \beta)\|\widehat{\mathcal{B}}^T\widehat{\mathcal{P}}\sum_{j=2}^{N}\mathcal{L}_{ij}\hat{\zeta}_j\|. \tag{6.55}$$

By noting that $\widehat{\mathcal{P}}\widehat{\mathcal{F}} = -\mathcal{I}$ and substituting (6.53), (6.54), and (6.55) into (6.52), we can get that

$$\dot{V}_4 \leq Z(\hat{\zeta}) - \sum_{i=2}^{N}\varphi_i(\tilde{d}_i^2 + \beta\tilde{d}_i) - (\beta - \gamma)\sum_{i=2}^{N}\|\widehat{\mathcal{B}}^T\widehat{\mathcal{P}}\sum_{j=2}^{N}\mathcal{L}_{ij}\hat{\zeta}_j\|$$

$$\leq Z(\hat{\zeta}) + \frac{1}{2}\sum_{i=2}^{N}\varphi_i(\beta^2 - \tilde{d}_i^2),$$

where

$$Z(\hat{\zeta}) \triangleq \frac{1}{2}\hat{\zeta}^T[\mathcal{L}_1 \otimes (\widehat{\mathcal{P}}\widehat{\mathcal{S}} + \widehat{\mathcal{S}}^T\widehat{\mathcal{P}}) - 2\beta\mathcal{L}_1^2 \otimes \mathcal{I}]\hat{\zeta}. \tag{6.56}$$

ii) $d_i\|\mathcal{O}\sum_{j=2}^{N}\mathcal{L}_{ij}\hat{\zeta}_j\| \leq \kappa_i$, $i = 2, \cdots, N$.
From the definitions of $r_i(\cdot)$ and $R(\cdot)$, in this case we get that

$$\hat{\zeta}^T(\mathcal{L}_1 D \otimes \widehat{\mathcal{P}}\widehat{\mathcal{B}})R(\hat{\zeta}) = -\sum_{i=2}^{N}\frac{(\tilde{d}_i + \beta)^2}{\kappa_i}\|\widehat{\mathcal{B}}^T\widehat{\mathcal{P}}\sum_{j=2}^{N}\mathcal{L}_{ij}\hat{\zeta}_j\|^2. \tag{6.57}$$

Then, it follows from (6.53), (6.54), (6.57), and (6.52) that

$$\dot{V}_4 \leq Z(\hat{\zeta}) - \sum_{i=2}^{N}\varphi_i(\tilde{d}_i^2 + \beta\tilde{d}_i) - \sum_{i=2}^{N}\frac{(\tilde{d}_i + \beta)^2}{\kappa_i}\|\widehat{\mathcal{B}}^T\widehat{\mathcal{P}}\sum_{j=2}^{N}\mathcal{L}_{ij}\hat{\zeta}_j\|^2$$

$$+ \sum_{i=2}^{N}(\tilde{d}_i + \beta)\|\widehat{\mathcal{B}}^T\widehat{\mathcal{P}}\sum_{j=2}^{N}\mathcal{L}_{ij}\hat{\zeta}_j\| \tag{6.58}$$

$$\leq Z(\hat{\zeta}) - \sum_{i=2}^{N}\varphi_i(\tilde{d}_i^2 + \beta\tilde{d}_i) + \frac{1}{4}\sum_{i=2}^{N}\kappa_i.$$

Note that to get the last inequality in (6.58), we have used the fact that for

$d_i\|\widehat{\mathcal{B}}^T\widehat{\mathcal{P}}\sum_{j=2}^N\mathcal{L}_{ij}\zeta_j\| \leq \kappa_i,\ -\frac{d_i^2}{\kappa_i}\|\widehat{\mathcal{B}}^T\widehat{\mathcal{P}}\sum_{j=2}^N\mathcal{L}_{ij}\hat{\zeta}_j\|^2 + d_i\|\widehat{\mathcal{B}}^T\widehat{\mathcal{P}}\sum_{j=2}^N\mathcal{L}_{ij}\hat{\zeta}_j\| \leq \frac{1}{4}\kappa_i.$

iii) ζ *satisfies neither Case i) nor Case ii).*

In this case, without loss of generality, assume that $d_i\|\mathcal{O}\sum_{j=2}^N\mathcal{L}_{ij}\hat{\zeta}_j\| > \kappa_i$, $i = 2,\cdots,l$, *and* $d_i\|\mathcal{O}\sum_{j=2}^N\mathcal{L}_{ij}\hat{\zeta}_j\| \leq \kappa_i$, $i = l+1,\cdots,N$, *where* $3 \leq l \leq N-1$. *Combining (6.55) and (6.57), we have*

$$\hat{\zeta}^T(\mathcal{L}_1 D \otimes \widehat{\mathcal{P}}\widehat{\mathcal{B}})R(\hat{\zeta}) = -\sum_{i=2}^l(\tilde{d}_i + \beta)\|\widehat{\mathcal{B}}^T\widehat{\mathcal{P}}\sum_{j=2}^N\mathcal{L}_{ij}\hat{\zeta}_j\|$$

$$-\sum_{i=l+1}^N\frac{(\tilde{d}_i + \beta)^2}{\kappa_i}\|\widehat{\mathcal{B}}^T\widehat{\mathcal{P}}\sum_{j=2}^N\mathcal{L}_{ij}\hat{\zeta}_j\|^2.$$

Then, by following the steps in the two cases above, it is not difficult to get that in this case

$$\dot{V}_4 \leq Z(\hat{\zeta}) - \sum_{i=2}^N\varphi_i(\tilde{d}_i^2 + \beta\tilde{d}_i) + \frac{1}{4}\sum_{i=2}^{N-l}\kappa_i.$$

Analyzing the above three cases, we get that $\dot{V}_4$ *satisfies (6.58) for all* $\zeta \in \mathbf{R}^{2Nn}$. *Note that*

$$\mathrm{diag}(I,P)[\widehat{\mathcal{P}}\widehat{\mathcal{S}} + \widehat{\mathcal{S}}^T\widehat{\mathcal{P}} - 2\beta\lambda_{\min}(\mathcal{L}_1)\mathcal{I}]\mathrm{diag}(I,P)$$

$$= \begin{bmatrix} \Xi & -C^TL^T \\ -LC & AP + PA^T - 2\beta\lambda_{\min}(\mathcal{L}_1)BB^T \end{bmatrix}, \tag{6.59}$$

where $\Xi = \vartheta[\hat{P}(A + LC) + (A + LC)^T\hat{P}]$. *Because* $\beta\lambda_{\min}(\mathcal{L}_1) \geq 1$, *it follows from (6.6) that* $AP + PA^T - 2\beta\lambda_{\min}(\mathcal{L}_1)BB^T < 0$. *Then, choosing* $\vartheta > 0$ *sufficiently large and in virtue of Lemma 19 (Schur Complement Lemma), we can obtain from (6.59) that* $\widehat{\mathcal{P}}\widehat{\mathcal{S}} + \widehat{\mathcal{S}}^T\widehat{\mathcal{P}} - 2\beta\lambda_{\min}(\mathcal{L}_1)\mathcal{I} < 0$. *Observe that*

$$(\mathcal{L}_1^{-\frac{1}{2}} \otimes I)[\mathcal{L}_1 \otimes (\widehat{\mathcal{P}}\widehat{\mathcal{S}} + \widehat{\mathcal{S}}^T\widehat{\mathcal{P}}) - 2\beta\mathcal{L}_1^2 \otimes \mathcal{I}](\mathcal{L}_1^{-\frac{1}{2}} \otimes I)$$

$$\leq I \otimes [\widehat{\mathcal{P}}\widehat{\mathcal{S}} + \widehat{\mathcal{S}}^T\widehat{\mathcal{P}} - 2\beta\lambda_{\min}(\mathcal{L}_1)\mathcal{I}].$$

Thus, we know that $\mathcal{L}_1 \otimes (\widehat{\mathcal{P}}\widehat{\mathcal{S}} + \widehat{\mathcal{S}}^T\widehat{\mathcal{P}}) - 2\beta\mathcal{L}_1^2 \otimes \mathcal{I} < 0$, *implying that* $Z(\hat{\zeta}) - \frac{1}{2}\sum_{i=2}^N\varphi_i\tilde{d}_i^2 < 0$, *where* $Z(\hat{\zeta})$ *is defined in (6.56). In virtue of Lemma 15, we get from (6.58) that the states* ζ *and* d_i *of (6.51) are uniformly ultimately bounded.*

Remark 51 *As pointed out in Proposition 1 of Chapter 2, a sufficient condition for the existence of a protocol (6.50) satisfying Theorem 45 is that* (A, B, C) *is stabilizable and detectable. The adaptive protocol (6.50) can be implemented by each agent in a fully distributed fashion requiring neither global information of the communication topology nor the upper bound for the leader's control input.*

Remark 52 *Note that the term $d_i r_i(\sum_{j=1}^{N} a_{ij} F(\hat{v}_i - \hat{v}_j))$ in (6.50) is used to tackle the effect of the leader's nonzero control input u_1 on consensus. For the special case where $u_1 = 0$, we can simplify (6.50) into the following form:*

$$\dot{\hat{v}}_i = A\hat{v}_i + Bu_i + \hat{L}(C\hat{v}_i - y_i),$$

$$\dot{d}_i = \tau_i [\sum_{j=1}^{N} a_{ij}(\hat{v}_i - \hat{v}_j)^T] \hat{\Gamma} [\sum_{j=1}^{N} a_{ij}(\hat{v}_i - \hat{v}_j)] \tag{6.60}$$

$$u_i = d_i \sum_{j=1}^{N} a_{ij} \hat{F}(\hat{v}_i - \hat{v}_j), \quad i = 2, \cdots, N.$$

where $\hat{v}_1$ satisfies $\dot{\hat{v}}_1 = A\hat{v}_1 + L(C\hat{v}_1 - y_1)$.

For the special case where $u_1 = 0$, the adaptive consensus protocols (5.25) and (5.26) based on the relative output information of neighboring agents in the previous chapter can also be extended to solve the distributed tracking problem. For instance, a node-based adaptive protocol can be proposed as

$$\dot{\tilde{v}}_i = (A + BF)\tilde{v}_i + \tilde{d}_i L \sum_{j=1}^{N} a_{ij} \left[C(\tilde{v}_i - \tilde{v}_j) - (y_i - y_j) \right],$$

$$\dot{\tilde{d}}_i = \epsilon_i \left(\sum_{j=1}^{N} a_{ij} \begin{bmatrix} y_i - y_j \\ C(\tilde{v}_i - \tilde{v}_j) \end{bmatrix} \right)^T \Gamma \left(\sum_{j=1}^{N} a_{ij} \begin{bmatrix} y_i - y_j \\ C(\tilde{v}_i - \tilde{v}_j) \end{bmatrix} \right), \tag{6.61}$$

$$u_i = F\tilde{v}_i, \quad i = 2, \cdots, N,$$

where $\tilde{v}_i \in \mathbf{R}^n$ is the protocol state, $i = 2, \cdots, N$, $\tilde{v}_1$ is generated by $\dot{\tilde{v}}_1 = (A + BF)\tilde{v}_1$.

Remark 53 *Comparing the protocols (6.60) and (6.61), to solve the distributed tracking problem with u_1, the latter one is more preferable due to two reasons. First, the protocol (6.61) generally requires a lighter communication load (i.e., a lower dimension of information to be transmitted) than (6.60). Note that the protocol (6.61) exchanges y_i and $C\tilde{v}_i$ between neighboring agents while (6.60) exchanges $\hat{v}_i$. The sum of the dimensions of y_i and $C\tilde{v}_i$ is generally lower than the dimension of $\hat{v}_i$, e.g., for the single output case with $n > 2$. Second, the protocol (6.60) requires the absolute measures of the agents' outputs, which might not be available in some circumstances, e.g., the case where the agents are equipped with only ultrasonic range sensors. On the other hand, an advantage of (6.60) is that it can be modified (specifically, the protocol (6.50)) to solve the distributed tracking problem for the case where the leader's control input is nonzero and time varying. In contrast, it is not an easy job to extend (6.61) to achieve consensus for the case of a leader with nonzero input.*

6.5   Simulation Examples

In this section, a numerical example is presented to illustrate the effectiveness of the proposed theoretical results.

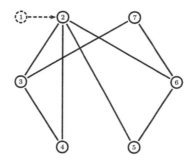

FIGURE 6.1: A leader-follower communication graph.

Example 21 *The dynamics of the agents are given by (6.1), with*

$$x_i = \begin{bmatrix} x_{i1} \\ x_{i2} \end{bmatrix}, \ A = \begin{bmatrix} 0 & 1 \\ -1 & 2 \end{bmatrix}, \ B = \begin{bmatrix} 0 \\ 1 \end{bmatrix}, \ C = \begin{bmatrix} 1 & 0 \end{bmatrix}.$$

The communication graph is given as in FIGURE 6.1, where the node indexed by 1 is the leader which is only accessible to the second node. Clearly all the agents are unstable without control. For the leader, design a first-order controller in the form of

$$\dot{v} = -3v - 2.5x_{11},$$
$$u_1 = -8v - 6x_{11}.$$

Then, the closed-loop dynamics of the leader have eigenvalues as -1 and $\pm \iota$. In this case, u_1 is clearly bounded. However, the bound γ, for which $\|u_1\| \leq \gamma$, depends on the initial state, which thereby might not be known to the followers. Here we use the adaptive controller (6.50) to solve the distributed tracking problem.

In (6.50), choose $\hat{L} = -\begin{bmatrix} 3 & 6 \end{bmatrix}^T$ such that $A + \hat{L}C$ is Hurwitz. Solving the LMI (6.6) gives the gain matrices $\hat{F}$ and $\hat{\Gamma}$ in (6.50) as

$$\hat{F} = -\begin{bmatrix} 1.2983 & 3.3878 \end{bmatrix}, \ \hat{\Gamma} = \begin{bmatrix} 1.6857 & 4.3984 \\ 4.3984 & 11.4769 \end{bmatrix}.$$

To illustrate Theorem 45, select $\kappa_i = 0.1$, $\varphi_i = 0.005$, and $\tau_i = 2$, $i = 2, \cdots, 7$, in (6.50). The tracking errors $x_i - x_1$, $i = 2, \cdots, 7$, of the agents under (6.50) designed as above are shown in FIGURE 6.2(a), implying that distributed tracking is indeed achieved. The coupling gains d_i associated with the followers are drawn in FIGURE 6.2(b), which are clearly bounded.

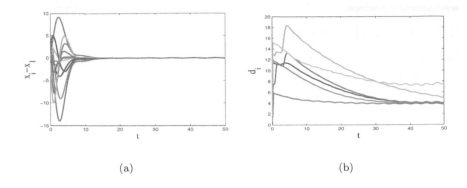

(a)   (b)

FIGURE 6.2: (a) The tracking errors $x_i - x_1$, $i = 2, \cdots, 7$; (b) the coupling gains d_i in (6.50).

6.6   Notes

The results in this chapter are mainly based on [92, 95]. In the literature, there exist other works on distributed tracking control of multi-agent systems with a leader of nonzero control input, e.g., [21, 109]. Specifically, in [21], discontinuous controllers were studied for multi-agent systems with first- and second-order integrator agent dynamics in the absence of velocity or acceleration measurements. The authors in [109] addressed a distributed coordinated tracking problem for multiple Euler-Lagrange systems with a dynamic leader. It is worthwhile to mention that the distributed tracking controllers proposed in [21, 109] are discontinuous, which will result in undesirable chattering phenomenon in real applications. In this chapter, to deal with the effect of the leader's nonzero control input on consensus, we actually borrow ideas from the sliding mode control literature and treat the leader's control input as an external disturbance satisfying the matching condition. Another possible approach is to design certain learning algorithm using the relative state information to estimate the leader's nonzero control input and then add the estimate into the consensus protocols. Related works along this line include [125, 192].

7

Containment Control of Linear Multi-Agent Systems with Multiple Leaders

CONTENTS

7.1   Containment of Continuous-Time Multi-Agent Systems with Leaders
of Zero Inputs 160
     7.1.1   Dynamic Containment Controllers 161
     7.1.2   Static Containment Controllers 164
7.2   Containment Control of Discrete-Time Multi-Agent Systems with
Leaders of Zero Inputs 164
     7.2.1   Dynamic Containment Controllers 165
     7.2.2   Static Containment Controllers 168
     7.2.3   Simulation Examples 168
7.3   Containment of Continuous-Time Multi-Agent Systems with Leaders
of Nonzero Inputs 170
     7.3.1   Distributed Continuous Static Controllers 171
     7.3.2   Adaptive Continuous Containment Controllers 176
     7.3.3   Simulation Examples 180
7.4   Notes 181

Roughly speaking, existing consensus algorithms can be categorized into two classes, namely, consensus without a leader and consensus with a leader. The case of consensus with a leader is also called leader-follower consensus or distributed tracking. Both the leaderless and leader-follower consensus problems have been addressed in the previous chapters. The distributed tracking problem deals with only one leader. However, in some practical applications, there might exist more than one leader in agent networks. In the presence of multiple leaders, the containment control problem arises where the followers are to be driven into a given geometric space spanned by the leaders [67]. The study of containment control has been motivated by many potential applications. For instance, a group of autonomous vehicles (designated as leaders) equipped with necessary sensors to detect the hazardous obstacles can be used to safely maneuver another group of vehicles (designated as followers) from one target to another, by ensuring that the followers are contained within the moving safety area formed by the leaders [19, 67].

     In this chapter, we extend the aforementioned papers to study the distributed containment control problem for multi-agent systems with general linear dynamics. Both the cases where the leaders have zero control inputs and have nonzero unknown inputs are considered. For the simple case where the leaders are of zero control inputs, the containment control problems for both continuous-time and discrete-time multi-agent systems with general linear dynamics under directed communication topologies are considered in Sections 7.1 and 7.2, respectively. Distributed dynamic containment controllers based on the relative outputs of neighboring agents are proposed for both the continuous-time and discrete-time cases. In the continuous-time case, a multistep algorithm is presented to construct a dynamic containment controller, under which the states of the followers will asymptotically converge to the convex hull formed by those of the leaders, if for each follower there exists at least one leader that has a directed path to that follower. In the discrete-time case, in light of the modified algebraic Riccati equation, an algorithm is given to design a dynamic containment controller that solves the containment control problem. Furthermore, as special cases of the dynamic controllers, static containment controllers relying on the relative states of neighboring agents are also discussed for both the continuous-time and discrete-time cases.

The containment control problem for the general case where the leaders have nonzero, bounded, and time-varying control inputs is further investigated in Section 7.3. Based on the relative states of neighboring agents, distributed continuous static and adaptive controllers are designed, under which the containment error is uniformly ultimately bounded and the upper bound of the containment error can be made arbitrarily small, if the subgraph associated with the followers is undirected and for each follower there exists at least one leader that has a directed path to that follower. Extensions to the case where only local output information is available are discussed. Based on the relative estimates of the states of neighboring agents, distributed observer-based containment controllers are proposed.

7.1   Containment of Continuous-Time Multi-Agent Systems with Leaders of Zero Inputs

Consider a group of N identical agents with general continuous-time linear dynamics, described by

$$
\begin{aligned}
\dot{x}_i &= Ax_i + Bu_i, \\
y_i &= Cx_i, \ i = 1, \cdots, N,
\end{aligned}
\tag{7.1}
$$

where $x_i \in \mathbf{R}^n$, $u_i \in \mathbf{R}^p$, and $y_i \in \mathbf{R}^q$ are, respectively, the state, the control input, and the measured output of the i-th agent, and A, B, C, are constant matrices with compatible dimensions.

We consider the case with multiple leaders. Suppose that there are M ($M < N$) followers and $N - M$ leaders. An agent is called a leader if the agent has no neighbor. An agent is called a follower if the agent has at least one neighbor. Without loss of generality, we assume that the agents indexed by $1, \cdots, M$, are followers, while the agents indexed by $M+1, \cdots, N$, are leaders whose control inputs are set to be zero. We use $\mathcal{R} \triangleq \{M+1, \cdots, N\}$ and $\mathcal{F} \triangleq \{1, \cdots, M\}$ to denote, respectively, the leader set and the follower set. In this section, we consider the case where the leaders have zero control inputs, i.e., $u_i = 0$, $i \in \mathcal{R}$. The communication topology among the N agents is represented by a directed graph $\mathcal{G}$. Note that here the leaders do not receive any information.

Assumption 7.1 *Suppose that for each follower, there exists at least one leader that has a directed path to that follower.*

The objective is to design distributed controllers to solve the containment control problem, defined as follows.

Definition 17 *The containment control problem is solved for the agents in (7.1) if the states of the followers under a certain distributed controller asymptotically converge to the convex hull formed by those of the leaders.*

Clearly the containment control problem will reduce to the distributed tracking problem discussed as in the previous chapters, when only one leader exists.

7.1.1 Dynamic Containment Controllers

In this subsection, it is assumed that each agent has access to the relative output measurements with respect to its neighbors. In this case, we propose the following distributed dynamic containment controller for each follower as

$$
\begin{aligned}
\dot{v}_i &= (A + BF)v_i + cL \sum_{j \in \mathcal{F} \cup \mathcal{R}} a_{ij}[C(v_i - v_j) - (y_i - y_j)], \\
u_i &= Fv_i, \ i \in \mathcal{F},
\end{aligned}
\tag{7.2}
$$

where $v_i \in \mathbf{R}^n$, $i \in \mathcal{F}$, is the state of the controller corresponding to the i-th follower, $v_j = 0 \in \mathbf{R}^n$ for $j \in \mathcal{R}$, a_{ij} is the (i,j)-th entry of the adjacency matrix associated with $\mathcal{G}$, $c > 0 \in \mathbf{R}$ denotes the coupling gain, and $L \in \mathbf{R}^{n \times q}$ and $F \in \mathbf{R}^{p \times n}$ are feedback gain matrices. In (7.2), the term $\sum_{j \in \mathcal{F} \cup \mathcal{R}} a_{ij} C(v_i - v_j)$ means that the agents need to transmit the virtual outputs of their corresponding controllers to their neighbors via the communication topology $\mathcal{G}$.

The above distributed controller is an extension of the observer-type consensus protocol (2.21) in Chapter 2 to the case with multiple leaders. The other observer-type protocol (2.33) can be similarly extended. The details are omitted here.

Denote by $\mathcal{L}$ the Laplacian matrix associated with $\mathcal{G}$. Because the leaders have no neighbors, $\mathcal{L}$ can be partitioned as

$$\mathcal{L} = \begin{bmatrix} \mathcal{L}_1 & \mathcal{L}_2 \\ 0_{(N-M)\times M} & 0_{(N-M)\times(N-M)} \end{bmatrix}, \tag{7.3}$$

where $\mathcal{L}_1 \in \mathbf{R}^{M\times M}$ and $\mathcal{L}_2 \in \mathbf{R}^{M\times(N-M)}$.

Lemma 30 ([111]) *Under Assumption 7.1, all the eigenvalues of $\mathcal{L}_1$ has positive real parts, each entry of $-\mathcal{L}_1^{-1}\mathcal{L}_2$ is nonnegative, and each row of $-\mathcal{L}_1^{-1}\mathcal{L}_2$ has a sum equal to one.*

Next, an algorithm is presented to select the control parameters in (7.2).

Algorithm 14 *Under Assumption 7.1, the containment controller (7.2) can be constructed as follows:*

1) *Choose the feedback gain matrix F such that $A + BF$ is Hurwitz.*

2) *Take $L = -Q^{-1}C^T$, where $Q > 0$ is a solution to the following linear matrix inequality (LMI):*

$$A^T Q + QA - 2C^T C < 0. \tag{7.4}$$

3) *Select the coupling gain $c \geq c_{th}$, with*

$$c_{th} = \frac{1}{\min\limits_{i=1,\cdots,M} \mathrm{Re}(\lambda_i)}, \tag{7.5}$$

where λ_i, $i = 1, \cdots, M$, are the nonzero eigenvalues of $\mathcal{L}_1$.

Remark 54 *As shown in Proposition 3 in Chapter 2, a necessary and sufficient condition on the existence of a positive-definite solution to the LMI (7.4) is that (A, C) is detectable. Therefore, a necessary and sufficient condition on the feasibility of Algorithm 14 is that (A, B, C) is stabilizable and detectable.*

Remark 55 *Similarly as in Chapter 2, the design method in Algorithm 14 essentially relies on the notion of consensus region. As discussed in Remark 6 in Chapter 2, Algorithm 14 has a favorable decoupling feature. Specifically, the first two steps deal with only the agent dynamics and the feedback gain matrices of (7.2), while the last step tackles the communication topology by adjusting the coupling gain c.*

Theorem 46 *Assume that (A, B, C) be stabilizable and detectable. For a directed communication graph $\mathcal{G}$ satisfying Assumption (7.1), the controller (7.2) constructed by Algorithm 14 solves the containment control problem for the agents described by (7.1). Specifically, $\lim_{t\to\infty}(x_f(t) - \varpi_x(t)) = 0$ and $\lim_{t\to\infty} v_f(t) = 0$, where $x_f = [x_1^T, \cdots, x_M^T]^T$, $v_f = [v_1^T, \cdots, v_M^T]^T$,*

$$\varpi_x(t) \triangleq (-\mathcal{L}_1^{-1}\mathcal{L}_2 \otimes e^{At}) \begin{bmatrix} x_{M+1}(0) \\ \vdots \\ x_N(0) \end{bmatrix}. \tag{7.6}$$

Proof 47 *Let* $z_i = [x_i^T, v_i^T]^T$, $z_f = [z_1^T, \cdots, z_M^T]^T$, *and* $z_l = [z_{M+1}^T, \cdots, z_N^T]^T$. *Then, the closed-loop network dynamics resulting from (7.1) and (7.2) can be written as*

$$\dot{z}_f = (I_M \otimes S + c\mathcal{L}_1 \otimes \mathcal{H}) z_f + c(\mathcal{L}_2 \otimes \mathcal{H}) z_l,$$
$$\dot{z}_l = (I_{N-M} \otimes S) z_l, \tag{7.7}$$

where

$$S = \begin{bmatrix} A & BF \\ 0 & A + BF \end{bmatrix}, \quad \mathcal{H} = \begin{bmatrix} 0 & 0 \\ -LC & LC \end{bmatrix}.$$

Let $\varrho_i = \sum_{j \in \mathcal{F} \cup \mathcal{R}} a_{ij}(z_i - z_j)$, $i \in \mathcal{F}$, *and* $\varrho = [\varrho_1^T, \cdots, \varrho_M^T]^T$. *Then, we have*

$$\varrho = (\mathcal{L}_1 \otimes I_{2n}) z_f + (\mathcal{L}_2 \otimes I_{2n}) z_l. \tag{7.8}$$

From (7.8), it follows that $\varrho = 0$ *if and only if* $z_f(t) = -(\mathcal{L}_1^{-1}\mathcal{L}_2 \otimes I_{2n}) z_l(t)$, *which is equivalent to that* $x_f = -(\mathcal{L}_1^{-1}\mathcal{L}_2 \otimes I_n) x_l$ *and* $v_f = -(\mathcal{L}_1^{-1}\mathcal{L}_2 \otimes I_n) v_l$. *Then, in virtue of Lemma 30, we can get that the containment control problem for the agents in (7.1) under the controller (7.2) is solved if* ϱ *asymptotically converges to zero. Hereafter, we refer to* ϱ *as the containment error for (7.1) under (7.2). Further, by noting that* $x_j(t) = e^{At} x_j(0)$ *and* $\lim_{t \to \infty} v_j(t) = 0$, $j \in \mathcal{R}$, *it then follows that if* $\lim_{t \to \infty} \varrho(t) = 0$, *then* $\lim_{t \to \infty}(x_f(t) - \varpi_x(t)) = 0$ *and* $\lim_{t \to \infty} v_f(t) = 0$.

Considering the special structure of $\mathcal{L}$ *as in (7.3), we can obtain from (7.7) and (7.8) that the containment error* ϱ *satisfies the following dynamics:*

$$\begin{aligned} \dot{\varrho} &= (\mathcal{L}_1 \otimes I_{2n}) \dot{z}_f + (\mathcal{L}_2 \otimes I_{2n}) \dot{z}_l \\ &= (\mathcal{L}_1 \otimes I_{2n})[(I_M \otimes S) z_f + c(\mathcal{L}_1 \otimes \mathcal{H}) z_f + c(\mathcal{L}_2 \otimes \mathcal{H}) z_l] \\ &\quad + (\mathcal{L}_2 \otimes I_{2n})(I_{N-M} \otimes S) z_l \\ &= (\mathcal{L}_1 \otimes S + c\mathcal{L}_1^2 \otimes \mathcal{H})[(\mathcal{L}_1^{-1} \otimes I_{2n})\varrho - (\mathcal{L}_1^{-1}\mathcal{L}_2 \otimes I_{2n}) z_l] \\ &\quad + (c\mathcal{L}_1\mathcal{L}_2 \otimes \mathcal{H} + \mathcal{L}_2 \otimes S) z_l \\ &= (I_M \otimes S + c\mathcal{L}_1 \otimes \mathcal{H})\varrho. \end{aligned} \tag{7.9}$$

Under Assumption 7.1, it follows from Lemma 30 that all the eigenvalues of $\mathcal{L}_1$ *have positive real parts. Let* $U \in \mathbf{C}^{M \times M}$ *be such a unitary matrix that* $U^H \mathcal{L}_1 U = \Lambda$, *where* Λ *is an upper-triangular matrix with* λ_i, $i = 1, \cdots, M$, *as its diagonal entries. Let* $\tilde{\varrho} \triangleq [\tilde{\varrho}_1^T, \cdots, \tilde{\varrho}_N^T]^T = (U^H \otimes I_{2n})\varrho$. *Then, it follows from (7.9) that*

$$\dot{\tilde{\varrho}} = (I_M \otimes S + c\Lambda \otimes \mathcal{H})\tilde{\varrho}. \tag{7.10}$$

By noting that Λ *is upper-triangular, it is clear that (7.10) is asymptotically stable if and only if the following* M *systems:*

$$\dot{\tilde{\varrho}}_i = (S + c\lambda_i \mathcal{H})\tilde{\varrho}_i, \quad i = 1, \cdots, M, \tag{7.11}$$

are simultaneously asymptotically stable. Multiplying the left and right sides of $S + c\lambda_i \mathcal{H}$ *by* $T = \begin{bmatrix} I & -I \\ 0 & I \end{bmatrix}$ *and* T^{-1}, *respectively, we get*

$$T(S + c\lambda_i \mathcal{H})T^{-1} = \begin{bmatrix} A + c\lambda_i LC & 0 \\ -c\lambda_i LC & A + BF \end{bmatrix}. \tag{7.12}$$

By steps 1) and 3) in Algorithm 14, it follows that there exists a $Q > 0$ satisfying

$$Q(A + c\lambda_i LC) + (A + c\lambda_i LC)^H Q = QA + A^T Q - 2c\mathrm{Re}(\lambda_i)C^T C$$
$$\leq QA + A^T Q - 2C^T C < 0.$$

That is, $A + c\lambda_i LC$, $i = 1, \cdots, M$, are Hurwitz. Therefore, using (7.10), (7.11), and (7.12), it follows that (7.9) is asymptotically stable, which implies that the containment problem is solved.

7.1.2   Static Containment Controllers

In this subsection, a special case where the relative states between neighboring agents are available is considered. For this case, a distributed static containment controller is proposed as

$$u_i = cK \sum_{j \in \mathcal{F} \cup \mathcal{R}} a_{ij}(x_i - x_j), \ i \in \mathcal{F}, \tag{7.13}$$

where c and a_{ij} are defined as in (7.2), and $K \in \mathbf{R}^{p \times n}$ is the feedback gain matrix to be designed. Let $x_f = [x_1^T, \cdots, x_M^T]^T$ and $x_l = [x_{M+1}^T, \cdots, x_N^T]^T$. Using (7.13) for (7.1) gives the closed-loop network dynamics as

$$\dot{x}_f = (I_M \otimes A + c\mathcal{L}_1 \otimes BK)x_f + c(\mathcal{L}_2 \otimes BK)x_l,$$
$$\dot{x}_l = (I_{N-M} \otimes A)x_l,$$

where $\mathcal{L}_1$ and $\mathcal{L}_2$ are defined in (7.3).

The following algorithm is given for designing the control parameters in (7.13).

Algorithm 15 *Under Assumption 7.1, the controller (7.13) can be constructed as follows:*

1) Choose the feedback gain matrix $K = -B^T P^{-1}$, where $P > 0$ is a solution to the following LMI:

$$AP + PA^T - 2BB^T < 0. \tag{7.14}$$

2) Select the coupling gain $c \geq c_{th}$, with c_{th} given by (7.5).

Note that the LMI (7.14) is dual to the LMI (7.4). Thus, a sufficient and necessary condition for the above algorithm is that (A, B) is stabilizable.

The following result can be derived by following similar steps as in the proof of Theorem 46.

Corollary 8 *Assume that (A, B) is stabilizable and the communication graph $\mathcal{G}$ satisfies Assumption 7.1. Then, the static controller (7.13) constructed by Algorithm 15 solves the containment control problem for the agents in (7.1). Specifically, $\lim_{t \to \infty}(x_f(t) - \varpi_x(t)) = 0$, where $x_f(t)$ and $\varpi_x(t)$ are defined in (7.6).*

7.2   Containment Control of Discrete-Time Multi-Agent Systems with Leaders of Zero Inputs

This section focuses on the discrete-time counterpart of the last section. Consider a group of N identical agents with general discrete-time linear dynamics, described by

$$
\begin{aligned}
x_i^+ &= Ax_i + Bu_i, \\
y_i &= Cx_i, \ i = 1, \cdots, N,
\end{aligned}
\tag{7.15}
$$

where $x_i = x_i(k) \in \mathbf{R}^n$, $x_i^+ = x_i(k+1)$, $u_i \in \mathbf{R}^p$, and $y_i \in \mathbf{R}^q$ are, respectively, the state at the current time instant, the state at the next time instant, the control input, the measured output of the i-th agent. Without loss of generality, it is assumed throughout this section that B is of full column rank and C has full row rank.

Similar to the last section, we assume that the agents indexed by $1, \cdots, M$, are followers, while the agents indexed by $M+1, \cdots, N$, are leaders. The leader set and the follower set are denoted, respectively, by $\mathcal{R} = \{M + 1, \cdots, N\}$ and $\mathcal{F} = \{1, \cdots, M\}$. We again let $u_i = 0$, $i \in \mathcal{R}$. The communication graph $\mathcal{G}$ among the N agents is assumed to be directed and satisfy Assumption 7.1.

7.2.1   Dynamic Containment Controllers

Similar to the continuous-time case, we propose the following distributed dynamic containment controller for each follower as

$$
\begin{aligned}
\hat{v}_i^+ &= (A + BF)\hat{v}_i + L \sum_{j \in \mathcal{F} \cup \mathcal{R}} d_{ij}[C(\hat{v}_i - \hat{v}_j) - (y_i - y_j)], \\
u_i &= F\hat{v}_i, \ i \in \mathcal{F},
\end{aligned}
\tag{7.16}
$$

where $\hat{v}_i \in \mathbf{R}^n$, $i \in \mathcal{F}$, is the state of the controller corresponding to the i-th follower, $\hat{v}_j = 0$, $j \in \mathcal{R}$, d_{ij} is the (i, j)-th entry of the row-stochastic matrix $\mathcal{D}$ associated with $\mathcal{G}$, and $L \in \mathbf{R}^{n \times q}$ and $F \in \mathbf{R}^{p \times n}$ are feedback gain matrices.

Because the last $N - M$ agents are leaders that have no neighbors, $\mathcal{D}$ can be partitioned as

$$
\mathcal{D} = \begin{bmatrix} \mathcal{D}_1 & \mathcal{D}_2 \\ 0_{(N-M) \times M} & I_{N-M} \end{bmatrix},
\tag{7.17}
$$

where $\mathcal{D}_1 \in \mathbf{R}^{M \times M}$ and $\mathcal{D}_2 \in \mathbf{R}^{M \times (N-M)}$.

Lemma 31 *Under Assumption 7.1, all the eigenvalues of $\mathcal{D}_1$ lie in the open unit disk, each entry of $(I_M - \mathcal{D}_1)^{-1}\mathcal{D}_2$ is nonnegative, and each row of $(I_M - \mathcal{D}_1)^{-1}\mathcal{D}_2$ has a sum equal to one.*

Proof 48 *Consider the following new row-stochastic matrix:*

$$
\overline{\mathcal{D}} = \begin{bmatrix} \mathcal{D}_1 & \mathcal{D}_2 \mathbf{1}_{N-M} \\ 0_{1 \times M} & 1 \end{bmatrix}.
$$

According to the definition of the directed spanning tree in Section 1.3.2 of Chapter 1, the graph associated with $\overline{\mathcal{D}}$ has a directed spanning tree if $\mathcal{G}$ satisfies Assumption 7.1. Therefore, by Lemma 3.4 in [133], if Assumption 7.1 holds, then 1 is a simple eigenvalue of $\overline{\mathcal{D}}$ and all the other eigenvalues of $\overline{\mathcal{D}}$ are in the open unit disk, which further implies that all the eigenvalues of $\mathcal{D}_1$ lie in the open unit disk.

Because every entry of $\mathcal{D}_1$ is nonnegative and the spectral radius of $\mathcal{D}_1$ is less than 1, $I_M - \mathcal{D}_1$ is a nonsingular M-matrix [3]. Therefore, each entry of $(I_M - \mathcal{D}_1)^{-1}$ is nonnegative [3]. Clearly, each entry of $(I_M - \mathcal{D}_1)^{-1}\mathcal{D}_2$ is also nonnegative. Note that $(I_N - D)\mathbf{1}_N = 0$, implying that $(I_M - \mathcal{D}_1)\mathbf{1}_M - \mathcal{D}_2\mathbf{1}_{N-M} = 0$. That is, $(I_M - \mathcal{D}_1)^{-1}\mathcal{D}_2\mathbf{1}_{N-M} = \mathbf{1}_M$, which shows that each row of $(I_M - \mathcal{D}_1)^{-1}\mathcal{D}_2$ has a sum equal to 1.

Next, an algorithm for determining the control parameters in (7.16) is presented.

Algorithm 16 *Under Assumption 7.1, the containment controller (7.16) can be constructed as follows:*

1) *The feedback gain matrix F is chosen such that $A + BF$ is Schur stable.*

2) *Select $L = -APC^T(CPC^T)^{-1}$, where $P > 0$ is the unique solution to the following modified algebraic Riccati equation (MARE):*

$$P = APA^T - (1 - \max_{i=1,\cdots,M}|\hat{\lambda}_i|^2)APC^T(CPC^T)^{-1}CPA^T + R, \qquad (7.18)$$

where $R > 0$ and $\hat{\lambda}_i$, $i = 1, \cdots, M$ are the eigenvalues of $\mathcal{D}_1$.

Remark 56 *The solvability of the MARE (7.18) can be found in Lemma 26 in Chapter 3. As shown in Lemma 26, for the case where A has at least eigenvalue outside the unit circle and C is of rank one, a sufficient condition for the existence of the controller (7.16) by using Algorithm 16 is that (A, B, C) is stabilizable and detectable, and $\max_{i=1,\cdots,M}|\hat{\lambda}_i|^2 < 1 - \frac{1}{\prod_i|\lambda_i^u(A)|}$. For the case where A has no eigenvalues with magnitude larger than 1, the sufficient condition is reduced to that (A, B, C) is stabilizable and detectable.*

The following theorem shows that the controller (7.16) given by Algorithm 16 indeed solves the containment problem.

Theorem 47 *Suppose that the directed communication graph $\mathcal{G}$ satisfies Assumption 7.1. Then, the controller (7.16) given by Algorithm 16 solves the containment control problem for the agents described by (7.15). Specifically, $\lim_{k\to\infty}(x_f(k) - \psi_x(k)) = 0$ and $\lim_{k\to\infty}\hat{v}_f(k) = 0$, where $x_f = [x_1^T, \cdots, x_M^T]^T$, and*

$$\psi_x(k) \triangleq \left[(I_M - \mathcal{D}_1)^{-1}\mathcal{D}_2 \otimes A^k\right]\begin{bmatrix} x_{M+1}(0) \\ \vdots \\ x_N(0) \end{bmatrix}. \qquad (7.19)$$

Proof 49 *Let* $\hat{z}_i = [x_i^T, \hat{v}_i^T]^T$, $\hat{z}_f = [\hat{z}_1^T, \cdots, \hat{z}_M^T]^T$, *and* $\hat{z}_l = [\hat{z}_{M+1}^T, \cdots, \hat{z}_N^T]^T$. *Then, we can obtain from (7.15) and (7.16) that the collective network dynamics can be written as*

$$\hat{z}_f^+ = [I_M \otimes \mathcal{S} + (I_M - \mathcal{D}_1) \otimes \mathcal{H}]\hat{z}_f - (\mathcal{D}_2 \otimes \mathcal{H})\hat{z}_l, \tag{7.20}$$
$$\hat{z}_l^+ = (I_{N-M} \otimes \mathcal{S})\hat{z}_l,$$

where $\mathcal{S}$ and $\mathcal{H}$ are defined in (7.7). Let $\zeta_i = \sum_{j \in \mathcal{F} \cup \mathcal{R}} d_{ij}(\hat{z}_i - \hat{z}_j)$, $i \in \mathcal{F}$, and $\zeta = [\zeta_1^T, \cdots, \zeta_M^T]^T$. Then, we have

$$\zeta = [(I_M - \mathcal{D}_1) \otimes I_{2n}]\hat{z}_f - (\mathcal{D}_2 \otimes I_{2n})\hat{z}_l. \tag{7.21}$$

Similarly as shown in the proof of Theorem 46, $\zeta = 0$ if and only if $z_f(k) = [(I_M - \mathcal{D}_1)^{-1}\mathcal{D}_2 \otimes I_{2n}]z_l(k)$, which implies that $\lim_{k \to \infty} (x_f(k) - \psi_x(k)) = 0$ and $\lim_{k \to \infty} \hat{v}_f(k) = 0$. In virtue of Lemma 31, we know that the containment problem of the agents in (7.15) under the controller (7.16) is solved if $\lim_{k \to \infty} \zeta = 0$. By following similar steps to those in the proof of Theorem 46, we can obtain from (7.20) and (7.21) that ζ satisfies the following dynamics:

$$\zeta^+ = [I_M \otimes \mathcal{S} + (I_M - \mathcal{D}_1) \otimes \mathcal{H}]\zeta. \tag{7.22}$$

Under Assumption 7.1, it follows from Lemma 31 that all the eigenvalues of $I_M - \mathcal{D}_1$ have positive real parts. Let $\hat{U} \in \mathbf{C}^{M \times M}$ be such a unitary matrix that $\hat{U}^H(I_M - \mathcal{D}_1)\hat{U} = \hat{\Lambda}$, where $\hat{\Lambda}$ is an upper-triangular matrix with $1 - \hat{\lambda}_i$, $i = 1, \cdots, M$, on the diagonal. Let $\tilde{\zeta} \triangleq [\tilde{\zeta}_1^T, \cdots, \tilde{\zeta}_N^T]^T = (\hat{U}^H \otimes I_{2n})\zeta$. Then, (7.22) can be rewritten as

$$\tilde{\zeta}^+ = (I_M \otimes \mathcal{S} + \hat{\Lambda} \otimes \mathcal{H})\tilde{\zeta}, \tag{7.23}$$

Clearly, (7.23) is asymptotically stable if and if the following M systems:

$$\tilde{\zeta}_i^+ = [\mathcal{S} + (1 - \hat{\lambda}_i)\mathcal{H}]\tilde{\zeta}_i, \quad i = 1, \cdots, M, \tag{7.24}$$

are simultaneously asymptotically stable. As shown in the proof of Theorem 46, $\mathcal{S} + (1 - \hat{\lambda}_i)\mathcal{H}$ is similar to $\begin{bmatrix} A+(1-\hat{\lambda}_i)LC & 0 \\ -(1-\hat{\lambda}_i)LC & A+BF \end{bmatrix}$, $i = 1, \cdots, M$. In light of step 2) in Algorithm 16, we can obtain

$$[A + (1 - \hat{\lambda}_i)LC]P[A + (1 - \hat{\lambda}_i)LC]^H - P$$
$$= APA^T - P + [-2\text{Re}(1 - \hat{\lambda}_i) + |1 - \hat{\lambda}_i|^2]APC^T(CPC^T)^{-1}CPA^T$$
$$= APA^T - P + (|\hat{\lambda}_i|^2 - 1)APC^T(CPC^T)^{-1}CPA^T$$
$$\leq APA^T - P - (1 - \max_{i=1,\cdots,M}|\hat{\lambda}_i|^2)APC^T(CPC^T)^{-1}CPA^T$$
$$= -R < 0. \tag{7.25}$$

Then, (7.25) implies that $A + (1 - \hat{\lambda}_i)LC$, $i = 1, \cdots, M$, are Schur stable. Therefore, considering (7.23) and (7.24), we obtain that (7.22) is asymptotically stable, i.e., the containment control problem is solved.

Remark 57 *Theorem 47 gives the discrete-time counterpart of the results in Theorem 46. In contrast to the continuous-time case where the Laplacian matrix $\mathcal{L}$ is used to represent the communication graph, the row-stochastic matrix $\mathcal{D}$ is utilized here in the discrete-time case. Different from Theorem 46 in the last subsection, Algorithm 16 for constructing discrete-time containment controllers and the proof of Theorem 47 rely on the modified algebraic Riccati equation. Furthermore, the eigenvalues of $\mathcal{D}$ in the discrete-time case have to further satisfy a constraint related to the unstable eigenvalues of the state matrix A, when A has eigenvalues outside the unit circle.*

7.2.2   Static Containment Controllers

In this subsection, a special case where the relative states between neighboring agents are available is considered. For this case, a distributed static containment controller is proposed as

$$u_i = K \sum_{j \in \mathcal{F} \cup \mathcal{R}} d_{ij}(x_i - x_j), \ i \in \mathcal{F}, \qquad (7.26)$$

where d_{ij} is defined as in (7.16) and $K \in \mathbf{R}^{p \times n}$ is the feedback gain matrix to be designed.

Let $x_f = [x_1^T, \cdots, x_M^T]^T$ and $x_l = [x_{M+1}^T, \cdots, x_N^T]^T$. Using (7.26) for (7.1) gives the closed-loop network dynamics as

$$x_f^+ = [I_N \otimes A + (I_M - \mathcal{D}_1) \otimes BK]x_f - (\mathcal{D}_2 \otimes BK)x_l,$$
$$x_l^+ = (I_{N-M} \otimes A)x_l,$$

where $\mathcal{D}_1$ and $\mathcal{D}_2$ are defined in (7.17).

Corollary 9 *Assume that the communication graph $\mathcal{G}$ satisfies Assumption 7.1. Then, the controller (7.26) with $K = -(B^T QB)^{-1}B^T QA$ solve the containment control problem, where $Q > 0$ is the unique solution to the following MARE:*

$$Q = A^T QA - (1 - \max_{i=1,\cdots,M}|\hat{\lambda}_i|^2)A^T QB(B^T QB)^{-1}B^T QA + R, \qquad (7.27)$$

with $R > 0$ and $\hat{\lambda}_i$, $i = 1, \cdots, M$, being the eigenvalues of $\mathcal{D}_1$. Specifically, $\lim_{k \to \infty}(x_f(k) - \psi_x(k)) = 0$, where $x_f(k)$ and $\psi_x(k)$ are defined in (7.19).

7.2.3   Simulation Examples

In this subsection, a simulation example is provided to validate the effectiveness of the theoretical results.

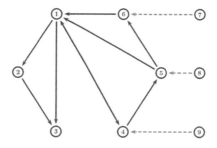

FIGURE 7.1: The communication topology with multiple leaders.

Example 22 *In this example, we take the agents in (7.1) as the Caltech multi-vehicle wireless testbed vehicles, whose linearized dynamics can be described by (7.1) [51], with*

$$
x_i = \begin{bmatrix} x_{i1} & x_{i2} & x_{i3} & \dot{x}_{i1} & \dot{x}_{i2} & \dot{x}_{i3} \end{bmatrix}^T,
$$

$$
A = \begin{bmatrix}
0 & 0 & 0 & 1 & 0 & 0 \\
0 & 0 & 0 & 0 & 1 & 0 \\
0 & 0 & 0 & 0 & 0 & 1 \\
0 & 0 & -0.2003 & -0.2003 & 0 & 0 \\
0 & 0 & 0.2003 & 0 & -0.2003 & 0 \\
0 & 0 & 0 & 0 & 0 & -1.6129
\end{bmatrix},
$$

$$
B = \begin{bmatrix}
0 & 0 \\
0 & 0 \\
0 & 0 \\
0.9441 & 0.9441 \\
0.9441 & 0.9441 \\
-28.7097 & 28.7097
\end{bmatrix},
$$

where x_{i1} and x_{i2} are, respectively, the positions of the i-th vehicle along the x and y coordinates, x_{i3} is the orientation of the i-th vehicle. The objective is to design a static containment controller in the form of (7.13) by using local state information of neighboring vehicles.

Solving the LMI (7.14) by using the LMI toolbox of MATLAB gives the feedback gain matrix of (7.13) as

$$
K = \begin{bmatrix}
-0.0089 & 0.0068 & 0.0389 & -0.0329 & 0.0180 & 0.0538 \\
0.0068 & -0.0089 & -0.0389 & 0.0180 & -0.0329 & -0.0538
\end{bmatrix}.
$$

For illustration, let the communication graph $\mathcal{G}$ be given by FIGURE 7.1, where nodes $7, 8, 9$ are three leaders and the others are followers. It can be verified that $\mathcal{G}$ satisfies Assumption 7.1. Correspondingly, the matrix $\mathcal{L}_1$ in

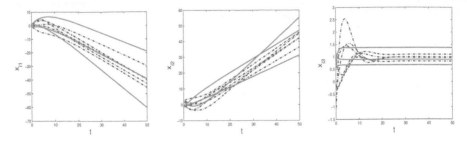

FIGURE 7.2: The positions and orientations of the nine vehicles under (7.13). The solid and dashdotted lines denote, respectively, the trajectories of the leaders and the followers.

(7.3) is

$$
\mathcal{L}_1 =
\begin{bmatrix}
3 & 0 & 0 & -1 & -1 & -1 \\
-1 & 1 & 0 & 0 & 0 & 0 \\
-1 & -1 & 2 & 0 & 0 & 0 \\
-1 & 0 & 0 & 2 & 0 & 0 \\
0 & 0 & 0 & -1 & 2 & 0 \\
0 & 0 & 0 & 0 & -1 & 2
\end{bmatrix},
$$

whose eigenvalues are 0.8213, 1, 2, 2.3329 ± ι0.6708, 3.5129. By Algorithm 15, we choose the coupling gain c ≥ 1.2176. The positions and orientations of the nine vehicles under the controller (7.13) with K as above and c = 2 are depicted in FIGURE 7.2, from which it can be observed that the containment control problem is indeed solved.

7.3   Containment of Continuous-Time Multi-Agent Systems with Leaders of Nonzero Inputs

In Section 7.1, the control inputs of the leaders are assumed to be zero, which is restrictive in many circumstances. In real applications, the leaders might need nonzero control actions to regulate their state trajectories, e.g., to avoid obstacles or to form a desirable safety area.

The focus of this section is to extend Section 7.1 to consider the general case where the leaders' control inputs are possibly nonzero and time varying. Note that due to the nonzero control inputs of the leaders which are available to at most a subset of follower, the containment control problem is more challenging.

Suppose that the following mild assumption holds.

Assumption 7.2 *The leaders' control inputs u_i, $i \in \mathcal{R}$, are bounded, i.e., $\|u_i\| \leq \gamma_i$, $i \in \mathcal{R}$, where γ_i are positive constants.*

The communication graph $\mathcal{G}$ among the N agents in this case is assumed to satisfy the following assumption, which is more stringent than Assumption (7.1).

Assumption 7.3 *The subgraph $\mathcal{G}_s$ associated with the M followers is undirected. For each follower, there exists at least one leader that has a directed path to that follower.*

Similarly as in (7.3), in this case $\mathcal{L}$ can be partitioned as

$$\mathcal{L} = \begin{bmatrix} \mathcal{L}_1 & \mathcal{L}_2 \\ 0_{(N-M)\times M} & 0_{(N-M)\times(N-M)} \end{bmatrix}, \tag{7.28}$$

where $\mathcal{L}_1 \in \mathbf{R}^{M\times M}$ is symmetric and $\mathcal{L}_2 \in \mathbf{R}^{M\times(N-M)}$.

The following result is a special case of Lemma 31.

Lemma 32 ([111]) *Under Assumption 7.3, all the eigenvalues of $\mathcal{L}_1$ are positive, each entry of $-\mathcal{L}_1^{-1}\mathcal{L}_2$ is nonnegative, and each row of $-\mathcal{L}_1^{-1}\mathcal{L}_2$ has a sum equal to one.*

7.3.1 Distributed Continuous Static Controllers

In order to solve the containment problem, based on the relative state information of neighboring agents, we propose a distributed continuous static controller for each follower as

$$u_i = c_1 K \sum_{j=1}^{N} a_{ij}(x_i - x_j) + c_2 g_i(K \sum_{j=1}^{N} a_{ij}(x_i - x_j)), \ i \in \mathcal{F}, \tag{7.29}$$

where $c_1 > 0$ and $c_2 > 0 \in \mathbf{R}$ are constant coupling gains, $K \in \mathbf{R}^{p\times n}$ is the feedback gain matrix, a_{ij} is the (i,j)-th entry of the adjacency matrix associated with $\mathcal{G}$, and the nonlinear functions $g_i(\cdot)$ are defined such that for $w \in \mathbf{R}^n$,

$$g_i(w) = \begin{cases} \frac{w}{\|w\|} & \text{if } \|w\| > \kappa_i \\ \frac{w}{\kappa_i} & \text{if } \|w\| \leq \kappa_i \end{cases} \tag{7.30}$$

with κ_i being small positive scalars.

Remark 58 *The above containment controller (7.29) is a direct extension of the tracking controller (6.22) in the previous chapter to the case with multiple leaders. If only one leader exists, (7.29) clearly reduces to (6.22). In (7.29), the information of the leaders's control inputs u_i, $i \in \mathcal{R}$ is actually not used. Comparing (7.29) to (7.13), it can observed that the term $c_2 g_i(K \sum_{j=1}^{N} a_{ij}(x_i - x_j))$*

in (7.29) is introduced to deal with the effect of the nonzero u_i, $i \in \mathcal{R}$. For the special case where $u_i = 0$, $i \in \mathcal{R}$, (7.29) can be reduced to (7.13) by removing $c_2 g_i(K \sum_{j=1}^{N} a_{ij}(x_i - x_j))$.

Remark 59 *Note that the nonlinear functions $g_i(\cdot)$ in (7.30) are continuous, which are actually continuous approximations, via the boundary layer approach [26, 71], of the discontinuous function:*

$$\hat{g}(w) = \begin{cases} \frac{w}{\|w\|} & \text{if } \|w\| \neq 0 \\ 0 & \text{if } \|w\| = 0. \end{cases} \tag{7.31}$$

The values of κ_i in (7.30) define the widths of the boundary layers. As $\kappa_i \to 0$, the continuous functions $g_i(\cdot)$ approach the discontinuous function $\hat{g}(\cdot)$.

Let $x_f = [x_1^T, \cdots, x_M^T]^T$, $x_l = [x_{M+1}^T, \cdots, x_N^T]^T$, $x = [x_f^T, x_l^T]^T$, and $u_l = [u_{M+1}^T, \cdots, u_N^T]^T$. Then, it follows from (7.1) and (7.29) that the closed-loop network dynamics can be written as

$$\dot{x}_f = (I_M \otimes A + c_1 \mathcal{L}_1 \otimes BK)x_f + c_1(\mathcal{L}_2 \otimes BK)x_l + c_2(I_M \otimes B)G(x),$$
$$\dot{x}_l = (I_{N-M} \otimes A)x_f + (I_{N-M} \otimes B)u_l, \tag{7.32}$$

where $\mathcal{L}_1$ and $\mathcal{L}_2$ are defined as in (7.28), and

$$G(x) \triangleq \begin{bmatrix} g_1(K \sum_{j=1}^{N} a_{1j}(x_1 - x_j)) \\ \vdots \\ g_N(K \sum_{j=1}^{N} a_{Mj}(x_M - x_j)) \end{bmatrix}.$$

Introduce the following variable:

$$\xi \triangleq x_f + (\mathcal{L}_1^{-1}\mathcal{L}_2 \otimes I_n)x_l, \tag{7.33}$$

where $\xi = [\xi_1^T, \cdots, \xi_M^T]^T$. From (7.33), it is easy to see that $\xi = 0$ if and only if $x_f = (-\mathcal{L}_1^{-1}\mathcal{L}_2 \otimes I_n)x_l$. In virtue of Lemma 32, we can get that the containment control problem of the agents in (7.1) under the controller (7.29) is solved if ξ converges to zero. Hereafter, we refer to ξ as the containment error for (7.1) under (7.29). By (7.33) and (7.32), it is not difficult to obtain that ξ satisfies the following dynamics:

$$\dot{\xi} = (I_M \otimes A + c_1 \mathcal{L}_1 \otimes BK)\xi + c_2(I_M \otimes B)\widehat{G}(\xi) + (\mathcal{L}_1^{-1}\mathcal{L}_2 \otimes B)u_l, \tag{7.34}$$

where

$$\widehat{G}(\xi) \triangleq \begin{bmatrix} g_1(K \sum_{j=1}^{M} \mathcal{L}_{1j}\xi_j) \\ \vdots \\ g_N(K \sum_{j=1}^{M} \mathcal{L}_{Mj}\xi_j) \end{bmatrix}, \tag{7.35}$$

with $\mathcal{L}_{ij}$ denoting the (i,j)-th entry of $\mathcal{L}_1$. Actually, $\widehat{G}(\xi) = G(x)$, which follows from the fact that $\sum_{j=1}^{N} a_{ij}(x_i - x_j) = \sum_{j=1}^{M} \mathcal{L}_{ij}\xi_j$, $i = 1, \cdots, M$.

The following theorem states the ultimate boundedness of the containment error ξ.

Theorem 48 *Assume that Assumptions 7.2 and 7.3 hold. The parameters in the containment controller (7.29) are designed as $c_1 \geq \frac{1}{\lambda_{\min}(\mathcal{L}_1)}$, $c_2 \geq \gamma_{\max}$, and $K = -B^T P^{-1}$, where $\gamma_{\max} \triangleq \max_{i \in \mathcal{R}} \gamma_i$ and $P > 0$ is a solution to the LMI (7.14). Then, the containment error ξ of (7.34) is uniformly ultimately bounded and exponentially converges to the residual set*

$$\mathcal{D}_1 \triangleq \{\xi : \|\xi\|^2 \leq \frac{2\lambda_{\max}(P)\gamma_{\max}}{\alpha\lambda_{\min}(\mathcal{L}_1)} \sum_{i=1}^{M} \kappa_i\}, \tag{7.36}$$

where

$$\alpha = \frac{-\lambda_{\max}(AP + PA^T - 2BB^T)}{\lambda_{\max}(P)}. \tag{7.37}$$

Proof 50 *Consider the following Lyapunov function candidate:*

$$V_1 = \frac{1}{2}\xi^T(\mathcal{L}_1 \otimes P^{-1})\xi. \tag{7.38}$$

It follows from Lemma 32 that $\mathcal{L}_1 > 0$, so V_1 is clearly positive definite. The time derivative of V_1 along the trajectory of (7.34) is given by

$$\begin{aligned}
\dot{V}_1 &= \xi^T(\mathcal{L}_1 \otimes P^{-1})\dot{\xi} \\
&= \xi^T(\mathcal{L}_1 \otimes P^{-1}A + c_1\mathcal{L}_1^2 \otimes P^{-1}BK)\xi \\
&\quad + c_2\xi^T(\mathcal{L}_1 \otimes P^{-1}B)\widehat{G}(\xi) + \xi^T(\mathcal{L}_2 \otimes P^{-1}B)u_l \\
&= \frac{1}{2}\xi^T\mathcal{X}\xi + c_2\xi^T(\mathcal{L}_1 \otimes P^{-1}B)\widehat{G}(\xi) + \xi^T(\mathcal{L}_2 \otimes P^{-1}B)u_l,
\end{aligned} \tag{7.39}$$

where

$$\mathcal{X} = \mathcal{L}_1 \otimes (P^{-1}A + A^T P^{-1}) - 2c_1\mathcal{L}_1^2 \otimes P^{-1}BB^T P^{-1}. \tag{7.40}$$

Let b_{ij} denote the (i,j)-th entry of $-\mathcal{L}_1^{-1}\mathcal{L}_2$, which by Lemma 32, satisfies that $b_{ij} \geq 0$ and $\sum_{j=1}^{N-M} b_{ij} = 1$ for $i = 1, \cdots, M$. In virtue of Assumption

7.2, we have

$$\xi^T(\mathcal{L}_2 \otimes P^{-1}B)u_l$$

$$= \xi^T(\mathcal{L}_1 \otimes I_n)(\mathcal{L}_1^{-1}\mathcal{L}_2 \otimes P^{-1}B)u_l$$

$$= -\left[\sum_{j=1}^{M}\mathcal{L}_{1j}\xi_j^T \quad \cdots \quad \sum_{j=1}^{M}\mathcal{L}_{Mj}\xi_j^T\right]\begin{bmatrix}\sum_{k=1}^{N-M}b_{1k}P^{-1}Bu_k\\ \vdots \\ \sum_{k=1}^{N-M}b_{Mk}P^{-1}Bu_k\end{bmatrix}$$

$$= -\sum_{i=1}^{M}\sum_{j=1}^{M}\mathcal{L}_{ij}\xi_j^TP^{-1}B\sum_{k=1}^{N-M}b_{jk}u_k$$

$$\leq \sum_{i=1}^{M}\|B^TP^{-1}\sum_{j=1}^{M}\mathcal{L}_{ij}\xi_j\|\sum_{k=1}^{N-M}b_{jk}\|u_k\|$$

$$\leq \gamma_{\max}\sum_{i=1}^{M}\|B^TP^{-1}\sum_{j=1}^{M}\mathcal{L}_{ij}\xi_j\|.$$

(7.41)

Next, consider the following three cases:
i) $\|K\sum_{j=1}^{M}\mathcal{L}_{ij}\xi_j\| > \kappa_i$, *i.e.,* $\|B^TP^{-1}\sum_{j=1}^{M}\mathcal{L}_{ij}\xi_j\| > \kappa_i$, $i = 1, \cdots, M$.
In this case, it follows from (7.30) and (7.35) that

$$\xi^T(\mathcal{L}_1 \otimes P^{-1}B)\widehat{G}(\xi) = -\sum_{i=1}^{M}\|B^TP^{-1}\sum_{j=1}^{M}\mathcal{L}_{ij}\xi_j\|. \tag{7.42}$$

Then, we can get from (7.39), (7.41), and (7.42) that

$$\dot{V}_1 \leq \frac{1}{2}\xi^T\mathcal{X}\xi - (c_2 - \gamma_{\max})\sum_{i=1}^{M}\|B^TP^{-1}\sum_{j=1}^{M}\mathcal{L}_{ij}\xi_j\|$$

$$\leq \frac{1}{2}\xi^T\mathcal{X}\xi.$$

ii) $\|B^TP^{-1}\sum_{j=1}^{M}\mathcal{L}_{ij}\xi_j\| \leq \kappa_i$, $i = 1, \cdots, M$.
From (7.41), we can obtain that

$$\xi^T(\mathcal{L}_2 \otimes P^{-1}B)u_l \leq \gamma_{\max}\sum_{i=1}^{M}\kappa_i. \tag{7.43}$$

Further, it follows from (7.30), (7.35), and (7.42) that

$$\xi^T(\mathcal{L}_1 \otimes P^{-1}B)\widehat{G}(\xi) = -\frac{1}{\kappa_i}\sum_{i=1}^{M}\|B^TP^{-1}\sum_{j=1}^{M}\mathcal{L}_{ij}\xi_j\|^2 \leq 0. \tag{7.44}$$

Thus, we get from (7.39), (7.43), and (7.44) that

$$\dot{V}_1 \le \frac{1}{2}\xi^T \mathcal{X} \xi + \gamma_{\max} \sum_{i=1}^{M} \kappa_i. \tag{7.45}$$

iii) ξ satisfies neither Case i) nor Case ii).

Without loss of generality, assume that $\|B^T P^{-1} \sum_{j=1}^{M} \mathcal{L}_{ij}\xi_j\| > \kappa_i$, $i = 1, \cdots, l$, and $\|B^T P^{-1} \sum_{j=1}^{M} \mathcal{L}_{ij}\xi_j\| \le \kappa_i$, $i = l+1, \cdots, M$. It is easy to see from (7.41), (7.30), and (7.42) that

$$\xi^T(\mathcal{L}_2 \otimes P^{-1}B)u_l \le \gamma_{\max}[\sum_{i=1}^{l} \|B^T P^{-1} \sum_{j=1}^{M} \mathcal{L}_{ij}\xi_j\| + \sum_{i=1}^{M-l} \kappa_i],$$

$$\xi^T(\mathcal{L}_1 \otimes P^{-1}B)\widehat{G}(\xi) \le -\sum_{i=1}^{l} \|B^T P^{-1} \sum_{j=1}^{M} \mathcal{L}_{ij}\xi_j\|.$$

Clearly, in this case we have

$$\dot{V}_1 \le \frac{1}{2}\xi^T \mathcal{X} \xi + \gamma_{\max} \sum_{i=1}^{M-l} \kappa_i.$$

Therefore, by analyzing the above three cases, we get that $\dot{V}_1$ satisfies (7.45) for all $\xi \in \mathbf{R}^{Mn}$. Note that (7.45) can be rewritten as

$$\dot{V}_1 \le -\alpha V_1 + \alpha V_1 + \frac{1}{2}\xi^T \mathcal{X} \xi + \gamma_{\max} \sum_{i=1}^{M} \kappa_i$$

$$= -\alpha V_1 + \frac{1}{2}\xi^T (\mathcal{X} + \alpha \mathcal{L}_1 \otimes P^{-1})\xi + \gamma_{\max} \sum_{i=1}^{M} \kappa_i. \tag{7.46}$$

Because $\alpha = \frac{-\lambda_{\max}(AP + PA^T - 2BB^T)}{\lambda_{\max}(P)}$, in light of (7.14), we can obtain that

$$(\mathcal{L}_1^{-\frac{1}{2}} \otimes P)(\mathcal{X} + \alpha \mathcal{L}_1 \otimes P^{-1})(\mathcal{L}_1^{-\frac{1}{2}} \otimes P)$$
$$\le I_N \otimes [AP + PA^T + \alpha P - 2BB^T] < 0.$$

That is, $\mathcal{X} + \alpha \mathcal{L}_1 \otimes P^{-1} < 0$. Then, it follows from (7.46) that

$$\dot{V}_1 \le -\alpha V_1 + \gamma_{\max} \sum_{i=1}^{M} \kappa_i. \tag{7.47}$$

By using Lemma 16 (the Comparison lemma), we can obtain from (7.47) that

$$V_1(\xi) \le [V_1(\xi(0)) - \frac{\gamma_{\max} \sum_{i=1}^{M} \kappa_i}{\alpha}]e^{-\alpha t} + \frac{\gamma_{\max} \sum_{i=1}^{M} \kappa_i}{\alpha}, \tag{7.48}$$

which implies that ξ exponentially converges to the residual set $\mathcal{D}_1$ in (7.36) with a convergence rate not less than $e^{-\alpha t}$.

Remark 60 *Note that the residual set $\mathcal{D}_1$ of the containment error ξ depends on the communication graph $\mathcal{G}$, the number of followers, the upper bounds of the leader's control inputs, and the widths κ_i of the boundary layers. By choosing sufficiently small κ_i, the containment error ξ under the continuous controller (7.29) can be arbitrarily small, which is acceptable in most circumstances. It can be shown that the containment error ξ can asymptotically converge to zero under the controller (7.29) with $g_i(\cdot)$ replaced by $\hat{g}(\cdot)$. The advantage of the continuous controller (7.29) is that it can avoid the undesirable the chattering effect caused by (7.29) with $g_i(\cdot)$ replaced by $\hat{g}(\cdot)$. The tradeoff is that the continuous controller (7.29) does not guarantee asymptotic stability.*

7.3.2   Adaptive Continuous Containment Controllers

In the last subsection, to design the controller (7.29) we have to use the minimal eigenvalue $\lambda_{\min}(\mathcal{L}_1)$ of $\mathcal{L}_1$ and the upper bound $\gamma_{\max}$ of the leaders' control inputs. However, $\lambda_{\min}(\mathcal{L}_1)$ is global information in the sense that each follower has to know the entire communication graph to compute it, and it is not practical to assume that the upper bound $\gamma_{\max}$ are explicitly known to all followers. In this subsection, we intend to design fully distributed controllers to solve the containment problem without requiring $\lambda_{\min}(\mathcal{L}_1)$ nor $\gamma_{\max}$.

Based on the relative states of neighboring agents, we propose the following distributed controller with an adaptive law for updating the coupling gain (weight) for each follower:

$$u_i = d_i K \sum_{j=1}^{N} a_{ij}(x_i - x_j) + d_i r_i(K \sum_{j=1}^{N} a_{ij}(x_i - x_j)),$$

$$\dot{d}_i = \tau_i[-\varphi_i d_i + (\sum_{j=1}^{N} a_{ij}(x_i - x_j)^T)\Gamma(\sum_{j=1}^{N} a_{ij}(x_i - x_j)) \tag{7.49}$$

$$+ \|K \sum_{j=1}^{N} a_{ij}(x_i - x_j)\|], \quad i = 1, \cdots, M,$$

where $d_i(t)$ denotes the time-varying coupling gain (weight) associated with the i-th follower, φ_i are small positive constants, $\Gamma \in \mathbf{R}^{n \times n}$ is the feedback gain matrix, τ_i are positive scalars, the nonlinear functions $r_i(\cdot)$ are defined such that for $w \in \mathbf{R}^n$,

$$r_i(w) = \begin{cases} \frac{w}{\|w\|} & \text{if } d_i\|w\| > \kappa_i \\ \frac{w}{\kappa_i}d_i & \text{if } d_i\|w\| \leq \kappa_i \end{cases} \tag{7.50}$$

and the rest of the variables are defined as in (7.29).

Let x_f, x_l, x, u_l, and ξ be defined as in (7.34) and (7.33). Let $D(t) =$

$\mathrm{diag}(d_1(t), \cdots, d_M(t))$. Then, it follows from (7.1) and (7.49) that the containment error ξ and the coupling gains $D(t)$ satisfy the following dynamics:

$$
\begin{aligned}
\dot{\xi} &= \dot{x}_f + (\mathcal{L}_1^{-1}\mathcal{L}_2 \otimes I_n)\dot{x}_l \\
&= (I \otimes A + D\mathcal{L}_1 \otimes BK)x_f + (D\mathcal{L}_2 \otimes BK)x_l + (D \otimes B)R(x) \\
&\quad + (\mathcal{L}_1^{-1}\mathcal{L}_2 \otimes A)x_l + (\mathcal{L}_1^{-1}\mathcal{L}_2 \otimes B)u_l \\
&= (I_M \otimes A + D\mathcal{L}_1 \otimes BK)\xi + (D \otimes B)R(\xi) + (\mathcal{L}_1^{-1}\mathcal{L}_2 \otimes B)u_l,
\end{aligned}
$$

$$
\dot{d}_i = \tau_i\left[-\varphi_i d_i + \left(\sum_{j=1}^{M}\mathcal{L}_{ij}\xi_j^T\right)\Gamma\left(\sum_{j=1}^{M}\mathcal{L}_{ij}\xi_j\right) + \|K\sum_{j=1}^{M}\mathcal{L}_{ij}\xi_j\|\right], i = 1, \cdots, M,
$$

$$(7.51)$$

where $R(\xi) = \begin{bmatrix} r_1(K\sum_{j=1}^{M}\mathcal{L}_{1j}\xi_j) \\ \vdots \\ r_M(K\sum_{j=1}^{M}\mathcal{L}_{Mj}\xi_j) \end{bmatrix}$.

The following theorem shows the ultimate boundedness of the states ξ and d_i of (7.51).

Theorem 49 *Suppose that Assumptions 7.2 and 7.3 hold. The feedback gain matrices of the adaptive controller (7.49) are designed as $K = -B^T P^{-1}$ and $\Gamma = P^{-1}BB^T P^{-1}$, where $P > 0$ is a solution to the LMI (7.14). Then, both the containment error ξ and the coupling gains d_i, $i = 1, \cdots, M$, in (7.51) are uniformly ultimately bounded. Furthermore, if φ_i are chosen such that $\varrho \triangleq \min_{i=1,\cdots,M}\varphi_i\tau_i < \alpha$, where α is defined as in (7.37), then ξ exponentially converges to the residual set*

$$
\mathcal{D}_2 \triangleq \left\{\xi : \|\xi\|^2 \leq \frac{\lambda_{\max}(P)}{\lambda_{\min}(\mathcal{L}_1)(\alpha - \varrho)}\sum_{i=1}^{M}(\beta^2\varphi_i + \frac{1}{2}\kappa_i)\right\}, \qquad (7.52)
$$

where $\beta = \max\{\gamma_{\max}, \frac{1}{\lambda_{\min}(\mathcal{L}_1)}\}$.

Proof 51 *Let $\tilde{d}_i = d_i - \beta$, $i = 1, \cdots, M$. Then, (7.51) can be rewritten as*

$$
\dot{\xi} = (I_M \otimes A + D\mathcal{L}_1 \otimes BK)\xi + (D \otimes B)R(\xi) + (\mathcal{L}_1^{-1}\mathcal{L}_2 \otimes B)u_l,
$$

$$
\dot{\tilde{d}}_i = \tau_i\left[-\varphi_i(\tilde{d}_i + \beta) + \left(\sum_{j=1}^{M}\mathcal{L}_{ij}\xi_j^T\right)\Gamma\left(\sum_{j=1}^{M}\mathcal{L}_{ij}\xi_j\right) + \|K\sum_{j=1}^{M}\mathcal{L}_{ij}\xi_j\|\right], i = 1, \cdots, M,
$$

$$(7.53)$$

where $D = \mathrm{diag}(\tilde{d}_1 + \beta, \cdots, \tilde{d}_M + \beta)$.

Consider the following Lyapunov function candidate

$$
V_2 = \frac{1}{2}\xi^T(\mathcal{L}_1 \otimes P^{-1})\xi + \sum_{i=1}^{M}\frac{\tilde{d}_i^2}{2\tau_i},
$$

As stated in the proof of Theorem 48, it is easy to see that V_2 is positive *definite. The time derivative of V_2 along (7.53) can be obtained as*

$$\dot{V}_2 = \xi^T(\mathcal{L}_1 \otimes P^{-1}A + \mathcal{L}_1 D\mathcal{L}_1 \otimes P^{-1}BK)\xi$$
$$+ \xi^T(\mathcal{L}_1 D \otimes P^{-1}B)R(\xi) + \xi^T(\mathcal{L}_2 \otimes P^{-1}B)u_l$$
$$+ \sum_{i=1}^{M} \tilde{d}_i[-\varphi_i(\tilde{d}_i + \beta) + (\sum_{j=1}^{M}\mathcal{L}_{ij}\xi_j^T)\Gamma(\sum_{j=1}^{M}\mathcal{L}_{ij}\xi_j) + \|K\sum_{j=1}^{M}\mathcal{L}_{ij}\xi_j\|].$$

$$(7.54)$$

By substituting $K = -BP^{-1}$, it is easy to get that

$$\xi^T(\mathcal{L}_1 D\mathcal{L}_1 \otimes P^{-1}BK)\xi = -\sum_{i=1}^{M}(\tilde{d}_i + \beta)[\sum_{j=1}^{M}\mathcal{L}_{ij}\xi_j]^T P^{-1}BB^T P^{-1}[\sum_{j=1}^{M}\mathcal{L}_{ij}\xi_j].$$

$$(7.55)$$

For the case where $d_i\|K\sum_{j=1}^{M}\mathcal{L}_{ij}\xi_j\| > \kappa_i$, $i = 1, \cdots, M$, we can get from (7.50) that

$$\xi^T(\mathcal{L}_1 D \otimes P^{-1}B)R(\xi) = -\sum_{i=1}^{M}(\tilde{d}_i + \beta)\|B^T P^{-1}\sum_{j=1}^{M}\mathcal{L}_{ij}\xi_j\|. \qquad (7.56)$$

Substituting (7.55), (7.56), and (7.41) into (7.54) yields

$$\dot{V}_2 \leq \xi^T(\mathcal{L}_1 \otimes P^{-1}A)\xi - \beta\sum_{i=1}^{M}\sum_{j=1}^{M}[\sum_{j=1}^{M}\mathcal{L}_{ij}\xi_j]^T P^{-1}BB^T P^{-1}[\sum_{j=1}^{M}\mathcal{L}_{ij}\xi_j]$$
$$- (\beta - \gamma_{\max})\sum_{i=1}^{M}\|B^T P^{-1}\sum_{j=1}^{M}\mathcal{L}_{ij}\xi_j\| - \sum_{i=1}^{M}\varphi_i(\tilde{d}_i^2 + \tilde{d}_i\beta)$$
$$\leq \frac{1}{2}\xi^T \mathcal{Z}\xi + \frac{1}{2}\sum_{i=1}^{M}\varphi_i(-\tilde{d}_i^2 + \beta^2),$$

where we have used the fact that $\beta \geq \max_{i\in\mathcal{R}}\gamma_i$ and $-\tilde{d}_i^2 - \tilde{d}_i\beta \leq -\frac{1}{2}\tilde{d}_i^2 + \frac{1}{2}\beta^2$ to get the last inequality and $\mathcal{Z} = \mathcal{L}_1 \otimes (P^{-1}A + A^T P^{-1}) - 2\beta\mathcal{L}_1^2 \otimes P^{-1}BBP^{-1}$.

For the case where $d_i\|K\sum_{j=1}^{M}\mathcal{L}_{ij}\xi_j\| \leq \kappa_i$, $i = 1, \cdots, M$, we can get from (7.50) that

$$\xi^T(\mathcal{L}_1 D \otimes P^{-1}B)R(\xi) = -\sum_{i=1}^{M}\frac{(\tilde{d}_i + \beta)^2}{\kappa_i}\|B^T P^{-1}\sum_{j=1}^{M}\mathcal{L}_{ij}\xi_j\|^2. \qquad (7.57)$$

Then, it follows from (7.55), (7.57), (7.41), and (7.54) that

$$\dot{V}_2 \leq \frac{1}{2}\xi^T \mathcal{Z}\xi + \sum_{i=1}^{M} \varphi_i(\tilde{d}_i^2 + \tilde{d}_i\beta) - \sum_{i=1}^{M} \frac{(\tilde{d}_i + \beta)^2}{\kappa_i} \|B^T P^{-1} \sum_{j=1}^{M} \mathcal{L}_{ij}\xi_j\|^2$$

$$+ \sum_{i=1}^{M}(\tilde{d}_i + \beta)\|B^T P^{-1} \sum_{j=1}^{M} \mathcal{L}_{ij}\xi_j\| \qquad (7.58)$$

$$\leq \frac{1}{2}\xi^T \mathcal{Z}\xi + \frac{1}{2}\sum_{i=1}^{M} \varphi_i(-\tilde{d}_i^2 + \beta^2) + \sum_{i=1}^{M} \frac{1}{4}\kappa_i.$$

Note that to get the last inequality in (7.58), we have used the following fact:

$$-\frac{(\tilde{d}_i + \beta)^2}{\kappa_i}\|B^T P^{-1} \sum_{j=1}^{M} \mathcal{L}_{ij}\xi_j\|^2 + (\tilde{d}_i + \beta)\|B^T P^{-1} \sum_{j=1}^{M} \mathcal{L}_{ij}\xi_j\| \leq \frac{1}{4}\kappa_i,$$

for $(\tilde{d}_i + \beta)\|K \sum_{j=1}^{M} \mathcal{L}_{ij}\xi_j\| \leq \kappa_i$, $i = 1, \cdots, M$.

For the case where $d_i\|K \sum_{j=1}^{M} \mathcal{L}_{ij}\xi_j\| \leq \kappa_i$, $i = 1, \cdots, l$, and $d_i\|K \sum_{j=1}^{M} \mathcal{L}_{ij}\xi_j\| > \kappa_i$, $i = l + 1, \cdots, M$. By following the steps in the two cases above, it is easy to get that

$$\dot{V}_2 \leq \frac{1}{2}\xi^T \mathcal{Z}\xi + \frac{1}{2}\sum_{i=1}^{M} \varphi_i(-\tilde{d}_i^2 + \beta^2) + \sum_{i=1}^{M-l} \frac{1}{4}\kappa_i.$$

Therefore, $\dot{V}_2$ satisfies (7.58) for all $\xi \in \mathbf{R}^{Nn}$. Because $\beta\lambda_{\min}(\mathcal{L}_1) \geq 1$, by following similar steps as in the proof of Theorem 48, it is easy to shown that $\mathcal{Z} < -\alpha\mathcal{L}_1 \otimes P^{-1} < 0$ and thereby $\xi^T \mathcal{Z}\xi - \sum_{i=1}^{M} \varphi_i\tilde{d}_i^2 < 0$. In virtue of Lemma 15, we get that the states ξ and d_i of (7.51) are uniformly ultimately bounded.

Next, we will derive the residual set for the containment error ξ. Rewrite (7.58) into

$$\dot{V}_2 \leq -\varrho V_2 + \frac{1}{2}\xi^T(\mathcal{Z} + \alpha\mathcal{L}_1 \otimes P^{-1})\xi - \frac{1}{2}\sum_{i=1}^{M}(\varphi_i - \frac{\varrho}{\tau_i})\tilde{d}_i^2$$

$$- \frac{\alpha - \varrho}{2}\xi^T(\mathcal{L}_1 \otimes P^{-1})\xi + \frac{1}{2}\sum_{i=1}^{M}(\beta^2\varphi_i + \frac{1}{2}\kappa_i) \qquad (7.59)$$

$$\leq -\varrho V_2 - \frac{\lambda_{\min}(\mathcal{L}_1)(\alpha - \varrho)}{2\lambda_{\max}(P)}\|\xi\|^2 + \frac{1}{2}\sum_{i=1}^{M}(\beta^2\varphi_i + \frac{1}{2}\kappa_i).$$

Obviously, it follows from (7.59) that $\dot{V}_2 \leq -\varrho V_2$ if $\|\xi\|^2 > \frac{\lambda_{\max}(P)}{\lambda_{\min}(\mathcal{L}_1)(\alpha-\varrho)}\sum_{i=1}^{M}$ $(\beta^2\varphi_i + \frac{1}{2}\kappa_i)$. Then, by noting $V_2 \geq \frac{\lambda_{\min}(\mathcal{L}_1)}{2\lambda_{\max}(P)}\|\xi\|^2$, we can get that if $\varrho \leq \alpha$ then ξ exponentially converges to the residual set $\mathcal{D}_2$ in (7.52) with a convergence rate faster than $e^{-\varrho t}$.

Remark 61 *As shown in Theorem 49, contrary to the static containment controller (7.29), the design of the adaptive controller (7.49) relies on only the agent dynamics and the relative state information, requiring neither the minimal eigenvalue $\lambda_{\min}(\mathcal{L}_1)$ nor the upper bounds of the leaders' control input. In (7.49), both the σ-modification technique in [63] and the boundary layer concept are used to guarantee the ultimate boundedness of the containment error ξ and the adaptive gains d_i. It can be observed that apart from the facts in (7.36), the residual set $\mathcal{D}_2$ in (7.52) also depends on the small constants φ_i. We can choose φ_i and κ_i to relatively small in order to guarantee a small containment error ξ.*

For the case where the relative state information is not available, the ideas in Section 6.4 can be extended to construct dynamic containment controllers using the relative estimates of the states of neighboring agents. The details are omitted here. Interested readers can do it as an exercise.

7.3.3   Simulation Examples

In this subsection, a simulation example is provided to validate the effectiveness of the theoretical results.

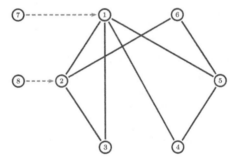

FIGURE 7.3: A communication graph with two leaders.

Consider a network of eight agents. For illustration, let the communication graph among the agents be given as in FIGURE 7.3, where nodes 7 and 8 are two leaders and the others are followers. The dynamics of the agents are given by (7.1), with

$$x_i = \begin{bmatrix} x_{i1} \\ x_{i2} \end{bmatrix}, \quad A = \begin{bmatrix} 0 & 1 \\ -1 & 1 \end{bmatrix}, \quad B = \begin{bmatrix} 0 \\ 1 \end{bmatrix},$$

Design the control inputs for the leaders as $u_7 = K_7 x_7 + 4\sin(2t)$ and $u_8 = K_8 x_8 + 2\cos(t)$, with $K_7 = -\begin{bmatrix} 0 & 2 \end{bmatrix}$ and $K_8 = -\begin{bmatrix} 1 & 3 \end{bmatrix}$. It is easy to see that in this case u_7 and u_8 are bounded. Here we use the adaptive control (7.49) to solve the containment control problem.

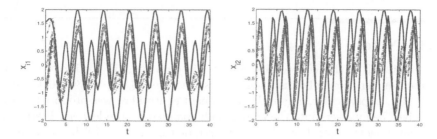

FIGURE 7.4: The state trajectories of the agents. The solid and dashdotted lines denote, respectively, the trajectories of the leaders and the followers.

Solving the LMI (7.14) by using the SeDuMi toolbox [158] gives the gain matrices K and Γ in (7.49) as

$$K = - \begin{bmatrix} 1.6203 & 4.7567 \end{bmatrix}, \ \Gamma = \begin{bmatrix} 2.6255 & 7.7075 \\ 7.7075 & 22.6266 \end{bmatrix}.$$

To illustrate Theorem 49, select $\kappa_i = 0.1$, $\varphi_i = 0.005$, and $\tau_i = 5$, $i = 1, \cdots, 6$, in (7.49). The state trajectories $x_i(t)$ of the agents under (7.49) designed as above are depicted in FIGURE 7.4, implying that the containment control problem is indeed solved. The coupling gains d_i associated with the followers are drawn in FIGURE 7.5, which are clearly bounded.

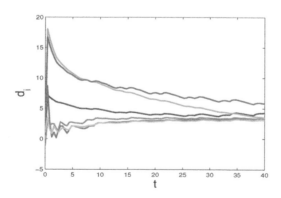

FIGURE 7.5: The coupling gains d_i in (7.49).

7.4   Notes

The materials of this chapter are mainly based on [88, 94]. For further results on containment control of multi-agent systems, see [16, 18, 19, 36, 46, 67, 104, 110, 111]. In particular, a hybrid containment control law was proposed in [67] to drive the followers into the convex hull spanned by the leaders. Distributed containment control problems were studied in [16, 18, 19] for a group of first-order and second-order integrator agents under fixed and switching directed communication topologies. The containment control was considered in [104] for second-order multi-agent systems with random switching topologies. A hybrid model predictive control scheme was proposed in [46] to solve the containment and distributed sensing problems in leader/follower multi-agent systems. The authors in [36, 110, 111] study the containment control problem for a collection of Euler-Lagrange systems. In particular, [36] discussed the case with multiple stationary leaders, [111] studied the case of dynamic leaders with finite-time convergence, and [110] considered the case with parametric uncertainties. In the above-mentioned works, the agent dynamics are assumed to be single, double integrators, or second-order Euler-Lagrange systems, which might be restrictive in some circumstances.

8

Distributed Robust Cooperative Control for Multi-Agent Systems with Heterogeneous Matching Uncertainties

CONTENTS

8.1	Distributed Robust Leaderless Consensus	184
	8.1.1 Distributed Static Consensus Protocols	185
	8.1.2 Distributed Adaptive Consensus Protocols	190
8.2	Distributed Robust Consensus with a Leader of Nonzero Control Input	197
8.3	Robustness with Respect to Bounded Non-Matching Disturbances .	201
8.4	Distributed Robust Containment Control with Multiple Leaders ...	205
8.5	Notes	205

The consensus problem of multi-agent systems with general discrete-time and continuous-time linear dynamics has been studied in the previous chapters, where various consensus protocols have been designed to achieve consensus. A common assumption in the previous chapters is that the dynamics of the agents are identical and precisely known, which might be restrictive and not practical in many circumstances. In practical applications, the agents may be subject to certain parameter uncertainties or unknown external disturbances.

In this chapter, we consider the distributed consensus problem of multi-agent systems with identical nominal linear dynamics but subject to different matching uncertainties. A typical example belonging to this scenario is a network of mass-spring systems with different masses or unknown spring constants. Due to the existence of the nonidentical uncertainties which may be time-varying, nonlinear and unknown, the multi-agent systems discussed in this chapter are essentially heterogeneous. The heterogeneous multi-agent systems in this chapter contain the homogeneous linear multi-agent systems studied in the previous chapters as a special case where the uncertainties do not exist. Note that because of the existence of the uncertainties, the consensus problem in this case becomes quite challenging to solve and the consensus algorithms given in in the previous chapters are not applicable any more.

In this chapter, we present a systematic procedure to address the distributed robust consensus problem of multi-agent systems with matching uncertainties for both the cases without and with a leader of possibly nonzero control input. The case where the communication graph is undirected and connected is considered in Section 8.1. By assuming that there exists time-varying upper bounds of the matching uncertainties, distributed continuous static consensus protocols based on the relative states of neighboring agents are designed, which includes a nonlinear term to deal with the effect of the uncertainties. It is shown that the consensus error under this static protocol is uniformly ultimately bounded and exponentially converges to a small residual set. Note that the design of this static protocol relies on the eigenvalues of the Laplacian matrix and the upper bounds of the matching uncertainties. In order to remove these requirements, a fully distributed adaptive protocol is further designed, under which the residual set of the consensus error is also given. One desirable feature is that for both the static and adaptive protocols, the residual sets of the consensus error can be made to be reasonably small by properly selecting the design parameters of the protocols and the convergence rates of the consensus error are explicitly given.

The robust consensus problem for the case where there exists a leader with nonzero control input is investigated in Section 8.2. Here we study the general case where the leader's control input is not available to any follower, which imposes additional difficulty. Distributed adaptive consensus protocols based on the relative state information are proposed and designed to ensure that the consensus error can converge to a small adjustable residual set.

The robustness issue of the proposed consensus protocols with respect to external disturbances which do not satisfy the matching condition is discussed in Section 8.3, where the proposed consensus protocols are redesigned to guarantee the boundedness of the consensus error and the adaptive gains in the presence of bounded external disturbances. Extensions of the results to the containment control problem with multiple leaders are discussed in Section 8.4.

8.1   Distributed Robust Leaderless Consensus

In this section, we consider a network of N autonomous agents with identical nominal linear dynamics but subject to heterogeneous uncertainties. The dynamics of the i-th agent are described by

$$\dot{x}_i = Ax_i + Bu_i + H_i(x_i, t) + \nu_i(t), \; i = 1, \cdots, N, \tag{8.1}$$

where $x_i \in \mathbf{R}^n$ is the state, $u_i \in \mathbf{R}^p$ is the control input, A and B are constant known matrices with compatible dimensions, and $H_i(x_i, t) \in \mathbf{R}^n$ and $\nu_i(t) \in \mathbf{R}^n$ denote, respectively, the parameter uncertainties and external disturbances associated with the i-th agent, which are assumed to satisfy the following standard matching condition [43, 183].

Assumption 8.1 *There exist functions $\hat{H}_i(x_i, t)$ and $\hat{\nu}_i(t)$ such that $H_i(x_i, t) = B\hat{H}_i(x_i, t)$ and $\nu_i(t) = B\hat{\nu}_i(t)$, $i = 1, \cdots, N$.*

By letting $f_i(x_i, t) = \hat{H}_i(x_i, t) + \hat{\nu}_i$ represent the lumped uncertainty of the i-th agent, (8.1) can be rewritten into

$$\dot{x}_i = Ax_i + B[u_i + f_i(x_i, t)], \ i = 1, \cdots, N. \tag{8.2}$$

In the previous chapters, the agents are identical linear systems and free of uncertainties. In contrast, the agents (8.2) considered in this chapter are subject to nonidentical uncertainties, which makes the resulting multi-agent systems are essentially heterogeneous. The agents (8.2) can recover the nominal linear agents in the previous chapters when the uncertainties $f_i(x_i, t)$ do not exist. Note that the existence of the uncertainties associated with the agents makes the consensus problem quite challenging to solve, as detailed in the sequel.

Regarding the bounds of the uncertainties $f_i(x_i, t)$, we introduce the following assumption.

Assumption 8.2 *There exist continuous scalar valued functions $\rho_i(x_i, t)$, $i = 1, \cdots, N$, such that $\|f_i(x_i, t)\| \leq \rho_i(x_i, t)$, $i = 1, \cdots, N$, for all $t \geq 0$ and $x_i \in \mathbf{R}^n$.*

The communication graph among the N agents is represented by a undirected graph $\mathcal{G}$, which is assumed to be connected throughout this section. The objective of this section is to solve the consensus problem for the agents in (8.2), i.e., to design distributed consensus protocols such that $\lim_{t \to \infty} \|x_i(t) - x_j(t)\| = 0, \ \forall i, j = 1, \cdots, N$.

8.1.1 Distributed Static Consensus Protocols

Based on the relative states of neighboring agents, the following distributed static consensus protocol is proposed:

$$u_i = cK \sum_{j=0}^{N} a_{ij}(x_i - x_j) + \rho_i(x_i, t)g_i(K \sum_{j=0}^{N} a_{ij}(x_i - x_j)), \ i = 1, \cdots, N,$$
$$\tag{8.3}$$

where $c > 0$ is the constant coupling gain, $K \in \mathbf{R}^{p \times n}$ is the feedback gain matrix, a_{ij} is the (i, j)-th entry of the adjacency matrix associated with $\mathcal{G}$, and the nonlinear functions $g_i(\cdot)$ are defined such that for $w \in \mathbf{R}^n$,

$$g_i(w) = \begin{cases} \frac{w}{\|w\|} & \text{if } \rho_i(x_i, t)\|w\| > \kappa_i \\ \frac{w}{\kappa_i}\rho_i(x_i, t) & \text{if } \rho_i(x_i, t)\|w\| \leq \kappa_i \end{cases} \tag{8.4}$$

where κ are small positive constants.

Let $x = [x_1^T, \cdots, x_N^T]^T$ and $\rho(x,t) = \text{diag}(\rho_1(x_1,t), \cdots, \rho_N(x_N,t))$. Using (8.3) for (8.2), we can obtain the closed-loop network dynamics as

$$\dot{x} = (I_N \otimes A + c\mathcal{L} \otimes BK)x + (I_N \otimes B)F(x,t) + [\rho(x,t) \otimes B]G(x), \qquad (8.5)$$

where $\mathcal{L}$ denotes the Laplacian matrix of $\mathcal{G}$, and

$$F(x,t) \triangleq \begin{bmatrix} f_1(x_1,t) \\ \vdots \\ f_N(x_N,t) \end{bmatrix}, \quad G(x) \triangleq \begin{bmatrix} g_1(K\sum_{j=1}^{N} \mathcal{L}_{1j}x_j) \\ \vdots \\ g_N(K\sum_{j=1}^{N} \mathcal{L}_{Nj}x_j) \end{bmatrix}. \qquad (8.6)$$

Let $\xi = (M \otimes I_n)x$, where $M = I_N - \frac{1}{N}\mathbf{1}\mathbf{1}^T$ and $\xi = [\xi_1^T, \cdots, \xi_N^T]^T$. It is easy to see that 0 is a simple eigenvalue of M with $\mathbf{1}$ as a corresponding right eigenvector and 1 is the other eigenvalue with multiplicity $N - 1$. Then, it follows that $\xi = 0$ if and only if $x_1 = \cdots = x_N$. Therefore, the consensus problem under the protocol (8.3) is solved if and only if ξ asymptotically converges to zero. Hereafter, we refer to ξ as the consensus error. By noting that $\mathcal{L}M = \mathcal{L}$, it is not difficult to obtain from (8.5) that the consensus error ξ satisfies

$$\begin{aligned} \dot{\xi} &= (I_N \otimes A + c\mathcal{L} \otimes BK)\xi + (M \otimes B)F(x,t) \\ &\quad + [M\rho(x,t) \otimes B]G(\xi). \end{aligned} \qquad (8.7)$$

The following result provides a sufficient condition to design the consensus protocol (8.3).

Theorem 50 *Suppose that the communication graph $\mathcal{G}$ is undirected and connected and Assumption 8.2 holds. The parameters in the distributed protocol (8.3) are designed as $c \geq \frac{1}{\lambda_2}$ and $K = -B^T P^{-1}$, where λ_2 is the smallest nonzero eigenvalue of $\mathcal{L}$ and $P > 0$ is a solution to the following linear matrix inequality (LMI):*

$$AP + PA^T - 2BB^T < 0, \qquad (8.8)$$

Then, the consensus error ξ of (8.7) is uniformly ultimately bounded and exponentially converges to the residual set

$$\mathcal{D}_1 \triangleq \{\xi : \|\xi\|^2 \leq \frac{2\lambda_{\max}(P)}{\alpha\lambda_2} \sum_{i=1}^{N} \kappa_i\}, \qquad (8.9)$$

where

$$\alpha = \frac{-\lambda_{\max}(AP + PA^T - 2BB^T)}{\lambda_{\max}(P)}. \qquad (8.10)$$

Proof 52 *Consider the following Lyapunov function candidate:*

$$V_1 = \frac{1}{2}\xi^T(\mathcal{L} \otimes P^{-1})\xi.$$

By the definition of ξ, it is easy to see that $(1^T \otimes I)\xi = 0$. For a connected graph $\mathcal{G}$, it then follows from Lemma 2 that

$$V_1(\xi) \geq \frac{1}{2}\lambda_2 \xi^T (I_N \otimes P^{-1})\xi \geq \frac{\lambda_2}{2\lambda_{\max}(P)}\|\xi\|^2. \tag{8.11}$$

The time derivative of V_1 along the trajectory of (8.5) is given by

$$\begin{aligned} \dot{V}_1 &= \xi^T (\mathcal{L} \otimes P^{-1}A + c\mathcal{L}^2 \otimes P^{-1}BK)\xi \\ &\quad + \xi^T (\mathcal{L} \otimes P^{-1}B)F(x,t) + \xi^T[\mathcal{L}\rho(x,t) \otimes P^{-1}B]G(\xi). \end{aligned} \tag{8.12}$$

By using Assumption 8.2, we can obtain that

$$\begin{aligned} \xi^T (\mathcal{L} \otimes P^{-1}B)F(x,t) &\leq \sum_{i=1}^{N} \|B^T P^{-1} \sum_{j=1}^{N} \mathcal{L}_{ij}\xi_j\| \|f_i(x_i,t)\| \\ &\leq \sum_{i=1}^{N} \rho_i(x_i,t)\|B^T P^{-1} \sum_{j=1}^{N} \mathcal{L}_{ij}\xi_j\|. \end{aligned} \tag{8.13}$$

Next, consider the following three cases.
i) $\rho_i(x_i,t)\|K\sum_{j=1}^{N} \mathcal{L}_{ij}\xi_j\| > \kappa_i$, $i = 1, \cdots, N$.
In this case, it follows from (8.4) and (8.6) that

$$\xi^T[\mathcal{L}\rho(x,t) \otimes P^{-1}B]G(\xi) = -\sum_{i=1}^{N} \rho_i(x_i,t)\|B^T P^{-1} \sum_{j=1}^{N} \mathcal{L}_{ij}\xi_j\|. \tag{8.14}$$

Substituting (8.14) and and (8.13) into (8.12) yields

$$\dot{V}_1 \leq \frac{1}{2}\xi^T \mathcal{X}\xi,$$

where $\mathcal{X} = \mathcal{L} \otimes (P^{-1}A + A^T P^{-1}) - 2c\mathcal{L}^2 \otimes P^{-1}BB^T P^{-1}$.
ii) $\rho_i(x_i,t)\|K\sum_{j=1}^{N} \mathcal{L}_{ij}\xi_j\| \leq \kappa_i$, $i = 1, \cdots, N$.
In this case, we can get from (8.4) and (8.6) that

$$\xi^T[\mathcal{L}\rho(x,t) \otimes P^{-1}B]G(\xi) = -\sum_{i=1}^{N} \frac{\rho_i(x_i,t)^2}{\kappa_i}\|B^T P^{-1} \sum_{j=1}^{N} \mathcal{L}_{ij}\xi_j\|^2 \leq 0. \tag{8.15}$$

Substituting (8.13) and (8.15) into (8.12) gives

$$\begin{aligned} \dot{V}_1 &\leq \frac{1}{2}\xi^T \mathcal{X}\xi + \sum_{i=1}^{N} \rho_i(x_i,t)\|B^T P^{-1} \sum_{j=1}^{N} \mathcal{L}_{ij}\xi_j\| \\ &\leq \frac{1}{2}\xi^T \mathcal{X}\xi + \sum_{i=1}^{N} \kappa_i. \end{aligned} \tag{8.16}$$

iii) ξ satisfies neither Case i) nor Case ii).

Without loss of generality, assume that $\rho_i(x_i,t)\|K\sum_{j=1}^N \mathcal{L}_{ij}\xi_j\| > \kappa_i$, $i = 1,\cdots,l$, and $\rho_i(x_i,t)\|K\sum_{j=1}^N \mathcal{L}_{ij}\xi_j\| \le \kappa_i$, $i = l+1,\cdots,N$, where $2 \le l \le N-1$. By combining (8.14) and (8.15), in this case we can get that

$$\xi^T[\mathcal{L}\rho(x,t) \otimes P^{-1}B]G(\xi) = -\sum_{i=1}^l \rho_i(x_i,t)\|B^T P^{-1}\sum_{j=1}^N \mathcal{L}_{ij}\xi_j\|$$
$$-\sum_{i=l+1}^N \frac{\rho_i(x_i,t)^2}{\kappa_i}\|B^T P^{-1}\sum_{j=1}^N \mathcal{L}_{ij}\xi_j\|^2 \quad (8.17)$$
$$\le -\sum_{i=1}^l \rho_i(x_i,t)\|B^T P^{-1}\sum_{j=1}^N \mathcal{L}_{ij}\xi_j\|.$$

Then, it follows from (8.12), (8.17), and (8.13) that

$$\dot{V}_1 \le \frac{1}{2}\xi^T \mathcal{X}\xi + \sum_{i=1}^{N-l} \kappa_i.$$

Therefore, by analyzing the above three cases, we get that $\dot{V}_1$ satisfies (8.16) for all $\xi \in \mathbf{R}^{Nn}$. Note that (8.16) can be rewritten as

$$\dot{V}_1 \le -\alpha V_1 + \alpha V_1 + \frac{1}{2}\xi^T \mathcal{X}\xi + \sum_{i=1}^N \kappa_i$$
$$= -\alpha V_1 + \frac{1}{2}\xi^T(\mathcal{X} + \alpha \mathcal{L} \otimes P^{-1})\xi + \sum_{i=1}^N \kappa_i, \quad (8.18)$$

where $\alpha > 0$.

Because $\mathcal{G}$ is connected, it follows from Lemma 1 that zero is a simple eigenvalue of $\mathcal{L}$ and all the other eigenvalues are positive. Let $U = [\frac{1}{\sqrt{N}} \ Y_1]$ and $U^T = \begin{bmatrix} \frac{1^T}{\sqrt{N}} \\ Y_2 \end{bmatrix}$, with $Y_1 \in \mathbf{R}^{N\times(N-1)}$, $Y_2 \in \mathbf{R}^{(N-1)\times N}$, be such unitary matrices that $U^T \mathcal{L} U = \Lambda \triangleq \mathrm{diag}(0,\lambda_2,\cdots,\lambda_N)$, where $\lambda_2 \le \cdots \le \lambda_N$ are the nonzero eigenvalues of $\mathcal{L}$. Let $\bar{\xi} \triangleq [\bar{\xi}_1^T,\cdots,\bar{\xi}_N^T]^T = (U^T \otimes P^{-1})\xi$. By the definitions of ξ and $\bar{\xi}$, it is easy to see that $\bar{\xi}_1 = (\frac{1^T}{\sqrt{N}} \otimes P^{-1})\xi = (\frac{1^T}{\sqrt{N}}M \otimes P^{-1})x = 0$. Then, it follows that

$$\xi^T(\mathcal{X} + \alpha \mathcal{L} \otimes P^{-1})\xi = \sum_{i=2}^N \lambda_i \bar{\xi}_i^T(AP + PA^T + \alpha P - 2c\lambda_i BB^T)\bar{\xi}_i$$
$$\le \sum_{i=2}^N \lambda_i \bar{\xi}_i^T(AP + PA^T + \alpha P - 2BB^T)\bar{\xi}_i. \quad (8.19)$$

Because $\alpha = \frac{-\lambda_{\max}(AP+PA^T-2BB^T)}{\lambda_{\max}(P)}$, *we can see from (8.19) that* $\xi^T(\mathcal{X}+\alpha\mathcal{L}\otimes P^{-1})\xi \leq 0$. *Then, we can get from (8.18) that*

$$\dot{V}_1 \leq -\alpha V_1 + \sum_{i=1}^{N} \kappa_i. \tag{8.20}$$

By using Lemma 16 (the Comparison lemma), we can obtain from (8.20) that

$$V_1(\xi) \leq [V_1(\xi(0)) - \sum_{i=1}^{N} \kappa_i/\alpha]e^{-\alpha t} + \sum_{i=1}^{N} \kappa_i/\alpha, \tag{8.21}$$

which, by (8.11), implies that ξ *exponentially converges to the residual set* $\mathcal{D}_1$ *in (8.9) with a convergence rate not less than* $e^{-\alpha t}$.

Remark 62 *The distributed consensus protocol (8.3) consists of a linear part and a nonlinear part, where the term* $\rho_i(x_i,t)g_i(K\sum_{j=1}^{N} a_{ij}(x_i - x_j))$ *is used to suppress the effect of the uncertainties* $f_i(x_i,t)$. *For the case where* $f_i(x_i,t) = 0$, *we can accordingly remove* $\rho_i(x_i,t)g_i(K\sum_{j=1}^{N} a_{ij}(x_i - x_j))$ *from (8.3), which can recover the static consensus protocol (2.2) in Chapter 2. As shown in Proposition 1 of Chapter 2, a necessary and sufficient condition for the existence of a* $P > 0$ *to the LMI (8.8) is that* (A, B) *is stabilizable. Therefore, a sufficient condition for the existence of (8.3) satisfying Theorem 50 is that* (A, B) *is stabilizable. Note that in Theorem 50 the parameters c and K of (8.3) are independently designed.*

Note that the nonlinear components $g_i(\cdot)$ in (8.4) are actually continuous approximations, via the boundary layer concept [43, 71], of the discontinuous function $\hat{g}(w) = \begin{cases} \frac{w}{\|w\|} & \text{if } \|w\| \neq 0 \\ 0 & \text{if } \|w\| = 0 \end{cases}$. The values of κ_i in (8.4) define the widths of the boundary layers. As $\kappa_i \to 0$, the continuous functions $g_i(\cdot)$ approach the discontinuous function $\hat{g}(\cdot)$.

Corollary 10 *Assume that* $\mathcal{G}$ *is connected and Assumption 8.2 holds. The consensus error* ξ *converges to zero under the discontinuous consensus protocol:*

$$u_i = cK \sum_{j=1}^{N} a_{ij}(x_i - x_j) + \rho_i(x_i,t)\hat{g}(K \sum_{j=1}^{N} a_{ij}(x_i - x_j)), \ i = 1, \cdots, N,$$

$$\tag{8.22}$$

where c and K are chosen as in Theorem 50.

Remark 63 *An inherent drawback of the discontinuous protocol (8.22) is that it will result in the undesirable chattering effect in real implementation, due to imperfections in switching devices [43, 187]. The effect of chattering is avoided* *by using the continuous protocol (8.3). The cast is that the protocol (8.3) does not guarantee asymptotic stability but rather uniform ultimate boundedness of the consensus error ξ. Note that the residual set $\mathcal{D}_1$ of ξ depends on the smallest nonzero eigenvalue of $\mathcal{L}$, the number of agents, the largest eigenvalue of P, and the widths κ_i of the boundary layers. By choosing sufficiently small κ_i, the consensus error ξ under the protocol (8.3) can converge to an arbitrarily small neighborhood of zero, which is acceptable in most applications.*

8.1.2 Distributed Adaptive Consensus Protocols

In the last subsection, the design of the distributed protocol (8.3) relies on the minimal nonzero eigenvalue λ_2 of $\mathcal{L}$ and the upper bounds $\rho_i(x_i, t)$ of the matching uncertainties $f_i(x_i, t)$. However, λ_2 is global information in the sense that each agent has to know the entire communication graph to compute it. Besides, the bounds $\rho_i(x_i, t)$ of the uncertainties $f_i(x_i, t)$ might not be easily obtained in some cases, e.g., $f_i(x_i, t)$ contains certain unknown external disturbances. In this subsection, we will implement some adaptive control ideas to compensate the lack of λ_2 and $\rho_i(x_i, t)$ and thereby to solve the consensus problem using only the local information available to the agents.

Before moving forward, we introduce a modified assumption regarding the bounds of the lumped uncertainties $f_i(x_i, t)$, $i = 1, \cdots, N$.

Assumption 8.3 *There are positive constants d_i and e_i such that $\|f_i(x_i, t)\| \leq d_i + e_i\|x_i\|$, $i = 1, \cdots, N$.*

Based on the local state information of neighboring agents, we propose the following distributed adaptive protocol to each agent:

$$u_i = \bar{d}_i K \sum_{j=1}^{N} a_{ij}(x_i - x_j) + r_i(K \sum_{j=1}^{N} a_{ij}(x_i - x_j)),$$

$$\dot{\bar{d}}_i = \tau_i[-\varphi_i\bar{d}_i + (\sum_{j=1}^{N} a_{ij}(x_i - x_j)^T)\Gamma(\sum_{j=1}^{N} a_{ij}(x_i - x_j)) + \|K\sum_{j=1}^{N} a_{ij}(x_i - x_j)\|],$$

$$\dot{\bar{e}}_i = \epsilon_i[-\psi_i\bar{e}_i + \|K\sum_{j=1}^{N} a_{ij}(x_i - x_j)\|\|x_i\|], \quad i = 1, \cdots, N,$$

$$(8.23)$$

where $\bar{d}_i(t)$ and $\bar{e}_i(t)$ are the adaptive coupling gains associated with the i-th agent, $\Gamma \in \mathbf{R}^{n \times n}$ is the feedback gain matrix, τ_i and ϵ_i are positive scalars, φ_i and ψ_i are small positive constants chosen by the designer, the nonlinear functions $r_i(\cdot)$ are defined such that for $w \in \mathbf{R}^n$,

$$r_i(w) = \begin{cases} \frac{w(\bar{d}_i + \bar{e}_i\|x_i\|)}{\|w\|} & \text{if } (\bar{d}_i + \bar{e}_i\|x_i\|)\|w\| > \kappa_i \\ \frac{w(\bar{d}_i + \bar{e}_i\|x_i\|)^2}{\kappa_i} & \text{if } (\bar{d}_i + \bar{e}_i\|x_i\|)\|w\| \leq \kappa_i \end{cases} \qquad (8.24)$$

and the rest of the variables are defined as in (8.3).

Let the consensus error ξ be defined as in (8.7) and $\overline{D} = \text{diag}(\bar{d}_1, \cdots, \bar{d}_N)$. Then, it is not difficult to get from (8.2) and (8.23) that the closed-loop network dynamics can be written as

$$
\begin{aligned}
\dot{\xi} &= (I_N \otimes A + M\overline{D}\mathcal{L} \otimes BK)\xi + (M \otimes B)[F(x,t) + R(\xi)], \\
\dot{\bar{d}}_i &= \tau_i[-\varphi_i \bar{d}_i + (\sum_{j=1}^N \mathcal{L}_{ij}\xi_j^T)\Gamma(\sum_{j=1}^N \mathcal{L}_{ij}\xi_j) + \|K\sum_{j=1}^N \mathcal{L}_{ij}\xi_j\|], \\
\dot{\bar{e}}_i &= \epsilon_i[-\psi_i \bar{e}_i + \|K\sum_{j=1}^N \mathcal{L}_{ij}\xi_j\|\|x_i\|], \quad i = 1, \cdots, N,
\end{aligned}
\tag{8.25}
$$

where

$$
R(\xi) \triangleq \begin{bmatrix} r_1(K\sum_{j=1}^N \mathcal{L}_{1j}\xi_j) \\ \vdots \\ r_N(K\sum_{j=1}^N \mathcal{L}_{Nj}\xi_j) \end{bmatrix},
\tag{8.26}
$$

and the rest of the variables are defined as in (8.5).

To establish the ultimate boundedness of the states ξ, $\bar{d}_i$, and $\bar{e}_i$ of (8.25), we use the following Lyapunov function candidate

$$
V_2 = \frac{1}{2}\xi^T(\mathcal{L} \otimes P^{-1})\xi + \sum_{i=1}^N \frac{\tilde{d}_i^2}{2\tau_i} + \sum_{i=1}^N \frac{\tilde{e}_i^2}{2\epsilon_i},
\tag{8.27}
$$

where $\tilde{e}_i = \bar{e}_i - e_i$, $\tilde{d}_i = \bar{d}_i - \beta$, $i = 1, \cdots, N$, and $\beta = \max_{i=1,\cdots,N}\{d_i, \frac{1}{\lambda_2}\}$.

The following theorem shows the ultimate boundedness of the states ξ, $\bar{d}_i$, and $\bar{e}_i$ of (8.25).

Theorem 51 *Suppose that $\mathcal{G}$ is connected and Assumption 8.3 holds. The feedback gain matrices of the distributed adaptive protocol (8.23) are designed as $K = -B^T P^{-1}$ and $\Gamma = P^{-1}BB^T P^{-1}$, where $P > 0$ is a solution to the LMI (8.8). Then, both the consensus error ξ and the adaptive gains $\bar{d}_i$ and $\bar{e}_i$, $i = 1, \cdots, N$, in (8.25) are uniformly ultimately bounded and the following statements hold.*

i) For any φ_i and ψ_i, the parameters ξ, $\tilde{d}_i$, and $\tilde{e}_i$ exponentially converge to the residual set

$$
\mathcal{D}_2 \triangleq \{\xi, \tilde{d}_i, \tilde{e}_i : V_2 < \frac{1}{2\delta}\sum_{i=1}^N (\beta^2 \varphi_i + e_i^2 \psi_i + \frac{\kappa_i}{2})\},
\tag{8.28}
$$

where $\delta \triangleq \min_{i=1,\cdots,N}\{\alpha, \varphi_i\tau_i, \psi_i\epsilon_i\}$ and α is defined as in (8.10).

ii) If small φ_i and ψ_i satisfy $\varrho \triangleq \min_{i=1,\cdots,N}\{\varphi_i\tau_i, \psi_i\epsilon_i\} < \alpha$, then in addition to i), ξ exponentially converges to the residual set

$$\mathcal{D}_3 \triangleq \{\xi : \|\xi\|^2 \leq \frac{\lambda_{\max}(P)}{\lambda_2(\alpha - \varrho)}\sum_{i=1}^{N}(\beta^2\varphi_i + e_i^2\psi_i + \frac{1}{2}\kappa_i)\}. \tag{8.29}$$

Proof 53 *The time derivative of V_2 along (8.25) can be obtained as*

$$\begin{aligned}
\dot{V}_2 = \xi^T[&(\mathcal{L} \otimes P^{-1}A + \mathcal{L}D\mathcal{L} \otimes P^{-1}BK)\xi \\
&+ (\mathcal{L} \otimes P^{-1}B)F(x,t) + (\mathcal{L} \otimes P^{-1}B)R(\xi)] \\
&+ \sum_{i=1}^{N}\tilde{d}_i[-\varphi_i(\tilde{d}_i + \beta) + (\sum_{j=1}^{N}\mathcal{L}_{ij}\xi_j^T)\Gamma(\sum_{j=1}^{N}\mathcal{L}_{ij}\xi_j) \\
&+ \|K\sum_{j=1}^{N}\mathcal{L}_{ij}\xi_j\|] + \sum_{i=1}^{N}\tilde{e}_i[-\psi_i(\tilde{e}_i + e_i) + \|K\sum_{j=1}^{N}\mathcal{L}_{ij}\xi_j\|\|x_i\|],
\end{aligned} \tag{8.30}$$

where $D = \operatorname{diag}(\tilde{d}_1 + \beta, \cdots, \tilde{d}_N + \beta)$.

By noting that $K = -BP^{-1}$, it is easy to get that

$$\begin{aligned}
\xi^T&(\mathcal{L}D\mathcal{L} \otimes P^{-1}BK)\xi \\
&= -\sum_{i=1}^{N}(\tilde{d}_i + \beta)(\sum_{j=1}^{N}\mathcal{L}_{ij}\xi_j)^T P^{-1}BB^T P^{-1}(\sum_{j=1}^{N}\mathcal{L}_{ij}\xi_j).
\end{aligned} \tag{8.31}$$

In light of Assumption 8.3, we can obtain that

$$\xi^T(\mathcal{L} \otimes P^{-1}B)F(x,t) \leq \sum_{j=1}^{N}(d_i + e_i\|x_i\|)\|B^T P^{-1}\sum_{j=1}^{N}\mathcal{L}_{ij}\xi_j\|. \tag{8.32}$$

In what follows, we consider three cases.
i) $(\bar{d}_i + \bar{e}_i\|x_i\|)\|K\sum_{j=1}^{N}\mathcal{L}_{ij}\xi_j\| > \kappa_i, i = 1, \cdots, N$.
In this case, we can get from (8.24) and (8.26) that

$$\xi^T(\mathcal{L} \otimes P^{-1}B)R(\xi) = -\sum_{i=1}^{N}[\bar{d}_i + \bar{e}_i\|x_i\|]\|B^T P^{-1}\sum_{j=1}^{N}\mathcal{L}_{ij}\xi_j\|. \tag{8.33}$$

Substituting (8.31), (8.32), and (8.33) into (8.30) yields

$$\begin{aligned}
\dot{V}_2 \leq &\frac{1}{2}\xi^T\mathcal{Y}\xi - \sum_{i=1}^{N}(\beta - d_i)\|B^T P^{-1}\sum_{j=1}^{N}\mathcal{L}_{ij}\xi_j\| \\
&- \frac{1}{2}\sum_{i=1}^{N}(\varphi_i\tilde{d}_i^2 + \psi_i\tilde{e}_i^2) + \frac{1}{2}\sum_{i=1}^{N}(\beta^2\varphi_i + e_i^2\psi_i) \\
\leq &\frac{1}{2}\xi^T\mathcal{Y}\xi - \frac{1}{2}\sum_{i=1}^{N}(\varphi_i\tilde{d}_i^2 + \psi_i\tilde{e}_i^2) + \frac{1}{2}\sum_{i=1}^{N}(\beta^2\varphi_i + e_i^2\psi_i),
\end{aligned}$$

where $\mathcal{Y} \triangleq \mathcal{L} \otimes (P^{-1}A + A^T P^{-1}) - 2\beta \mathcal{L}^2 \otimes P^{-1} B B^T P^{-1}$ *and we have used the facts that* $\beta \geq \max_{i=1,\cdots,N} d_i$ *and* $-\tilde{d}_i^2 - \tilde{d}_i \beta \leq -\frac{1}{2}\tilde{d}_i^2 + \frac{1}{2}\beta^2$.

ii) $(\bar{d}_i + \bar{e}_i \|x_i\|) \|K \sum_{j=1}^{N} \mathcal{L}_{ij} \xi_j\| \leq \kappa_i$, $i = 1, \cdots, N$.

In this case, we can get from (8.24) and (8.26) that

$$\xi^T (\mathcal{L} \otimes P^{-1} B) R(\xi) = -\sum_{i=1}^{N} \frac{(\bar{d}_i + \bar{e}_i \|x_i\|)^2}{\kappa_i} \|B^T P^{-1} \sum_{j=1}^{N} \mathcal{L}_{ij} \xi_j\|^2. \qquad (8.34)$$

Then, it follows from (8.31), (8.32), (8.34), and (8.30) that

$$\dot{V}_2 \leq \frac{1}{2}\xi^T \mathcal{Y}\xi - \frac{1}{2}\sum_{i=1}^{N}(\varphi_i \tilde{d}_i^2 + \psi_i \tilde{e}_i^2)$$
$$+ \frac{1}{2}\sum_{i=1}^{N}(\beta^2 \varphi_i + e_i^2 \psi_i + \frac{1}{2}\kappa_i), \qquad (8.35)$$

where we have used the fact that $-\frac{(\bar{d}_i + \bar{e}_i \|x_i\|)^2}{\kappa_i} \|B^T P^{-1} \sum_{j=1}^{N} \mathcal{L}_{ij} \xi_j\|^2 + (\bar{d}_i + \bar{e}_i \|x_i\|) \|B^T P^{-1} \sum_{j=1}^{N} \mathcal{L}_{ij} \xi_j\| \leq \frac{1}{4}\kappa_i$, *for* $(\bar{d}_i + \bar{e}_i \|x_i\|) \|K \sum_{j=1}^{N} \mathcal{L}_{ij} \xi_j\| \leq \kappa_i$, $i = 1, \cdots, N$.

iii) $(\bar{d}_i + \bar{e}_i \|x_i\|) \|K \sum_{j=1}^{N} \mathcal{L}_{ij} \xi_j\| > \kappa_i$, $i = 1, \cdots, l$, *and* $(\bar{d}_i + \bar{e}_i \|x_i\|) \|K \sum_{j=1}^{N} \mathcal{L}_{ij} \xi_j\| \leq \kappa_i$, $i = l+1, \cdots, N$, *where* $2 \leq l \leq N - 1$.

By following similar steps in the two cases above, it is not difficult to get that

$$\dot{V}_2 \leq \frac{1}{2}\xi^T \mathcal{Y}\xi - \frac{1}{2}\sum_{i=1}^{N}(\varphi_i \tilde{d}_i^2 + \psi_i \tilde{e}_i^2)$$
$$+ \frac{1}{2}\sum_{i=1}^{N}(\beta^2 \varphi_i + e_i^2 \psi_i) + \frac{1}{4}\sum_{i=1}^{N-l}\kappa_i.$$

Therefore, based on the above three cases, we can get that $\dot{V}_2$ *satisfies (8.35) for all* $\xi \in \mathbf{R}^{Nn}$. *Note that (8.35) can be rewritten into*

$$\dot{V}_2 \leq -\delta V_2 + \delta V_2 + \frac{1}{2}\xi^T \mathcal{Y}\xi - \frac{1}{2}\sum_{i=1}^{N}(\varphi_i \tilde{d}_i^2 + \psi_i \tilde{e}_i^2)$$
$$+ \frac{1}{2}\sum_{i=1}^{N}(\beta^2 \varphi_i + e_i^2 \psi_i + \frac{1}{2}\kappa_i)$$
$$= -\delta V_2 + \frac{1}{2}\xi^T (\mathcal{Y} + \delta \mathcal{L} \otimes P^{-1})\xi - \frac{1}{2}\sum_{i=1}^{N}[(\varphi_i - \frac{\delta}{\tau_i})\tilde{d}_i^2$$
$$+ (\psi_i - \frac{\delta}{\epsilon_i})\tilde{e}_i^2)] + \frac{1}{2}\sum_{i=1}^{N}(\beta^2 \varphi_i + e_i^2 \psi_i + \frac{1}{2}\kappa_i). \qquad (8.36)$$

Because $\beta \lambda_2 \geq 1$ *and* $0 < \delta \leq \alpha$, *by following similar steps in the proof of*

Theorem 50, we can show that $\xi^T(\mathcal{Y} + \delta\mathcal{L} \otimes P^{-1})\xi \leq 0$. Further, by noting that $\delta \leq \min_{i=1,\cdots,N}\{\varphi_i\tau_i, \psi_i\epsilon_i\}$, it follows from (8.36) that

$$\dot{V}_2 \leq -\delta V_2 + \frac{1}{2}\sum_{i=1}^{N}(\beta^2\varphi_i + e_i^2\psi_i + \frac{1}{2}\kappa_i), \tag{8.37}$$

which implies that

$$V_2 \leq [V_2(0) - \frac{1}{2\delta}\sum_{i=1}^{N}(\beta^2\varphi_i + e_i^2\psi_i + \frac{1}{2}\kappa_i)]e^{-\delta t}$$
$$+ \frac{1}{2\delta}\sum_{i=1}^{N}(\beta^2\varphi_i + e_i^2\psi_i + \frac{1}{2}\kappa_i). \tag{8.38}$$

Therefore, V_2 exponentially converges to the residual set $\mathcal{D}_2$ in (8.28) with a convergence rate faster than $e^{-\delta t}$, which, in light of $V_2 \geq \frac{\lambda_2}{2\lambda_{\max}(P)}\|\xi\|^2 + \sum_{i=1}^{N}\frac{\tilde{d}_i^2}{2\tau_i} + \sum_{i=1}^{N}\frac{\tilde{e}_i^2}{2\epsilon_i}$, implies that ξ, $\bar{d}_i$, and $\bar{e}_i$ are uniformly ultimately bounded.

Next, if $\varrho \triangleq \min_{i=1,\cdots,N}\{\varphi_i\tau_i, \psi_i\epsilon_i\} < \alpha$, then we can choose $\delta = \varrho$ and it is easy to see that $\mathcal{D}_2$ increase as $\min\{\varphi_i\tau_i, \psi_i\epsilon_i\}$ decreases. However, for the case where $\varrho < \alpha$, we can obtain a smaller residual set for ξ by rewriting (8.36) into

$$\dot{V}_2 \leq -\varrho V_2 + \frac{1}{2}\xi^T(\mathcal{Y} + \alpha\mathcal{L} \otimes P^{-1})\xi$$
$$- \frac{\alpha - \varrho}{2}\xi^T(\mathcal{L} \otimes P^{-1})\xi + \frac{1}{2}\sum_{i=1}^{N}(\beta^2\varphi_i + e_i^2\psi_i + \frac{1}{2}\kappa_i) \tag{8.39}$$
$$\leq -\varrho V_2 - \frac{\lambda_2(\alpha - \varrho)}{2\lambda_{\max}(P)}\|\xi\|^2 + \frac{1}{2}\sum_{i=1}^{N}(\beta^2\varphi_i + e_i^2\psi_i + \frac{1}{2}\kappa_i).$$

Obviously, it follows from (8.39) that $\dot{V}_2 \leq -\varrho V_2$ if $\|\xi\|^2 > \frac{\lambda_{\max}(P)}{\lambda_2(\alpha-\varrho)}\sum_{i=1}^{N}(\beta^2\varphi_i + e_i^2\psi_i + \frac{1}{2}\kappa_i)$. Then, by noting $V_2 \geq \frac{\lambda_2}{2\lambda_{\max}(P)}\|\xi\|^2$, we can get that if $\varrho \leq \alpha$ then ξ exponentially converges to the residual set $\mathcal{D}_3$ in (8.29) with a convergence rate faster than $e^{-\varrho t}$.

Remark 64 *It is worth mentioning that both the σ-modification technique in [63, 183] and the boundary layer concept play a vital role to guarantee the ultimate boundedness of the consensus error ξ and the adaptive gains $\bar{d}_i$ and $\bar{e}_i$. From (8.28) and (8.29), we can observe that the residual sets $\mathcal{D}_2$ and $\mathcal{D}_3$ decrease as κ_i decrease. Given κ_i, smaller φ_i and ψ_i give a smaller bound for ξ and at the same time yield a larger bound for $\bar{d}_i$ and $\bar{e}_i$. For the case where $\varphi_i = 0$ or $\psi_i = 0$, $\bar{d}_i$ and $\bar{e}_i$ will tend to infinity. In real implementations, if large $\bar{d}_i$ and $\bar{e}_i$ are acceptable, we can choose φ_i, ψ_i, and κ_i to be relatively small in order to guarantee a small consensus error ξ.*

Remark 65 *Contrary to the static protocol (8.3), the design of the adaptive protocol (8.23) relies on only the agent dynamics and the local state information of each agent and its neighbors, requiring neither the minimal nonzero eigenvalue of $\mathcal{L}$ nor the upper bounds of of the uncertainties $f_i(x_i, t)$. Thus, the adaptive protocol (8.23) can be implemented by each agent in a fully distributed fashion without requiring any global information.*

Remark 66 *A special case of the uncertainties $f_i(x_i, t)$ satisfying Assumptions 8.2 and 8.3 is that there exist positive constants d_i such that $\| f_i(x_i, t)\| \leq d_i$, $i = 1, \cdots, N$. For this case, the proposed protocols (8.3) and (8.23) can be accordingly simplified. For (8.3), we can simply replace $\rho_i(x_i, t)$ by d_i. The adaptive protocol (8.23) in this case can be modified into*

$$u_i = \bar{d}_i K \sum_{j=1}^{N} a_{ij}(x_i - x_j) + \bar{r}_i(K \sum_{j=1}^{N} a_{ij}(x_i - x_j)),$$

$$\dot{\bar{d}}_i = \tau_i[-\varphi_i \bar{d}_i + (\sum_{j=1}^{N} a_{ij}(x_i - x_j)^T)\Gamma(\sum_{j=1}^{N} a_{ij}(x_i - x_j)) \tag{8.40}$$

$$+ \|K \sum_{j=1}^{N} a_{ij}(x_i - x_j)\|], \ i = 1, \cdots, N,$$

where the nonlinear functions $\bar{r}_i(\cdot)$ are defined such that $\bar{r}_i(w) = \begin{cases} \frac{w \bar{d}_i}{\|w\|} & \text{if } \bar{d}_i \|w\| > \kappa \\ \frac{w \bar{d}_i^2}{\kappa} & \text{if } \bar{d}_i \|w\| \leq \kappa \end{cases}$ and the rest of the variables are defined as in (8.23).

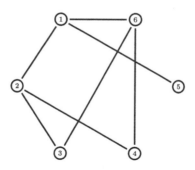

FIGURE 8.1: The leaderless communication graph.

Example 23 *Consider a network of mass-spring systems with a common mass m but different unknown spring constants, described by*

$$m\ddot{y}_i + k_i y_i = u_i, \ i = 1, \cdots, N, \tag{8.41}$$

where y_i are the displacements from certain reference positions and k_i, $i = 1, \cdots, N$, are the bounded unknown spring constants. Denote by $x_i = [\, y_i \; \dot{y}_i \,]^T$ the state of the i-th agent. Then, (8.41) can be rewritten as

$$\dot{x}_i = A x_i + B(u_i + k_i E x_i), \quad i = 1, \cdots, N, \tag{8.42}$$

with

$$A = \begin{bmatrix} 0 & 1 \\ 0 & 0 \end{bmatrix}, \quad B = \begin{bmatrix} 0 \\ \frac{1}{m} \end{bmatrix}, \quad E = \begin{bmatrix} -1 & 0 \end{bmatrix}.$$

It is easy to see that $k_i E x_i$, $i = 1, \cdots, N$, satisfy Assumption 8.3, i.e., $\| k_i E x_i \| \le k_i \| x_i \|$, $i = 1, \cdots, N$.

Because the spring constants k_i are unknown, we will use the adaptive protocol (8.23) to solve the consensus problem. Let $m = 2.5 kg$ and k_i be randomly chosen. Solving the LMI (8.8) by using the SeDuMi toolbox [158] gives the feedback gain matrices of (8.23) as

$$K = -\begin{bmatrix} 0.6693 & 2.4595 \end{bmatrix}, \quad \Gamma = \begin{bmatrix} 0.4480 & 1.6462 \\ 1.6462 & 6.0489 \end{bmatrix}.$$

Assume that the communication topology is given in FIGURE 8.1. In (8.23), select $\kappa = 0.5$, $\varphi_i = \psi_i = 0.05$, and $\tau_i = \epsilon_i = 10$, $i = 1, \cdots, 6$, in (8.23). The state trajectories $x_i(t)$ of (8.42) under (8.23) designed as above are depicted in FIGURE 8.2, which implies that consensus is indeed achieved. The adaptive gains $\bar{d}_i$ and $\bar{e}_i$ in (8.23) are shown in FIGURE 8.3, from which it can be observed that $\bar{d}_i$ and $\bar{e}_i$ tend to be quite small.

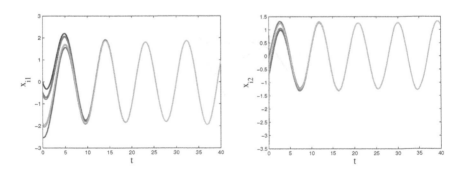

FIGURE 8.2: The state trajectories of the mass-spring systems.

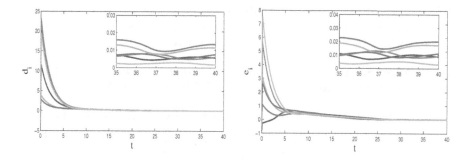

FIGURE 8.3: The adaptive gains $\bar{d}_i$ and $\bar{e}_i$ in (8.23).

8.2 Distributed Robust Consensus with a Leader of Nonzero Control Input

For the leaderless consensus problem in the previous section where the communication graph is undirected, the final consensus values reached by the agents under the protocols (8.3) and (8.23) are generally difficult to be explicitly obtained. The main difficulty lies in that the agents are subject to uncertainties and the protocols (8.3) and (8.23) are essentially nonlinear. In this section, we consider the leader-follower consensus (distributed tracking) problem, for which case the agents' states are required to converge onto a reference trajectory.

Consider a network of $N + 1$ agents consisting of N followers and one leader. Without loss of generality, let the agent indexed by 0 be the leader and the agents indexed by $1, \cdots, N$, be the followers. The dynamics of the followers are described by (8.2). For simplicity, we assume that the leader has the nominal linear dynamics, given by

$$\dot{x}_0 = Ax_0 + Bu_0, \qquad (8.43)$$

where $x_0 \in \mathbf{R}^n$ is the state and $u_0 \in \mathbf{R}^p$ is the control input of the leader, respectively. In some applications, the leader might need its own control action to achieve certain objectives, e.g., to reach a desirable consensus value. In this section, we consider the general case where u_0 is possibly nonzero and time varying and not accessible to any follower, which is much harder to solve than the case with $u_0 = 0$.

Before moving forward, the following mild assumption is needed.

Assumption 8.4 *The leader's control input u_0 is bounded, i.e., there exists a positive scalar γ such that $\|u_0\| \leq \gamma$.*

It is assumed that the leader receives no information from any follower and the state of the leader is available to only a subset of the followers. The communication graph among the $N + 1$ agents is represented by a directed graph $\widehat{\mathcal{G}}$, which satisfies the following assumption.

Assumption 8.5 $\widehat{\mathcal{G}}$ *contains a directed spanning tree with the leader as the root and the subgraph associated with the N followers is undirected.*

Denote by $\widehat{\mathcal{L}}$ the Laplacian matrix associated with $\widehat{\mathcal{G}}$. Because the leader has no neighbors, $\widehat{\mathcal{L}}$ can be partitioned as $\widehat{\mathcal{L}} = \begin{bmatrix} 0 & 0_{1 \times N} \\ \mathcal{L}_2 & \mathcal{L}_1 \end{bmatrix}$, where $\mathcal{L}_2 \in \mathbf{R}^{N \times 1}$ and $\mathcal{L}_1 \in \mathbf{R}^{N \times N}$. By Lemma 1 and Assumption 8.5, it is clear that $\mathcal{L}_1 > 0$.

The objective of this section is to solve the leader-follower consensus problem for the agents in (8.2) and (8.43), i.e., to design distributed protocols under which the states of the N followers converge to the state of the leader.

Based on the relative states of neighboring agents, we propose the following distributed adaptive protocol to each follower as

$$u_i = \hat{d}_i K \sum_{j=0}^{N} a_{ij}(x_i - x_j) + r_i(K \sum_{j=0}^{N} a_{ij}(x_i - x_j)),$$

$$\dot{\hat{d}}_i = \tau_i[-\varphi_i \hat{d}_i + (\sum_{j=0}^{N} a_{ij}(x_i - x_j)^T)\Gamma(\sum_{j=0}^{N} a_{ij}(x_i - x_j)) + \|K \sum_{j=0}^{N} a_{ij}(x_i - x_j)\|],$$

$$\dot{\hat{e}}_i = \epsilon_i[-\psi_i \hat{e}_i + \|K \sum_{j=0}^{N} a_{ij}(x_i - x_j)\| \|x_i\|], \quad i = 1, \cdots, N,$$

$$(8.44)$$

where $\hat{d}_i(t)$ and $\hat{e}_i(t)$ are the adaptive gains associated with the i-th follower, a_{ij} is the (i,j)-th entry of the adjacency matrix associated with $\widehat{\mathcal{G}}$, and the rest of the variables are defined as in (8.23).

Let $x = [x_1^T, \cdots, x_N^T]^T$, $\zeta_i = x_i - x_0$, $i = 1, \cdots, N$, and $\zeta = [\zeta_1^T, \cdots, \zeta_N^T]^T$. Then, it follows from (8.2), (8.43), and (8.44) that the closed-loop network dynamics can be obtained as

$$\dot{\zeta} = (I_N \otimes A + \widehat{D}\mathcal{L}_1 \otimes BK)\zeta + (I_N \otimes B)F(x, t)$$
$$+ (I_N \otimes B)R(\zeta) - (\mathbf{1} \otimes B)u_0,$$

$$\dot{\hat{d}}_i = \tau_i[-\varphi_i \hat{d}_i + (\sum_{j=1}^{N} \mathcal{L}_{ij}\zeta_j^T)\Gamma(\sum_{j=1}^{N} \mathcal{L}_{ij}\zeta_j) + \|K \sum_{j=1}^{N} \mathcal{L}_{ij}\zeta_j\|], \qquad (8.45)$$

$$\dot{\hat{e}}_i = \epsilon_i[-\psi_i \hat{e}_i + \|K \sum_{j=1}^{N} \mathcal{L}_{ij}\zeta_j\| \|x_i\|], \quad i = 1, \cdots, N,$$

where $\widehat{D} = \mathrm{diag}(\hat{d}_1, \cdots, \hat{d}_N)$, $R(\cdot)$ remains the same as in (8.26), and the rest of the variables are defined as in (8.25).

To present the following theorem, we use a Lyapunov function in the form of

$$V_3 = \frac{1}{2}\zeta^T(\mathcal{L}_1 \otimes P^{-1})\zeta + \sum_{i=1}^{N} \frac{\breve{d}_i^2}{2\tau_i} + \sum_{i=1}^{N} \frac{\breve{e}_i^2}{2\epsilon_i},$$

where $\breve{e}_i = \hat{e}_i - e_i$, $\breve{d}_i = \hat{d}_i - \hat{\beta}$, $i = 1, \cdots, N$, and $\hat{\beta} = \max_{i=1,\cdots,N}\{d_i + \gamma, \frac{1}{\lambda_{\min}(\mathcal{L}_1)}\}$.

Theorem 52 *Supposing that Assumptions 8.3, 8.4, and 8.5 hold, the leader-follower consensus error ζ and the adaptive gains $\hat{d}_i$ and $\hat{e}_i$, $i = 1, \cdots, N$, in (8.45) are uniformly ultimately bounded under the distributed adaptive protocol (8.44) with K and Γ designed as in Theorem 51. Moreover, the following two assertions hold.*

i) For any φ_i and ψ_i, the parameters ξ, $\breve{d}_i$, and $\breve{e}_i$ exponentially converge to the residual set

$$\mathcal{D}_4 \triangleq \{\zeta, \breve{d}_i, \breve{e}_i : V_3 < \frac{1}{2\delta}\sum_{i=1}^{N}(\hat{\beta}^2\varphi_i + e_i^2\psi_i + \frac{1}{2}\kappa_i)\}, \qquad (8.46)$$

where δ is defined as in Theorem 51.

ii) If $\varrho < \alpha$, where ϱ and α are defined as in Theorem 51 and (8.10), respectively, then in addition to i), ζ exponentially converges to the residual set

$$\mathcal{D}_5 \triangleq \{\zeta : \|\zeta\|^2 \leq \frac{\lambda_{\max}(P)}{\lambda_{\min}(\mathcal{L}_1)(\alpha - \varrho)}\sum_{i=1}^{N}(\hat{\beta}^2\varphi_i + e_i^2\psi_i + \frac{1}{2}\kappa_i)\}. \qquad (8.47)$$

Proof 54 *It can be completed by following similar steps as in the proof of Theorem 51 and by using the following assertion:*

$$-\zeta^T(\mathcal{L}_1\mathbf{1} \otimes PB)u_0 \leq \sum_{i=1}^{N}\|B^TP\sum_{j=1}^{N}\mathcal{L}_{ij}\zeta_j\|\|u_0\|$$

$$\leq \gamma\sum_{i=1}^{N}\|B^TP\sum_{j=1}^{N}\mathcal{L}_{ij}\zeta_j\|.$$

The details are omitted here for brevity.

The following example is presented to illustrate Theorem 52.

Example 24 *Consider a group of Chua's circuits, whose dynamics in the dimensionless form are given by [107]*

$$\begin{aligned}
\dot{x}_{i1} &= a[-x_{i1} + x_{i2} - h(x_{i1})] + u_i, \\
\dot{x}_{i2} &= x_{i1} - x_{i2} + x_{i3}, \\
\dot{x}_{i2} &= -bx_{i2}, \quad i = 0, \cdots, N,
\end{aligned} \qquad (8.48)$$

where $a > 0$, $b > 0$, and $h(x_{i1})$ is a nonlinear function represented by $h(x_{i1}) = m_i^1 x_{i1} + \frac{1}{2}(m_i^2 - m_i^1)(|x_{i1}+1| - |x_{i1}-1|)$, where $m_i^1 < 0$ and $m_i^2 < 0$. The circuit indexed by 0 is the leader and the other circuits are the followers. It is assumed that the Chua's circuits have nonidentical nonlinear components, i.e., m_i^1 and m_i^2 are different for different Chua's circuits. By letting $x_i = [x_{i1}, x_{i2}, x_{i3}]^T$, then (8.48) can be rewritten in a compact form as

$$\dot{x}_i = Ax_i + B[u_i + f_i(x_i)], \quad i = 0, \cdots, N, \tag{8.49}$$

where

$$A = \begin{bmatrix} -m_0^1(a+1) & a & 0 \\ 1 & -1 & 1 \\ 0 & -b & 0 \end{bmatrix}, \quad B = \begin{bmatrix} 1 \\ 0 \\ 0 \end{bmatrix},$$

$$f_0(x_0) = \frac{a}{2}(m_0^1 - m_0^2)(|x_{01}+1| - |x_{01}-1|),$$

$$f_i(x_i) = a(m_0^1 - m_i^1)x_{i1} + \frac{a}{2}(m_i^1 - m_i^2)(|x_{i1}+1| - |x_{i1}-1|), i = 1, \cdots, N.$$

For simplicity, we let $u_0 = 0$ and take $f_0(x_0)$ as the virtual control input of the leader, which clearly satisfies $\|f_0(x_0)\| \leq \frac{a}{2}|m_0^1 - m_0^2|$. Let $a = 9$, $b = 18$, $m_0^1 = -\frac{3}{4}$, and $m_0^2 = -\frac{4}{3}$. In this case, the leader displays a double-scroll chaotic attractor [107]. The parameters m_i^1 and m_i^2, $i = 1, \cdots, N$, are randomly chosen within the interval $[-6, 0)$. It is easy to see that

$$\|f_i(x_i)\| \leq a|m_0^1 - m_i^1| \|x_i\| + a|m_i^1 - m_i^2|$$

$$= \frac{189}{4}\|x_i\| + 54, \quad i = 1, \cdots, N.$$

Note that m_0^1 is a parameter of the leader, which might not be available to the followers. Therefore, although $f_i(x_i)$ satisfy the above condition, the upper bound of $m_i^1 - m_0^1$ might be not explicitly known for the followers. Hence, we will use the adaptive protocol (8.44) to solve the leader-follower consensus problem.

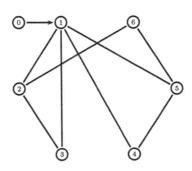

FIGURE 8.4: A leader-follower communication graph.

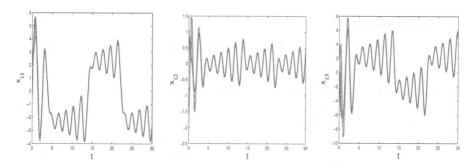

FIGURE 8.5: The state trajectories of the Chua's circuits.

The communication graph is given as in FIGURE 8.4, where the node indexed by 0 is the leader. Solving the LMI (8.8) gives the feedback gain matrices of (8.44) as

$$K = -\begin{bmatrix} 16.9070 & 16.5791 & 1.8297 \end{bmatrix}, \Gamma = \begin{bmatrix} 285.8453 & 280.3016 & 30.9344 \\ 280.3016 & 274.8654 & 30.3344 \\ 30.9344 & 30.3344 & 3.3477 \end{bmatrix}.$$

To illustrate Theorem 52, select $\kappa_i = 0.5$, $\varphi_i = \psi_i = 0.05$, and $\tau_i = \epsilon_i = 5$, $i = 1, \cdots, 6$, in (8.44). The state trajectories $x_i(t)$ of the circuits under (8.44) designed as above are depicted in FIGURE 8.5, implying that leader-follower consensus is indeed achieved. The adaptive gains $\hat{d}_i$ and $\hat{e}_i$ in (8.44) are shown in FIGURE 8.6, which are clearly bounded.

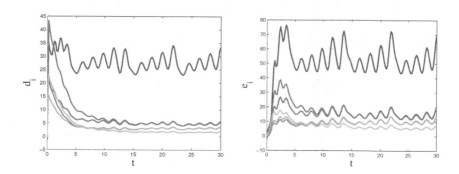

FIGURE 8.6: The adaptive gains $\hat{d}_i$ and $\hat{e}_i$ in (8.23).

8.3 Robustness with Respect to Bounded Non-Matching Disturbances

In the preceding sections, the external disturbances in (8.1) are assumed to satisfy the matching condition, i.e., Assumption 8.1. In this section, we examine the case where the agents are subject to external disturbances which do not necessarily satisfy the matching condition and investigate whether the proposed protocols in the preceding sections still ensure the boundedness of the consensus error. For conciseness, we consider here only the case of the leaderless consensus problem. The case of the leader-follower consensus problem can also be similarly discussed.

Consider a network of N agents whose communication graph is represented by an undirected graph $\mathcal{G}$. The dynamics of the i-th agent is described by

$$\dot{x}_i = Ax_i + B[u_i + f_i(x_i, t)] + \omega_i, \tag{8.50}$$

where $f_i(x_i, t)$ is the lumped matching uncertainty defined as in (8.2) and $\omega_i \in \mathbf{R}^n$ is the non-matching external disturbance, satisfying

Assumption 8.6 *There exist positive constants υ_i such that $\|\omega_i\| \le \upsilon_i$, $i = 1, \cdots, N$.*

Next, we will redesign the distributed adaptive protocol (8.23) to ensure the boundedness of the consensus error for the agents in (8.50). Using (8.23) for (8.50), we can obtain the closed-loop dynamics of the network as

$$\dot{\xi} = (I_N \otimes A + M\overline{D}\mathcal{L} \otimes BK)\xi + (M \otimes B)[F(x, t) + R(\xi)] + (M \otimes I)\omega,$$

$$\dot{\bar{d}}_i = \tau_i[-\varphi_i \bar{d}_i + (\sum_{j=1}^{N} \mathcal{L}_{ij}\xi_j^T)\Gamma(\sum_{j=1}^{N} \mathcal{L}_{ij}\xi_j) + \|K \sum_{j=1}^{N} \mathcal{L}_{ij}\xi_j\|],$$

$$\dot{\bar{e}}_i = \epsilon_i[-\psi_i \bar{e}_i + \|K \sum_{j=1}^{N} \mathcal{L}_{ij}\xi_j\|\|x_i\|], \quad i = 1, \cdots, N,$$

$$\tag{8.51}$$

where $\omega = [\omega_1^T, \cdots, \omega_N^T]^T$ and the variables are defined as in (8.25).

Theorem 53 *Suppose that $\mathcal{G}$ is connected and Assumptions 8.3, 8.5, and 8.6 hold. Then, both the consensus error ξ and the adaptive gains $\bar{d}_i$ and $\bar{e}_i$, $i = 1, \cdots, N$, in (8.25) are uniformly ultimately bounded under the distributed adaptive protocol (8.23) with $K = -B^T Q^{-1}$ and $\Gamma = Q^{-1}BB^T Q^{-1}$, where $Q > 0$ is a solution to the following LMI:*

$$AQ + QA^T + \varepsilon Q - 2BB^T < 0, \tag{8.52}$$

where $\varepsilon > 1$. Moreover, we have

i) For any φ_i and ψ_i, the parameters ξ, $\tilde{d}_i$, and $\tilde{e}_i$ exponentially converge to the residual set

$$\mathcal{D}_6 \triangleq \{\xi, \tilde{d}_i, \tilde{e}_i : V_6 < \frac{1}{2\sigma} \sum_{i=1}^{N} (\beta^2 \varphi_i + e_i^2 \psi_i + \frac{1}{2}\kappa_i) + \frac{\lambda_{\max}(\mathcal{L})}{2\sigma\lambda_{\min}(Q)} \sum_{i=1}^{N} v_i^2\},$$

(8.53)

where $\sigma \triangleq \min_{i=1,\cdots,N}\{\varepsilon - 1, \varphi_i\tau_i, \psi_i\epsilon_i\}$ *and*

$$V_4 = \frac{1}{2}\xi^T(\mathcal{L} \otimes Q^{-1})\xi + \sum_{i=1}^{N} \frac{\tilde{d}_i^2}{2\tau_i} + \sum_{i=1}^{N} \frac{\tilde{e}_i^2}{2\epsilon_i},$$

(8.54)

where the variables are defined as in (8.27).

ii) If φ_i and ψ_i satisfy $\varrho < \varepsilon - 1$, where ϱ is defined in Theorem 51, then in addition to *i)*, ξ exponentially converges to the residual set

$$\mathcal{D}_7 \triangleq \{\xi : \|\xi\|^2 \leq \frac{\lambda_{\max}(Q)}{\lambda_2(\varepsilon - 1 - \varrho)}[\sum_{i=1}^{N}(\beta^2\varphi_i + e_i^2\psi_i + \frac{1}{2}\kappa_i) + \frac{\lambda_{\max}(\mathcal{L})}{\lambda_{\min}(Q)} \sum_{i=1}^{N} v_i^2]\}.$$

(8.55)

Proof 55 *Choose the Lyapunov function candidate as in (8.54). By following similar steps as in the proof of Theorem 51, it is not difficult to get that the time derivative of V_4 along the trajectory of (8.51) can be obtained as*

$$\dot{V}_4 \leq \frac{1}{2}\xi^T\mathcal{W}\xi + \xi^T(\mathcal{L} \otimes Q^{-1})\omega$$
$$- \frac{1}{2}\sum_{i=1}^{N}(\varphi_i\tilde{d}_i^2 + \psi_i\tilde{e}_i^2) + \frac{1}{2}\sum_{i=1}^{N}(\beta^2\varphi_i + e_i^2\psi_i + \frac{1}{2}\kappa_i),$$

(8.56)

where $\mathcal{W} \triangleq \mathcal{L} \otimes (Q^{-1}A + A^TQ^{-1}) - 2\beta\mathcal{L}^2 \otimes Q^{-1}BB^TQ^{-1}$. *Using the following fact:*

$$\frac{1}{2}\xi^T(\mathcal{L} \otimes Q^{-1})\xi - \xi^T(\mathcal{L} \otimes Q^{-1})\omega + \frac{1}{2}\omega^T(\mathcal{L} \otimes Q^{-1})\omega$$
$$= \frac{1}{2}(\xi - \omega)^T(\mathcal{L} \otimes Q^{-1})(\xi - \omega) \geq 0,$$

we can get from (8.56) that

$$\dot{V}_4 \leq \frac{1}{2}\xi^T(\mathcal{W} + \mathcal{L} \otimes Q^{-1})\xi + \frac{1}{2}\omega^T(\mathcal{L} \otimes Q^{-1})\omega$$
$$- \frac{1}{2}\sum_{i=1}^{N}(\varphi_i\tilde{d}_i^2 + \psi_i\tilde{e}_i^2) + \frac{1}{2}\sum_{i=1}^{N}(\beta^2\varphi_i + e_i^2\psi_i + \frac{1}{2}\kappa_i),$$

(8.57)

Note that (8.57) can be rewritten into

$$\dot{V}_4 \leq -\sigma V_4 + \frac{1}{2}\xi^T[\mathcal{W} + (\sigma+1)\mathcal{L} \otimes Q^{-1}]\xi$$

$$+ \frac{1}{2}\omega^T(\mathcal{L} \otimes Q^{-1})\omega - \frac{1}{2}\sum_{i=1}^{N}[(\varphi_i - \frac{\sigma}{\tau_i})\tilde{d}_i^2$$

$$+ (\psi_i - \frac{\sigma}{\epsilon_i})\tilde{e}_i^2)] + \frac{1}{2}\sum_{i=1}^{N}(\beta^2\varphi_i + e_i^2\psi_i + \frac{1}{2}\kappa_i) \qquad (8.58)$$

$$\leq -\sigma V_4 + \frac{1}{2}\xi^T[\mathcal{W} + (\sigma+1)\mathcal{L} \otimes Q^{-1}]\xi$$

$$+ \frac{\lambda_{\max}(\mathcal{L})}{2\lambda_{\min}(Q)}\sum_{i=1}^{N}v_i^2 + \frac{1}{2}\sum_{i=1}^{N}(\beta^2\varphi_i + e_i^2\psi_i + \frac{1}{2}\kappa_i),$$

where we have used the fact that $\sigma \leq \min_{i=1,\cdots,N}\{\varphi_i\tau_i, \psi_i\epsilon_i\}$ to get that last inequality. Let $\hat{\xi} \triangleq [\hat{\xi}_1^T, \cdots, \hat{\xi}_N^T]^T = (U^T \otimes Q^{-1})\xi$, where U is defined as in the proof of Theorem 50. By the definition of $\hat{\xi}$, it is easy to see that $\hat{\xi}_1 = 0$. Since $\sigma \leq \varepsilon - 1$ and $\beta \geq \frac{1}{\lambda_2}$, we then have

$$\xi^T[\mathcal{W} + (\sigma+1)\mathcal{L} \otimes Q^{-1}]\xi \leq \sum_{i=2}^{N}\lambda_i\hat{\xi}_i^T[AQ + QA^T + \varepsilon Q - 2\beta\lambda_i BB^T]\hat{\xi}_i$$

$$\leq \sum_{i=2}^{N}\lambda_i\hat{\xi}_i^T[AQ + QA^T + \varepsilon Q - 2BB^T]\hat{\xi}_i \leq 0.$$

$$(8.59)$$

Then, it follows from (8.58) and (8.59) that

$$\dot{V}_4 \leq -\sigma V_4 + \frac{\lambda_{\max}(\mathcal{L})}{2\lambda_{\min}(Q)}\sum_{i=1}^{N}v_i^2 + \frac{1}{2}\sum_{i=1}^{N}(\beta^2\varphi_i + e_i^2\psi_i + \frac{1}{2}\kappa_i), \qquad (8.60)$$

which implies that V_4 exponentially converges to the residual set $\mathcal{D}_6$ in (8.53) with a convergence rate faster than $e^{-\sigma t}$. By following similar steps as in the last part of the proof of Theorem 51, it is not difficult to show that for the case where φ_i and ψ_i satisfy $\varrho < \varepsilon - 1$, ξ exponentially converges to the residual set $\mathcal{D}_7$. The details are omitted here for conciseness.

Remark 67 *As shown in Proposition 2 in Chapter 2, there exists a $Q > 0$ satisfying (8.52) if and only if (A, B) is controllable. Thus, a sufficient condition for the existence of (8.3) satisfying Theorem 52 is that (A, B) is controllable, which, compared to the existence condition of (8.3) satisfying Theorem 51, is more stringent. It is worth mentioning that large ε in (8.52) yields a faster convergence rate of the consensus error ξ, but meanwhile generally implies a high-gain K in the protocol (8.23). In implementation, a tradeoff has to be made when choosing ε.*

8.4 Distributed Robust Containment Control with Multiple Leaders

For the case where there exist multiple leaders, the robust containment control problem arises. The dynamics of the followers are described by (8.2) and the leaders have identical nominal linear dynamics, given by

$$\dot{x}_i = Ax_i + Bu_i, \ i = 1, \cdots, M, \tag{8.61}$$

where $x_i \in \mathbf{R}^n$ is the state and $u_i \in \mathbf{R}^p$ is the control input of the i-th leader. The communication graph of the agents consisting of the N followers and the M leaders is assumed to satisfy Assumption 7.3 in Section 7.3.

Distributed static and adaptive containment controllers can be designed by combing the ideas in the previous section and Section 7.3. Details of the derivations are omitted here for brevity. Interested readers can complete the derivations as an exercise.

8.5 Notes

The materials of this chapter are mainly based on [77]. The matching uncertainties associated with the agents are assumed to satisfy certain upper bounds in this chapter. Another kind of commonly encountered uncertainties is the uncertainties which can be linearly parameterized. The distributed cooperative control problems of multi-agent systems with linearly parameterized uncertainties have been considered in, e.g., [30, 31, 126, 193, 196]. Distributed control of multi-agent systems with norm-bounded uncertainties which do not necessarily satisfy the matching condition has been addressed in [89, 169], where the distributed robust stabilization problem of multi-agent systems with parameter uncertainties was concerned in [89] and the robust synchronization problem was investigated in [169] for uncertainties in the form of additive perturbations of the transfer matrices of the nominal dynamics.

9

Global Consensus of Multi-Agent Systems with Lipschitz Nonlinear Dynamics

CONTENTS

9.1   Global Consensus of Nominal Lipschitz Nonlinear Multi-Agent
Systems 208
     9.1.1   Global Consensus without Disturbances 208
     9.1.2   Global H_∞ Consensus Subject to External Disturbances .. 211
     9.1.3   Extensions to Leader-Follower Graphs 214
     9.1.4   Simulation Example 216
9.2   Robust Consensus of Lipschitz Nonlinear Multi-Agent Systems with
Matching Uncertainties 218
     9.2.1   Distributed Static Consensus Protocols 219
     9.2.2   Distributed Adaptive Consensus Protocols 224
     9.2.3   Adaptive Protocols for the Case without Uncertainties 230
     9.2.4   Simulation Examples 231
9.3   Notes 232

In the previous chapters, the agent dynamics or the nominal agent dynamics are assumed to be linear. This assumption might be restrictive in some circumstances. In this chapter, we intend to address the global consensus problem for a class of nonlinear multi-agent systems, the Lipschitz nonlinear multi-agent systems where each agent contains a nonlinear function satisfying a global Lipschitz condition. The Lipschitz nonlinear multi-agent systems will reduce to general linear multi-agent systems when the nonlinearity does not exist.

In Section 9.1, the global consensus problem of high-order multi-agent systems with Lipschitz nonlinearity and directed communication graphs is formulated. A distributed consensus protocol is proposed, based on the relative states of neighboring agents. A two-step algorithm is presented to construct one such protocol, under which a Lipschitz multi-agent system without disturbances can reach global consensus for a strongly connected directed communication graph. The existence condition of the proposed consensus protocol is also discussed. For the case where the agents are subject to external disturbances, the global H_∞ consensus problem is formulated and another algorithm is then given to design the protocol which achieves global consensus with a guaranteed H_∞ performance for a strongly connected balanced communication graph. It is worth mentioning that in these two algorithms the feedback gain design of the consensus protocol is decoupled from the communication graph. We further extend the results to the case with a leader-follower communication graph which contains a directed spanning tree with the leader as the root. In the final part of this section, a simulation example is presented for illustration.

In Section 9.2, we intend to consider the distributed robust consensus problem of a class of Lipschitz nonlinear multi-agent systems subject to different matching uncertainties. Contrary to Section 9.1 where the agents have identical nominal dynamics, the multi-agent systems considered in this section are subject to nonidentical matching uncertainties, so the multi-agent systems discussed in this section are essentially heterogeneous, for which case the consensus problem becomes quite challenging to solve. For the case where the communication graph is undirected and connected, a distributed continuous static consensus protocol based on the relative states of neighboring agents is designed, under which the consensus error is uniformly ultimately bounded and exponentially converges to an adjustable residual set. In order to remove the requirement of the nonzero eigenvalues of the Laplacian matrix and the upper bounds of the matching uncertainties, a fully distributed adaptive protocol is further designed, under which the residual set of the consensus error is also given. Note that the upper bound of the consensus error can be made to be reasonably small by properly selecting the design parameters of the proposed protocols.

9.1   Global Consensus of Nominal Lipschitz Nonlinear Multi-Agent Systems

9.1.1   Global Consensus without Disturbances

Consider a group of N identical nonlinear agents, described by

$$\dot{x}_i = Ax_i + g(x_i) + Bu_i, \quad i = 1, \cdots, N, \tag{9.1}$$

where $x_i \in \mathbf{R}^n$ and $u_i \in \mathbf{R}^p$ are the state and the control input of the i-th agent, respectively, A and B are constant matrices with compatible dimensions, and the nonlinear function $g(x_i)$ is assumed to satisfy the Lipschitz condition with a Lipschitz constant $\gamma > 0$, i.e.,

$$\|g(x) - g(y)\| \leq \gamma \|x - y\|, \quad \forall x, y \in \mathbf{R}^n. \tag{9.2}$$

The communication graph among the N agents is represented by a directed graph $\mathcal{G}$. It is assumed that $\mathcal{G}$ is strongly connected throughout this subsection and each agent has access to the relative states with respect to its neighbors. In order to achieve consensus, the following distributed consensus protocol is proposed:

$$u_i = cK \sum_{j=1}^{N} a_{ij}(x_i - x_j), \quad i = 1, \cdots, N, \tag{9.3}$$

where $c > 0 \in \mathbf{R}$ denotes the coupling gain, $K \in \mathbf{R}^{p \times n}$ is the feedback gain matrix, and a_{ij} is (i, j)-th entry of the adjacency matrix associated with $\mathcal{G}$.

   The objective is to design a consensus protocol (9.3) such that the N agents in (9.1) can achieve global consensus in the sense of $\lim_{t \to \infty} \|x_i(t) - x_j(t)\| = 0$, $\forall i, j = 1, \cdots, N$.

   Let $r = [r_1, \cdots, r_N]^T \in \mathbf{R}^N$ be the left eigenvector of $\mathcal{L}$ associated with the zero eigenvalue, satisfying $r^T \mathbf{1} = 1$ and $r_i > 0$, $i = 1, \cdots, N$. Define $e = (W \otimes I_n)x$, where $W = I_N - \mathbf{1}r^T$ and $e = [e_1^T, \cdots, e_N^T]^T$. It is easy to verify that $(r^T \otimes I_n)e = 0$. By the definition of r^T, it is easy to see that 0 is a simple eigenvalue of W with $\mathbf{1}$ as a right eigenvector and 1 is the other eigenvalue with multiplicity $N - 1$. Then, it follows that $e = 0$ if and only if $x_1 = \cdots = x_N$. Therefore, the consensus problem under the protocol (9.3) can be reduced to the asymptotical stability of e. Using (9.3) for (9.1), it can be verified that e satisfies the following dynamics:

$$\dot{e}_i = Ae_i + g(x_i) - \sum_{j=1}^{N} r_j g(x_j) + c \sum_{j=1}^{N} \mathcal{L}_{ij}BKe_j, \quad i = 1, \cdots, N, \tag{9.4}$$

where $\mathcal{L}_{ij}$ is the (i, j)-th entry of the Laplacian matrix associated with $\mathcal{G}$.

   Next, an algorithm is presented to select the control parameters in (9.3).

Algorithm 17 *For the agents in (9.1), a consensus protocol (9.3) can be constructed as follows:*

1) Solve the following linear matrix inequality (LMI):

$$\begin{bmatrix} AP + PA^T - \tau BB^T + \mu I & P \\ P & -(\gamma^{-1})^2 \mu I \end{bmatrix} < 0 \tag{9.5}$$

to get a matrix $P > 0$ and scalars $\tau > 0$ and $\mu > 0$. Then, choose $K = -\frac{1}{2}B^T P^{-1}$.

2) Select the coupling gain $c \geq \frac{\tau}{a(\mathcal{L})}$, where $a(\mathcal{L})$ is the generalized algebraic connectivity of $\mathcal{G}$ (defined in Lemma 4 in Chapter 1).

Remark 68 *It is worth noting that Algorithm 17 has a favorable decoupling feature. Specifically, the first step deals only with the agent dynamics in (9.1) while the second step tackles the communication graph by adjusting the coupling gain c. The generalized algebraic connectivity $a(\mathcal{L})$ for a given graph can be obtained by using Lemma 8 in [189]. By Lemma 4, $a(\mathcal{L})$ can be replaced by $\lambda_2(\frac{\mathcal{L} + \mathcal{L}^T}{2})$ in step 2), if the directed graph $\mathcal{L}$ is balanced and strongly connected.*

Remark 69 *By using using Lemma 20 (Finsler's Lemma), it is not difficult to get that there exist a $P > 0$, a $\tau > 0$ and a $\mu > 0$ such that (9.5) holds if and only if there exists a K such that $(A+BK)P+P(A+BK)^T+\mu I+\alpha^2\mu^{-1}P^2 < 0$, which, with $\mu = 1$, is the existence condition for the observer design problem of a single Lipschitz system in [128, 129]. According to Theorem 2 in [129], the LMI (9.5) is feasible if the distance to unobservability of (A^T, B^T) is larger than γ and A is stable. Besides, a scalar $\mu > 0$ is introduced into (9.5) so as to reduce conservatism.*

The following theorem presents a sufficient condition for the global consensus of (9.4).

Theorem 54 *Assume that the directed graph $\mathcal{G}$ is strongly connected and there exists a solution to (9.5). Then, the N agents in (9.1) can reach global consensus under the protocol (9.3) constructed by Algorithm 17.*

Proof 56 *Consider the Lyapunov function candidate*

$$V_1 = \sum_{i=1}^{N} r_i e_i^T P^{-1} e_i. \tag{9.6}$$

Clearly, V_1 is positive definite. The time derivative of V_1 along the trajectory of (9.4) is given by

$$\dot{V}_1 = 2\sum_{i=1}^{N} r_i e_i^T P^{-1}[Ae_i + g(x_i) - \sum_{j=1}^{N} r_i g(x_j) + c\sum_{j=1}^{N} \mathcal{L}_{ij} BKe_j]$$

$$= 2\sum_{i=1}^{N} r_i e_i^T P^{-1}[Ae_i + c\sum_{j=1}^{N} \mathcal{L}_{ij} BKe_j] \tag{9.7}$$

$$+ 2\sum_{i=1}^{N} r_i e_i^T P^{-1}[g(x_i) - g(\bar{x}) + g(\bar{x}) - \sum_{j=1}^{N} r_i g(x_j)],$$

where $\bar{x} = \sum_{j=1}^{N} r_j x_j$.
Using the Lipschitz condition (9.2) gives

$$2e_i^T P^{-1}(g(x_i) - g(\bar{x})) \leq 2\alpha\|P^{-1}e_i\|\|e_i\|$$
$$\leq e_i^T[\mu(P^{-1})^2 + \mu^{-1}\gamma^2 I]e_i, \tag{9.8}$$

where μ is a positive scalar. Since $\sum_{i=1} r_i e_i = 0$, we have

$$\sum_{i=1}^{N} r_i e_i^T P^{-1}[g(\bar{x}) - \sum_{j=1}^{N} r_j g(x_j)] = 0. \tag{9.9}$$

Let $\xi_i = P^{-1}e_i$, $\xi = [\xi_1^T, \cdots, \xi_N^T]^T$. Substituting $K = -B^T P^{-1}$ in (9.7), it follows from (9.7) by using (9.8) and (9.9) that

$$\dot{V}_1 \leq \sum_{i=1}^{N} r_i \xi_i^T [(AP + PA^T + \mu I + \mu^{-1}\gamma^2 P^2)\xi_i + 2c\sum_{j=1}^{N} \mathcal{L}_{ij} BKP\xi_j]$$

$$= \xi^T [R \otimes (AP + PA^T + \mu I + \mu^{-1}\gamma^2 P^2) - \frac{c}{2}(R\mathcal{L} + \mathcal{L}^T R) \otimes BB^T]\xi,$$

$$(9.10)$$

where $R = \text{diag}(r_1, \cdots, r_N)$.

Since $(r^T \otimes I_n)e = 0$, i.e., $(r^T \otimes I_n)\xi = 0$, we can get from Lemma 4 that

$$\xi^T[(R\mathcal{L} + \mathcal{L}^T R) \otimes I_n]\xi \geq 2a(\mathcal{L})\xi^T(R \otimes I_n)\xi, \qquad (9.11)$$

where $a(\mathcal{L}) > 0$. In light of (9.11), it then follows from (9.10) that

$$\dot{V}_1 \leq \xi^T[R \otimes (AP + PA^T + \mu I + \mu^{-1}\gamma^2 P^2 - ca(\mathcal{L})BB^T)]\xi. \qquad (9.12)$$

By steps 1) and 2) in Algorithm 17, we can obtain that

$$AP + PA^T + \mu I + \mu^{-1}\gamma^2 P^2 - ca(\mathcal{L})BB^T$$
$$\leq AP + PA^T + \mu I + \mu^{-1}\gamma^2 P^2 - \tau BB^T < 0, \qquad (9.13)$$

where the last inequality follows from (9.5) by using Lemma 19 (the Schur complement lemma). Therefore, it follows from (9.12) that $\dot{V}_1 < 0$, implying that $e(t) \to 0$, as $t \to \infty$. That is, the global consensus of the network (9.4) is achieved.

Remark 70 *Theorem 54 converts the global consensus of the N Lipschitz agents in (9.1) under the protocol (9.3) to the feasibility of a low-dimensional linear matrix inequality. The effect of the communication graph on consensus is characterized by the generalized algebraic connectivity of the communication graph.*

9.1.2 Global H_∞ Consensus Subject to External Disturbances

This subsection continues to consider a network of N identical nonlinear agents subject to external disturbances, described by

$$\dot{x}_i = Ax_i + g(x_i) + Bu_i + D\omega_i, \quad i = 1, \cdots, N, \qquad (9.14)$$

where $x_i \in \mathbf{R}^n$ is the state of the i-th agent, $u_i \in \mathbf{R}^p$ is the control input, $\omega_i \in \mathcal{L}_2^{m_1}[0, \infty)$ is the external disturbance, and $g(x_i)$ is a nonlinear function satisfying (9.2). The communication graph $\mathcal{G}$ is assumed to balanced and strongly connected in this subsection.

The objective in this subsection is to design a protocol (9.3) for the agents in (9.14) to reach global consensus and meanwhile maintain a desirable disturbance rejection performance. To this end, define the performance variable z_i, $i = 1, \cdots, N$, as the average of the weighted relative states of the agents:

$$z_i = \frac{1}{N} \sum_{j=1}^{N} C(x_i - x_j), \quad i = 1, \cdots, N, \tag{9.15}$$

where $z_i \in \mathbf{R}^{m_2}$ and $C \in \mathbf{R}^{m_2 \times n}$ is a constant matrix.

With (9.14), (9.3), and (9.15), the closed-loop network dynamics can be obtained as

$$\dot{x}_i = Ax_i + g(x_i) + c \sum_{j=1}^{N} \mathcal{L}_{ij} BKx_j + Dw_i,$$

$$z_i = \frac{1}{N} \sum_{j=1}^{N} C(x_i - x_j), \quad i = 1, \cdots, N, \tag{9.16}$$

The global H_∞ consensus problem for (9.14) under the protocol (9.3) is first defined.

Definition 18 *Given a positive scalar η, the protocol (9.3) is said to achieve global consensus with a guaranteed H_∞ performance η for the agents in (9.14), if the following two requirements hold:*

(1) The network (9.16) with $\omega_i = 0$ can reach global consensus in the sense of $\lim_{t \to \infty} \|x_i - x_j\| = 0$, $\forall i, j = 1, \cdots, N$.

(2) Under the zero-initial condition, the performance variable z satisfies

$$J = \int_0^\infty [z^T(t)z(t) - \eta^2 \omega^T(t)\omega(t)]dt < 0, \tag{9.17}$$

where $z = [z_1^T, \cdots, z_N^T]^T$, $\omega = [\omega_1^T, \cdots, \omega_N^T]^T$.

Next, an algorithm for the protocol (9.3) is presented.

Algorithm 18 *For a given scalar $\gamma > 0$ and the agents in (9.14), a consensus protocol (9.3) can be constructed as follows:*

1) Solve the following LMI:

$$\begin{bmatrix} AQ + QA^T - \varsigma BB^T + \mu I & Q & QC^T & D \\ Q & -(\gamma^{-1})^2 \mu & 0 & 0 \\ CQ & 0 & -I & 0 \\ D^T & 0 & 0 & \eta^2 I \end{bmatrix} < 0 \tag{9.18}$$

to get a matrix $Q > 0$ and scalars $\varsigma > 0$ and $\mu > 0$. Then, choose $K = -\frac{1}{2}B^T Q^{-1}$.

2) Select the coupling gain $c \geq \varsigma/\lambda_2(\frac{\mathcal{L}+\mathcal{L}^T}{2})$, where $\lambda_2(\frac{\mathcal{L}+\mathcal{L}^T}{2})$ denotes the smallest nonzero eigenvalue of $\frac{\mathcal{L}+\mathcal{L}^T}{2}$.

Theorem 55 *Assume that $\mathcal{G}$ is balanced and strongly connected, and there exists a solution to (9.18). Then, the protocol (9.3) constructed by Algorithm 18 can achieve global consensus with a guaranteed H_∞ performance $\eta > 0$ for the N agents in (9.14).*

Proof 57 *Let $e_i = x_i - \frac{1}{N}\sum_{j=1}^{N} x_j$, and $e = [e_1^T, \cdots, e_N^T]^T$. As shown in the last subsection, we know that $e = 0$ if and only if $x_1 = \cdots = x_N$. From (9.16), it is easy to get that e satisfies the following dynamics:*

$$\dot{e}_i = Ae_i + g(x_i) - \frac{1}{N}\sum_{j=1}^{N} g(x_j)$$

$$+ c\sum_{j=1}^{N} \mathcal{L}_{ij}BKe_j + \frac{1}{N}\sum_{j=1}^{N} D(\omega_i - \omega_j), \qquad (9.19)$$

$$z_i = Ce_i, \quad i = 1, \cdots, N.$$

Therefore, the protocol (9.3) solves the global H_∞ consensus problem, if the system (9.19) is asymptotically stable and satisfies (9.17).

Consider the Lyapunov function candidate

$$V_2 = \sum_{i=1}^{N} e_i^T Q^{-1} e_i.$$

By following similar steps to those in Theorem 54, we can obtain the time derivative of V_2 along the trajectory of (9.19) as

$$\dot{V}_2 \leq \sum_{i=1}^{N} e_i^T[(2Q^{-1}A + \gamma^2\mu^{-1}I + \mu(Q^{-1})^2)e_i + 2c\sum_{j=1}^{N} \mathcal{L}_{ij}Q^{-1}BKe_j$$

$$+ \frac{2}{N}\sum_{j=1}^{N} Q^{-1}D(\omega_i - \omega_j)] \qquad (9.20)$$

$$= \hat{e}^T[I_N \otimes (AQ + QA^T + \gamma^2\mu^{-1}Q^2 + \mu I) - \frac{c}{2}(\mathcal{L} + \mathcal{L}^T) \otimes BB^T]\hat{e}$$

$$+ 2\hat{e}^T(M \otimes D)\omega,$$

where $\hat{e}_i = Q^{-1}e_i$, $\hat{e} = [\hat{e}_1^T, \cdots, \hat{e}_N^T]^T$, and $M = I_N - \frac{1}{N}\mathbf{1}\mathbf{1}^T$.

Next, for any nonzero ω, we have

$$J = \int_0^\infty [z^T(t)z(t) - \eta^2\omega^T(t)\omega(t) + \dot{V}_2]dt - V_2(\infty) + V_2(0)$$

$$\leq \int_0^\infty \hat{e}^T[I_N \otimes (AQ + QA^T + \gamma^2\mu^{-1}Q^2 + \mu I) - \frac{c}{2}(\mathcal{L} + \mathcal{L}^T) \otimes BB^T$$

$$+ I_N \otimes QC^T CQ + \eta^{-2}M^2 \otimes DD^T]\hat{e}\,dt$$

$$- \int_0^\infty \eta^2[\omega - \eta^{-2}(M \otimes D^T)\hat{e}]^T[\omega - \eta^{-2}(M \otimes D^T)\hat{e}]dt$$

$$- V_2(\infty) + V_2(0).$$

$$\tag{9.21}$$

By noting that $\mathbf{1}$ is the right and also the left eigenvectors of $\mathcal{L} + \mathcal{L}^T$ and M associated with the zero eigenvalue, respectively, we have

$$M(\mathcal{L} + \mathcal{L}^T) = \mathcal{L} + \mathcal{L}^T = (\mathcal{L} + \mathcal{L}^T)M.$$

Thus there exists a unitary matrix $\hat{U} \in \mathbf{R}^{N \times N}$ such that $\hat{U}^T M \hat{U}$ and $\hat{U}^T(\mathcal{L} + \mathcal{L}^T)\hat{U}$ are both diagonal [58]. Moreover, we can choose $\hat{U} = [\frac{1}{\sqrt{N}} \; Y]$, $\hat{U}^T = \left[\begin{smallmatrix} \frac{1^T}{\sqrt{N}} \\ W \end{smallmatrix}\right]$, with $Y \in \mathbf{R}^{N \times (N-1)}$, $W \in \mathbf{R}^{(N-1) \times N}$, satisfying

$$\hat{U}^T M \hat{U} = \Pi = \begin{bmatrix} 0 & 0 \\ 0 & I_{N-1} \end{bmatrix},$$

$$\frac{1}{2}\hat{U}^T(\mathcal{L} + \mathcal{L}^T)\hat{U} = \Lambda = \text{diag}(0, \lambda_2, \cdots, \lambda_N),$$

$$\tag{9.22}$$

where λ_i, $i = 2, \cdots, N$, are the nonzero eigenvalues of $\frac{\mathcal{L} + \mathcal{L}^T}{2}$. Let $\zeta \triangleq [\zeta_1^T, \cdots, \zeta_N^T]^T = (\hat{U}^T \otimes I_n)\hat{e}$. Clearly, $\zeta_1 = (\frac{1^T}{\sqrt{N}} \otimes Q^{-1})e = 0$. By using (9.22), we can obtain that

$$\hat{e}^T[I_N \otimes (AQ + QA^T + \gamma^2\mu^{-1}Q^2 + \mu I) - \frac{c}{2}(\mathcal{L} + \mathcal{L}^T) \otimes BB^T$$

$$+ I_N \otimes QC^T CQ + \eta^{-2}M^2 \otimes DD^T]\hat{e}$$

$$= \zeta^T[I_N \otimes (AQ + QA^T + \gamma^2\mu^{-1}Q^2 + \mu I) - c\Lambda \otimes BB^T$$

$$+ I_N \otimes QC^T CQ + \eta^{-2}\Pi^2 \otimes DD^T]\zeta \tag{9.23}$$

$$= \sum_{i=2}^N \zeta_i^T[AQ + QA^T + \gamma^2\mu^{-1}Q^2 + \mu I$$

$$- c\lambda_i BB^T + \eta^{-2}DD^T + QC^T CQ]\zeta_i.$$

In light of steps 1) and 2) in Algorithm 18, we have

$$AQ + QA^T + \gamma^2\mu^{-1}Q^2 + \mu I - c\lambda_i BB^T + \eta^{-2}DD^T + QC^T CQ$$

$$\leq AQ + QA^T + \gamma^2\mu^{-1}Q^2 + \mu I - \varsigma BB^T + \eta^{-2}DD^T + QC^T CQ < 0.$$

$$\tag{9.24}$$

By comparing Algorithm 18 with Algorithm 17, it follows from Theorem 54 that the first condition in Definition 18 holds. Since $x(0) = 0$, it is clear that $V_2(0) = 0$. Considering (9.23) and (9.24), we can obtain from (9.21) that $J < 0$. Therefore, the global H_∞ consensus problem is solved.

Remark 71 *Theorem 55 and Algorithm 18 extend Theorem 54 and Algorithm 17 to evaluate the performance of a multi-agent network subject to external disturbances. The decoupling property of Algorithm 17 still holds for Algorithm 18.*

9.1.3 Extensions to Leader-Follower Graphs

In the above subsections, the communication graph is assumed to be strongly connected, where the final consensus value reached by the agents is generally not explicitly known. In many practical cases, it is desirable that the agents' states asymptotically approach a reference state. In this subsection, we intend to consider the case where a network of N agents in (9.1) maintains a leader-follower communication graph $\mathcal{G}$. An agent is called a leader if the agent has no neighbor, i.e., it does not receive any information. An agent is called a follower if the agent has at least one neighbor. Without loss of generality, assume that the agent indexed by 1 is the leader with $u_1 = 0$ and the rest $N - 1$ agents are followers. The following distributed consensus protocol is proposed for each follower:

$$u_i = cK \sum_{j=1}^{N} a_{ij}(x_i - x_j), \quad i = 2, \cdots, N, \tag{9.25}$$

where $c > 0$, K, and a_{ij} are the same as defined in (9.3).

The objective in this subsection is to solve the global leader-follower consensus problem, i.e., to design a consensus protocol (9.25) under which the states of the followers asymptotically approach the state of the leader in the sense of $\lim_{t \to \infty} \|x_i(t) - x_1(t)\| = 0$, $\forall i = 2, \cdots, N$.

In the sequel, we make the following assumption.

Assumption 9.1 *The communication graph $\mathcal{G}$ contains a directed spanning tree with the leader as the root.*

Because the leader has no neighbors, the Laplacian matrix $\mathcal{L}$ associated with $\mathcal{G}$ can be partitioned as $\mathcal{L} = \begin{bmatrix} 0 & 0_{1 \times (N-1)} \\ \mathcal{L}_2 & \mathcal{L}_1 \end{bmatrix}$, where $\mathcal{L}_2 \in \mathbf{R}^{(N-1) \times 1}$ and $\mathcal{L}_1 \in \mathbf{R}^{(N-1) \times (N-1)}$. Under Assumption 9.1, it follows from Lemma 1 that all the eigenvalues of $\mathcal{L}_1$ have positive real parts and $\mathcal{L}_1$ is a nonsingular M-matrix. By Lemma 29, we have

$$G \triangleq \mathrm{diag}(q_2, \cdots, q_N) > 0 \quad \text{and} \quad H \triangleq \frac{1}{2}(G\mathcal{L}_1 + \mathcal{L}_1^T G) > 0, \tag{9.26}$$

where $q = [q_2, \cdots, q_N]^T = (\mathcal{L}_1^T)^{-1}\mathbf{1}$.

Algorithm 19 *For the agents in (9.1) satisfying Assumption 9.1, a consensus protocol (9.25) can be constructed as follows:*

1) *Solve the LMI (9.5) to get a matrix $P > 0$ and scalars $\tau > 0$ and $\mu > 0$. Then, choose $K = -\frac{1}{2}B^T P^{-1}$.*

2) *Select the coupling gain $c \geq \dfrac{\tau}{\lambda_{\min}(H) \min\limits_{i=2,\cdots,N} q_i}$, where H and q_i are defined in (9.26).*

Remark 72 *For the case where the subgraph associated with the followers is balanced and strongly connected, $\mathcal{L}_1 + \mathcal{L}_1^T > 0$ [86]. Then, by letting $G = I$, step 2) can be simplified to $c \geq \tau/\lambda_{\min}(\frac{\mathcal{L}_1 + \mathcal{L}_1^T}{2})$.*

Theorem 56 *Suppose that Assumption 9.1 holds and there exists a solution to (9.5). Then, the consensus protocol (9.25) given by Algorithm 19 solves the leader-follower global consensus problem for the N agents described by (9.1).*

Proof 58 *Denote the consensus errors by $v_i = x_i - x_1$, $i = 2, \cdots, N$. Then, we can obtain from (9.1) and (9.25) the closed-loop network dynamics as*

$$\dot{v}_i = Av_i + g(x_i) - g(x_1) + cBK \sum_{j=1}^{N} a_{ij}(v_i - v_j), \quad i = 2, \cdots, N, \quad (9.27)$$

Clearly, the leader-follower global consensus problem can be reduced to the asymptotical stability of (9.27).

  Consider the Lyapunov function candidate

$$V_3 = \sum_{i=2}^{N} q_i v_i^T P^{-1} v_i,$$

where q_i is defined as in (9.26). Clearly V_3 is positive definite. Following similar steps in proving Theorem 54, the time derivative of V_3 along the trajectory of (9.27) is obtained as

$$\dot{V}_3 = 2 \sum_{i=2}^{N} q_i v_i^T P^{-1}[Av_i + g(x_i) - g(x_1) + cBK \sum_{j=1}^{N} a_{ij}(v_i - v_j)]$$

$$\leq \sum_{i=2}^{N} q_i \tilde{v}_i^T[(AP + PA^T + \mu I + \gamma^2 \mu^{-1} P^2)\tilde{v}_i - cBB^T \sum_{j=1}^{N} a_{ij}(\tilde{v}_i - \tilde{v}_j)]$$

$$= \tilde{v}^T[G \otimes (AP + PA^T + \mu I + \gamma^2 \mu^{-1} P^2) - cH \otimes BB^T]\tilde{v}$$

$$= \tilde{v}^T[G \otimes (AP + PA^T + \mu I + \gamma^2 \mu^{-1} P^2 - \tau BB^T)]\tilde{v}$$

$$\quad + \tilde{v}^T[(\tau G - cH) \otimes BB^T]\tilde{v},$$

$$(9.28)$$

where $\tilde{v}_i = P^{-1}v_i$, $\tilde{v} = [\tilde{v}_2^T, \cdots, \tilde{v}_N^T]^T$, G and H are defined in (9.26), and we have used (9.8) to get the first inequality.

Since G and H are positive definite, we can get that $\tau G - cH < 0$ if $\tau G < c\lambda_{\min}(H)I$, which is true in light of step 2) in Algorithm 19. Thus, $(\tau G - cH) \otimes BB^T \leq 0$. Then, it follows from (9.13) and (9.28) that $\dot{V}_3 < 0$, implying that (9.27) is asymptotically stable. This completes the proof.

The global H_∞ consensus for the agents in (9.14) with a leader-follower communication graph can be discussed similarly, which is omitted here for brevity.

9.1.4 Simulation Example

In this subsection, a simulation example is provided for illustration.

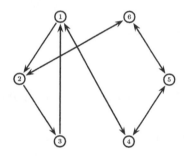

FIGURE 9.1: A directed balanced communication graph.

Example 25 *Consider a network of six single-link manipulators with revolute joints actuated by a DC motor. The dynamics of the i-th manipulator is described by (9.14), with [129, 201]*

$$x_i = \begin{bmatrix} x_{i1} \\ x_{i2} \\ x_{i3} \\ x_{i4} \end{bmatrix}, \quad A = \begin{bmatrix} 0 & 1 & 0 & 0 \\ -48.6 & -1.26 & 48.6 & 0 \\ 0 & 0 & 0 & 10 \\ 1.95 & 0 & -1.95 & 0 \end{bmatrix}, \quad B = \begin{bmatrix} 0 \\ 21.6 \\ 0 \\ 0 \end{bmatrix},$$

$$D = \begin{bmatrix} 0 & 1 & 0.4 & 0 \end{bmatrix}^T, \quad C = \begin{bmatrix} 1 & 0 & 0 & 0 \end{bmatrix},$$

$$g(x_i) = \begin{bmatrix} 0 & 0 & 0 & -0.333\sin(x_{i1}) \end{bmatrix}^T.$$

Clearly, $g(x_i)$ here satisfies (9.2) with a Lipschitz constant $\alpha = 0.333$. The external disturbance here is $\omega = [w, -w, 1.5w, 3w, -0.6w, 2w]^T$, where $w(t)$ is a one-period square wave starting at $t = 0$ with width 2 and height 1.

Choose the H_∞ performance index $\eta = 2$. Solving the LMI (9.18) by using

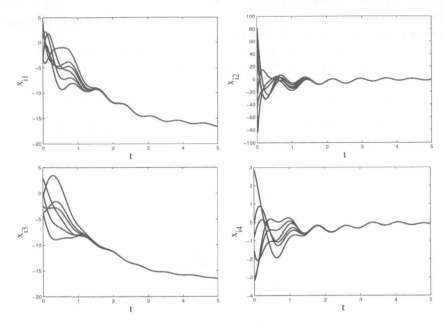

FIGURE 9.2: The six manipulators reach global consensus.

the LMI toolbox of MATLAB gives a feasible solution:

$$Q = \begin{bmatrix} 0.4060 & -0.9667 & 0.3547 & -0.0842 \\ -0.9667 & 67.6536 & 0.0162 & -0.0024 \\ 0.3547 & 0.0162 & 0.4941 & -0.0496 \\ -0.0842 & -0.0024 & -0.0496 & 0.0367 \end{bmatrix}, \quad \varsigma = 29.6636.$$

Thus, by Algorithm 18, the feedback gain matrix of (9.3) is chosen as

$$K = \begin{bmatrix} -2.4920 & -0.1957 & 1.4115 & -3.8216 \end{bmatrix}.$$

For illustration, let the communication graph $\mathcal{G}$ be given as in FIGURE 9.1. It is easy to verify that $\mathcal{G}$ is balanced and strongly connected. The corresponding Laplacian matrix is

$$\mathcal{L} = \begin{bmatrix} 2 & 0 & -1 & -1 & 0 & 0 \\ -1 & 2 & 0 & 0 & 0 & -1 \\ 0 & -1 & 1 & 0 & 0 & 0 \\ -1 & 0 & 0 & 2 & -1 & 0 \\ 0 & 0 & 0 & -1 & 2 & -1 \\ 0 & -1 & 0 & 0 & -1 & 2 \end{bmatrix}.$$

The smallest nonzero eigenvalue of $\frac{\mathcal{L}+\mathcal{L}^T}{2}$ is equal to 0.8139. By Theorem 55 and Algorithm 18, the protocol (9.3) with K chosen as above achieves global *consensus with a H_∞ performance $\gamma = 2$, if the coupling gain $c \geq 36.4462$. For the case without disturbances, the state trajectories of the six manipulators under the protocol (9.3) with K given as above and $c = 37$ are depicted in FIGURE 9.2, from which it can be observed that the global consensus is indeed achieved. With the zero-initial condition and external disturbances ω, the trajectories of the performance variables z_i, $i = 1, \cdots, 6$, are shown in FIGURE 9.3.*

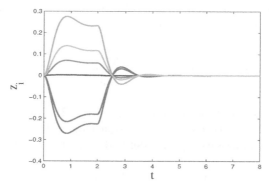

FIGURE 9.3: The performance variables z_i, $i = 1, \cdots, 6$.

9.2 Robust Consensus of Lipschitz Nonlinear Multi-Agent Systems with Matching Uncertainties

In this section, we consider a network of N autonomous agents with Lipschitz nonlinear dynamics subject to heterogeneous uncertainties. The dynamics of the i-th agent are described by

$$\dot{x}_i = Ax_i + g(x_i) + B[u_i + f_i(x_i, t)], \; i = 1, \cdots, N, \tag{9.29}$$

where $x_i \in \mathbf{R}^n$ is the state, $u_i \in \mathbf{R}^p$ is the control input, A and B are constant known matrices with compatible dimensions, and $f_i(x_i, t) \in \mathbf{R}^n$ denotes the lumped matching uncertainty (including the parameter uncertainties and external disturbances) associated with the i-th agent, and the nonlinear function $g(x_i)$ is assumed to satisfy the Lipschitz condition (9.2) with a Lipschitz constant $\gamma > 0$.

Regarding the bounds of the uncertainties $f_i(x_i, t)$, we introduce the following assumption.

Assumption 9.2 *There exist continuous scalar valued functions $\rho_i(x_i, t)$, $i = 1, \cdots, N$, such that $\|f_i(x_i, t)\| \leq \rho_i(x_i, t)$, $i = 1, \cdots, N$, for all $t \geq 0$ and $x_i \in \mathbf{R}^n$.*

The communication graph among the N agents is represented by a undirected graph $\mathcal{G}$, which is assumed to be connected throughout this section. The objective of this section is to solve the consensus problem, i.e., to design distributed protocols such that the agents in (9.29) achieve consensus in the sense of $\lim_{t \to \infty} \|x_i(t) - x_j(t)\| = 0$, $\forall i, j = 1, \cdots, N$.

9.2.1   Distributed Static Consensus Protocols

Based on the relative states of neighboring agents, the following distributed static consensus protocol is proposed:

$$u_i = cK \sum_{j=0}^{N} a_{ij}(x_i - x_j) + \rho_i(x_i, t) r_i (K \sum_{j=0}^{N} a_{ij}(x_i - x_j)), \ i = 1, \cdots, N,$$

(9.30)

where $c > 0$ is the constant coupling gain, $K \in \mathbf{R}^{p \times n}$ is the feedback gain matrix, a_{ij} is the (i, j)-th entry of the adjacency matrix associated with $\mathcal{G}$, and the nonlinear functions $r_i(\cdot)$ are defined such that for $w \in \mathbf{R}^n$,

$$r_i(w) = \begin{cases} \frac{w}{\|w\|} & \text{if } \rho_i(x_i, t)\|w\| > \kappa_i \\ \frac{w}{\kappa_i} \rho_i(x_i, t) & \text{if } \rho_i(x_i, t)\|w\| \leq \kappa_i \end{cases}$$

(9.31)

where κ_i are small positive values

Define $\xi = (M \otimes I_n)x$, where $M = I_N - \frac{1}{N}\mathbf{1}\mathbf{1}^T$ and $\xi = [\xi_1^T, \cdots, \xi_N^T]^T$. It is easy to see that 0 is a simple eigenvalue of M with $\mathbf{1}$ as a corresponding eigenvector and 1 is the other eigenvalue with multiplicity $N - 1$. Then, it follows that $\xi = 0$ if and only if $x_1 = \cdots = x_N$. Therefore, the states of (9.29) under (9.49) reach agreement if and only if ξ asymptotically converges to zero. Hereafter, we refer to ξ as the consensus error. By noting that $\mathcal{L}M = \mathcal{L}$, it is not difficult to obtain that the closed-loop network dynamics can be written in terms of ξ as

$$\dot{\xi} = (I_N \otimes A + c\mathcal{L} \otimes BK)\xi + (M \otimes I)G(x)$$
$$+ (M \otimes B)F(x, t) + [M\rho(x, t) \otimes B]R(\xi),$$

(9.32)

where $\mathcal{L}$ denotes the Laplacian matrix of $\mathcal{G}$, $x = [x_1^T, \cdots, x_N^T]^T$, $\rho(x, t) = \text{diag}(\rho_1(x_1, t), \cdots, \rho_N(x_N, t))$, $F(x, t) = [f_1(x_1, t)^T, \cdots, f_N(x_N, t)^T]^T$, $G(x) = [g(x_1)^T, \cdots, g(x_N)^T]^T$, and

$$R(\xi) = \begin{bmatrix} r_1(K \sum_{j=1}^{N} \mathcal{L}_{1j}\xi_j) \\ \vdots \\ r_N(K \sum_{j=1}^{N} \mathcal{L}_{Nj}\xi_j) \end{bmatrix}.$$

(9.33)

The following result provides a sufficient condition to design the consensus protocol (9.30).

Theorem 57 *Suppose that the communication graph $\mathcal{G}$ is undirected and connected and Assumption 9.2 holds. Solve the LMI (9.5) to get a matrix $P > 0$, scalars $\tau > 0$ and $\mu > 0$. The parameters in the distributed protocol (9.30) are designed as $c \geq \frac{\tau}{\lambda_2}$ and $K = -B^T P^{-1}$, where λ_2 is the smallest nonzero eigenvalue of $\mathcal{L}$. Then, the consensus error ξ of (9.32) is uniformly ultimately bounded and exponentially converges to the residual set*

$$\mathcal{D}_1 \triangleq \{\xi : \|\xi\|^2 \leq \frac{2\lambda_{\max}(P)}{\alpha\lambda_2} \sum_{i=1}^{N} \kappa_i\}, \tag{9.34}$$

where

$$\alpha = -\lambda_{\max}(AP + PA^T - 2\tau BB^T + \mu I + \mu^{-1}\gamma^2 P^2)/\lambda_{\max}(P). \tag{9.35}$$

Proof 59 *Consider the following Lyapunov function candidate:*

$$V_5 = \frac{1}{2}\xi^T(\mathcal{L} \otimes P^{-1})\xi.$$

By the definition of ξ, it is easy to see that $(\mathbf{1}^T \otimes I)\xi = 0$. For a connected graph $\mathcal{G}$, it then follows from Lemma 2 that

$$V_5(\xi) \geq \frac{1}{2}\lambda_2\xi^T(I \otimes P^{-1})\xi \geq \frac{\lambda_2}{2\lambda_{\max}(P)}\|\xi\|^2. \tag{9.36}$$

The time derivative of V_5 along the trajectory of (9.32) is given by

$$\begin{aligned}
\dot{V}_5 &= \xi^T(\mathcal{L} \otimes P^{-1})\dot{\xi} \\
&= \xi^T(\mathcal{L} \otimes P^{-1}A + c\mathcal{L}^2 \otimes P^{-1}BK)\xi \\
&\quad + \xi^T(\mathcal{L} \otimes P^{-1})G(x) + \xi^T(\mathcal{L} \otimes P^{-1}B)F(x,t) \\
&\quad + \xi^T[\mathcal{L}\rho(x,t) \otimes P^{-1}B]R(\xi).
\end{aligned} \tag{9.37}$$

By using Assumption 9.2, we can obtain that

$$\begin{aligned}
\xi^T(\mathcal{L} \otimes P^{-1}B)F(x,t) &\leq \sum_{i=1}^{N}\|B^T P^{-1}\sum_{j=1}^{N}\mathcal{L}_{ij}\xi_j\|\|f_i(x_i,t)\| \\
&\leq \sum_{i=1}^{N}\rho_i(x_i,t)\|B^T P^{-1}\sum_{j=1}^{N}\mathcal{L}_{ij}\xi_j\|.
\end{aligned} \tag{9.38}$$

Observe that

$$\begin{aligned}
\xi^T(\mathcal{L} \otimes P^{-1})G(x) &= \xi^T(\mathcal{L} \otimes P^{-1})[(M \otimes I)G(x)] \\
&= \xi^T(\mathcal{L} \otimes P^{-1})[G(x) - \mathbf{1} \otimes g(\bar{x}) \\
&\quad + \mathbf{1} \otimes g(\bar{x}) - (\frac{1}{N}\mathbf{1}\mathbf{1}^T \otimes I)G(x)] \\
&= \xi^T(\mathcal{L} \otimes P^{-1})[G(x) - \mathbf{1} \otimes g(\bar{x})],
\end{aligned} \tag{9.39}$$

where $\bar{x} = \frac{1}{N}(1^T \otimes I)x$ and we have used $\mathcal{L}1 = 0$ to get the last equation.
Next, consider the following three cases.
i) $\rho_i(x_i, t)\|K \sum_{j=1}^{N} \mathcal{L}_{ij}\xi_j\| > \kappa_i$, $i = 1, \cdots, N$.
In this case, it follows from (9.31) and (9.33) that

$$\xi^T[\mathcal{L}\rho(x,t) \otimes P^{-1}B]R(\xi) = -\sum_{i=1}^{N} \rho_i(x_i, t)\|B^T P^{-1} \sum_{j=1}^{N} \mathcal{L}_{ij}\xi_j\|. \qquad (9.40)$$

Substituting (9.38), (9.39), and (9.40) into (9.37) yields

$$\dot{V}_5 \le \frac{1}{2}\xi^T[\mathcal{X}\xi + 2(\mathcal{L} \otimes P^{-1})(G(x) - 1 \otimes g(\bar{x}))],$$

where $\mathcal{X} = \mathcal{L} \otimes (P^{-1}A + A^T P^{-1}) - 2c\mathcal{L}^2 \otimes P^{-1}BB^T P^{-1}$.
ii) $\rho_i(x_i, t)\|K \sum_{j=1}^{N} \mathcal{L}_{ij}\xi_j\| \le \kappa_i$, $i = 1, \cdots, N$.
In this case, we can get from (9.31) and (9.33) that

$$\xi^T[\mathcal{L}\rho(x,t) \otimes P^{-1}B]R(\xi)$$
$$= -\sum_{i=1}^{N} \frac{\rho_i(x_i, t)^2}{\kappa_i}\|B^T P^{-1} \sum_{j=1}^{N} \mathcal{L}_{ij}\xi_j\|^2 \le 0. \qquad (9.41)$$

Substituting (9.38), (9.39), and (9.41) into (9.37) gives

$$\dot{V}_5 \le \frac{1}{2}\xi^T[\mathcal{X}\xi + 2(\mathcal{L} \otimes P^{-1})(G(x) - 1 \otimes g(\bar{x}))] + \sum_{i=1}^{N} \kappa_i. \qquad (9.42)$$

iii) ξ satisfies neither Case i) nor Case ii).
Without loss of generality, assume that $\rho_i(x_i, t)\|K \sum_{j=1}^{N} \mathcal{L}_{ij}\xi_j\| > \kappa_i$, $i = 1, \cdots, l$, and $\rho_i(x_i, t)\|K \sum_{j=1}^{N} \mathcal{L}_{ij}\xi_j\| \le \kappa_i$, $i = l + 1, \cdots, N$, where $2 \le l \le N - 1$. By combing (9.40) and (9.41), in this case we can get that

$$\xi^T[\mathcal{L}\rho(x,t) \otimes P^{-1}B]R(\xi) \le -\sum_{i=1}^{l} \rho_i(x_i, t)\|B^T P^{-1} \sum_{j=1}^{N} \mathcal{L}_{ij}\xi_j\|. \qquad (9.43)$$

Then, it follows from (9.37), (9.40), (9.43), (9.38), and (9.39) that

$$\dot{V}_5 \le \frac{1}{2}\xi^T[\mathcal{X}\xi + 2(\mathcal{L} \otimes P^{-1})(G(x) - 1 \otimes g(\bar{x}))] + \sum_{i=1}^{N-l} \kappa_i.$$

Therefore, by analyzing the above three cases, we get that $\dot{V}_5$ satisfies (9.42) *for all $\xi \in \mathbf{R}^{Nn}$. Note that (9.42) can be rewritten as*

$$\dot{V}_5 \leq -\alpha V_5 + \alpha V_5 + \frac{1}{2}\xi^T[\mathcal{X}\xi + 2(\mathcal{L} \otimes P^{-1})(G(x) - \mathbf{1} \otimes g(\bar{x}))] + \sum_{i=1}^{N} \kappa_i$$

$$= -\alpha V_5 + \frac{1}{2}\xi^T[\mathcal{X} + \alpha\mathcal{L} \otimes P^{-1} + 2(\mathcal{L} \otimes P^{-1})(G(x) - \mathbf{1} \otimes g(\bar{x}))]$$

$$+ \sum_{i=1}^{N} \kappa_i,$$

(9.44)

where $\alpha > 0$.

Because $\mathcal{G}$ is connected, it follows from Lemma 1 that zero is a simple eigenvalue of $\mathcal{L}$ and all the other eigenvalues are positive. Let $U = [\frac{1}{\sqrt{N}} \; Y_1]$ and $U^T = \begin{bmatrix} \frac{1^T}{\sqrt{N}} \\ Y_2 \end{bmatrix}$, with $Y_1 \in \mathbf{R}^{N \times (N-1)}$, $Y_2 \in \mathbf{R}^{(N-1) \times N}$, be such unitary matrices that $U^T \mathcal{L} U = \Lambda \triangleq \mathrm{diag}(0, \lambda_2, \cdots, \lambda_N)$, where $\lambda_2 \leq \cdots \leq \lambda_N$ are the nonzero eigenvalues of $\mathcal{L}$. Let $\bar{\xi} \triangleq [\bar{\xi}_1^T, \cdots, \bar{\xi}_N^T]^T = (U^T \otimes I)\xi$. By the definitions of ξ and $\bar{\xi}$, it is easy to see that $\bar{\xi}_1 = (\frac{1^T}{\sqrt{N}} \otimes I)\xi = 0$. In light of Lemma 18, it then follows that*

$$2\xi^T(\mathcal{L} \otimes P^{-1})(G(x) - \mathbf{1} \otimes g(\bar{x}))$$

$$= 2\bar{\xi}^T(\Lambda \otimes P^{-1})(U^T \otimes I)(G(x) - \mathbf{1} \otimes g(\bar{x}))$$

$$\leq \mu\bar{\xi}^T(\Lambda \otimes P^{-1})(\Pi^{-1} \otimes I)(\Lambda \otimes P^{-1})\bar{\xi} + \frac{1}{\mu}[G(x) - \mathbf{1} \otimes g(\bar{x})]^T$$

$$\times (U \otimes I)(\Pi \otimes I)(U^T \otimes I)[G(x) - \mathbf{1} \otimes g(\bar{x})] \quad (9.45)$$

$$= \mu\bar{\xi}^T(\Lambda\Pi^{-1}\Lambda \otimes (P^{-1})^2)\bar{\xi} + \frac{1}{\mu}\sum_{i=1}^{N} \pi_i\|g(x_i) - g(\bar{x})\|^2$$

$$= \mu\bar{\xi}^T[\Lambda\Pi^{-1}\Lambda \otimes (P^{-1})^2 + \frac{1}{\mu}\gamma^2\Pi]\bar{\xi},$$

where $\Pi = \mathrm{diag}(\pi_1, \cdots, \pi_N)$ is a positive-definite diagonal matrix and $\mu > 0$, and we have used the Lipschitz condition (9.2) to get the last inequality. By choosing $\Pi = \mathrm{diag}(1, \lambda_2, \cdots, \lambda_N)$ and letting $\hat{\xi} = (I \otimes P^{-1})\bar{\xi} = (U^T \otimes P^{-1})\xi$, it then follows from (9.45) and (9.44) that

$$\xi^T[\mathcal{X} + \alpha\mathcal{L} \otimes P^{-1} + 2(\mathcal{L} \otimes P^{-1})(G(x) - \mathbf{1} \otimes g(\bar{x}))]$$

$$= \sum_{i=2}^{N} \lambda_i\hat{\xi}_i^T(AP + PA^T - 2c\lambda_i BB^T + \alpha P + \mu I + \mu^{-1}\gamma^2 P^2)\hat{\xi}_i$$

(9.46)

$$\leq \sum_{i=2}^{N} \lambda_i\hat{\xi}_i^T(AP + PA^T - 2\tau BB^T + \alpha P + \mu I + \mu^{-1}\gamma^2 P^2)\hat{\xi}_i.$$

By using Lemma 19 (Schur complement lemma), it follows from (9.5) that

$AP + PA^T - 2\tau BB^T + \mu I + \mu^{-1}\gamma^2 P^2 < 0$. Because $\alpha = -\lambda_{\max}(AP + PA^T - 2\tau BB^T + \mu I + \mu^{-1}\gamma^2 P^2)/\lambda_{\max}(P)$, we can see from (9.46) that

$$\xi^T[\mathcal{X} + \alpha\mathcal{L} \otimes P^{-1} + 2(\mathcal{L} \otimes P^{-1})(G(x) - 1 \otimes g(\bar{x}))] \leq 0.$$

Then, we can get from (9.44) that

$$\dot{V}_5 \leq -\alpha V_5 + \sum_{i=1}^{N} \kappa_i. \tag{9.47}$$

By using Lemma 16 (the Comparison lemma), we can obtain from (9.47) that

$$V_5(\xi) \leq [V_5(\xi(0)) - \frac{1}{\alpha}\sum_{i=1}^{N} \kappa_i]e^{-\alpha t} + \frac{1}{\alpha}\sum_{i=1}^{N} \kappa_i, \tag{9.48}$$

which, by (9.36), implies that ξ exponentially converges to the residual set $\mathcal{D}_1$ in (9.34) with a convergence rate not less than $e^{-\alpha t}$.

Remark 73 *Note that the nonlinear functions $r_i(\cdot)$ in (9.31) are actually continuous approximations, via the boundary layer concept [43, 71], of the discontinuous function $\hat{g}(w) = \begin{cases} \frac{w}{\|w\|} & if \|w\| \neq 0 \\ 0 & if \|w\| = 0 \end{cases}$. The values of κ_i in (9.31) define the widths of the boundary layers. From (9.34) we can observe that the residual set $\mathcal{D}_1$ of the consensus error ξ depends on the smallest nonzero eigenvalue of $\mathcal{L}$, the number of agents, the largest eigenvalue of P, and the values of κ_i. By choosing sufficiently small κ_i, the consensus error ξ under the protocol (9.30) can converge to an arbitrarily small neighborhood of zero.*

9.2.2 Distributed Adaptive Consensus Protocols

In the last subsection, the design of the protocol (9.30) relies on the minimal nonzero eigenvalue λ_2 of $\mathcal{L}$ which is global information in the sense that each agent has to know the entire communication graph to compute it and the upper bounds $\rho_i(x_i, t)$ of the matching uncertainties $f_i(x_i, t)$, which might not be easily obtained in some cases, In this subsection, we will develop some fully distributed consensus protocols without using either λ_2 or $\rho_i(x_i, t)$.

Before moving forward, we introduce a modified assumption regarding the bounds of the lumped uncertainties $f_i(x_i, t)$, $i = 1, \cdots, N$.

Assumption 9.3 *There are positive constants a_i and b_i such that $\|f_i(x_i, t)\| \leq a_i + b_i\|x_i\|$, $i = 1, \cdots, N$.*

Based on the local state information of neighboring agents, we propose the following distributed adaptive protocol to each agent:

$$u_i = d_i K \sum_{j=1}^{N} a_{ij}(x_i - x_j) + \bar{r}_i (K \sum_{j=1}^{N} a_{ij}(x_i - x_j)),$$

$$\dot{d}_i = \nu_i [-\varphi_i d_i + (\sum_{j=1}^{N} a_{ij}(x_i - x_j)^T) \Gamma (\sum_{j=1}^{N} a_{ij}(x_i - x_j))$$

$$+ \| K \sum_{j=1}^{N} a_{ij}(x_i - x_j) \|],$$

(9.49)

$$\dot{e}_i = \epsilon_i [-\psi_i e_i + \| K \sum_{j=1}^{N} a_{ij}(x_i - x_j) \| \|x_i\|], \quad i = 1, \cdots, N,$$

where $d_i(t)$ and $e_i(t)$ are the adaptive gains associated with the i-th agent, $K \in \mathbf{R}^{p \times n}$ and $\Gamma \in \mathbf{R}^{n \times n}$ are the feedback gain matrices to be designed, ν_i and ϵ_i are positive scalars, φ_i and ψ_i are small positive constants, the nonlinear functions $\bar{r}_i(\cdot)$ are defined such that for $w \in \mathbf{R}^n$,

$$\bar{r}_i(w) = \begin{cases} \frac{w(d_i + e_i \|x_i\|)}{\|w\|} & \text{if } (d_i + e_i \|x_i\|) \|w\| > \kappa_i \\ \frac{w(d_i + e_i \|x_i\|)^2}{\kappa_i} & \text{if } (d_i + e_i \|x_i\|) \|w\| \le \kappa_i \end{cases}$$

(9.50)

and the rest of the variables are defined as in (9.30).

Let the consensus error ξ be defined as in (9.32) and $D(t) = \text{diag}(d_1(t), \cdots, d_N(t))$. Then, it is not difficult to get from (9.29) and (9.49) that the closed-loop network dynamics can be written as

$$\dot{\xi} = (I_N \otimes A + MD\mathcal{L} \otimes BK)\xi + (M \otimes I)G(x)$$
$$+ (M \otimes B)[F(x,t) + \bar{R}(\xi)],$$

$$\dot{d}_i = \nu_i [-\varphi_i d_i + (\sum_{j=1}^{N} \mathcal{L}_{ij} \xi_j^T) \Gamma (\sum_{j=1}^{N} \mathcal{L}_{ij} \xi_j) + \| K \sum_{j=1}^{N} \mathcal{L}_{ij} \xi_j \|],$$

(9.51)

$$\dot{e}_i = \epsilon_i [-\psi_i e_i + \| K \sum_{j=1}^{N} \mathcal{L}_{ij} \xi_j \| \|x_i\|], \quad i = 1, \cdots, N,$$

where

$$\bar{R}(\xi) = \begin{bmatrix} \bar{r}_i(K \sum_{j=1}^{N} \mathcal{L}_{1j} \xi_j) \\ \vdots \\ \bar{r}_i(K \sum_{j=1}^{N} \mathcal{L}_{Nj} \xi_j) \end{bmatrix},$$

(9.52)

and the rest of the variables are defined as in (9.32).

In order to establish the ultimate boundedness of the states ξ, d_i, and e_i of (9.51), we use the following Lyapunov function candidate

$$V_6 = \frac{1}{2} \xi^T (\mathcal{L} \otimes P^{-1}) \xi + \sum_{i=1}^{N} \frac{\tilde{d}_i^2}{2\nu_i} + \sum_{i=1}^{N} \frac{\tilde{e}_i^2}{2\epsilon_i},$$

(9.53)

where $\tilde{e}_i = e_i - b_i$, $\tilde{d}_i = d_i - \beta$, $i = 1, \cdots, N$, and $\beta = \max_{i=1,\cdots,N}\{a_i, \frac{\tau}{\lambda_2}\}$, and $P > 0$ and τ satisfy (9.5).

The following result provides a sufficient condition to design the consensus protocol (9.49).

Theorem 58 *Suppose that $\mathcal{G}$ is connected and Assumption 9.3 holds. Then, both the consensus error ξ and the adaptive gains d_i and e_i, $i = 1, \cdots, N$, in (9.51) are uniformly ultimately bounded, under the distributed adaptive protocol (9.49) with $K = -B^T P^{-1}$ and $\Gamma = P^{-1}BB^T P^{-1}$, where $P > 0$ is a solution to the LMI (9.5). Moreover, the following statements hold.*

i) For any φ_i and ψ_i, the parameters ξ, $\tilde{d}_i$, and $\tilde{e}_i$ exponentially converge to the residual set

$$\mathcal{D}_2 \triangleq \{\xi, \tilde{d}_i, \tilde{e}_i : V_6 < \frac{1}{2\delta}\sum_{i=1}^{N}(\beta^2\varphi_i + b_i^2\psi_i + \frac{1}{2}\kappa_i)\}, \qquad (9.54)$$

where $\delta \triangleq \min_{i=1,\cdots,N}\{\alpha, \varphi_i\nu_i, \psi_i\epsilon_i\}$ and α is defined as in (9.35).

ii) If φ_i and ψ_i satisfy $\varrho \triangleq \min_{i=1,\cdots,N}\{\varphi_i\nu_i, \psi_i\epsilon_i\} < \alpha$, then in addition to i), ξ exponentially converges to the residual set

$$\mathcal{D}_3 \triangleq \{\xi : \|\xi\|^2 \leq \frac{\lambda_{\max}(P)}{\lambda_2(\alpha - \varrho)}\sum_{i=1}^{N}(\beta^2\varphi_i + b_i^2\psi_i + \frac{1}{2}\kappa_i)\}. \qquad (9.55)$$

Proof 60 *Let V_6 in (9.53) be the Lyapunov function candidate. The time derivative of V_6 along (9.51) can be obtained as*

$$
\begin{aligned}
\dot{V}_6 &= \xi^T(\mathcal{L} \otimes P^{-1})\dot{\xi} + \sum_{i=1}^{N}\frac{\tilde{d}_i}{\nu_i}\dot{\tilde{d}}_i + \sum_{i=1}^{N}\frac{\tilde{e}_i}{\epsilon_i}\dot{\tilde{e}}_i \\
&= \xi^T[(\mathcal{L} \otimes P^{-1}A + \mathcal{L}D\mathcal{L} \otimes P^{-1}BK)\xi + (\mathcal{L} \otimes P^{-1})]G(x) \\
&\quad + \xi^T(\mathcal{L} \otimes P^{-1}B)[F(x,t) + \bar{R}(\xi)] \\
&\quad + \sum_{i=1}^{N}\tilde{d}_i[-\varphi_i(\tilde{d}_i + \beta) + (\sum_{j=1}^{N}\mathcal{L}_{ij}\xi_j^T)\Gamma(\sum_{j=1}^{N}\mathcal{L}_{ij}\xi_j) \\
&\quad + \|K\sum_{j=1}^{N}\mathcal{L}_{ij}\xi_j\|] + \sum_{i=1}^{N}\tilde{e}_i[-\psi_i(\tilde{e}_i + b_i) + \|K\sum_{j=1}^{N}\mathcal{L}_{ij}\xi_j\|\|x_i\|],
\end{aligned}
\qquad (9.56)
$$

where $D = \mathrm{diag}(\tilde{d}_1 + \beta, \cdots, \tilde{c}_N + \beta)$.

By noting that $K = -BP^{-1}$, it is easy to get that

$$
\begin{aligned}
&\xi^T(\mathcal{L}D\mathcal{L} \otimes P^{-1}BK)\xi \\
&= -\sum_{i=1}^{N}(\tilde{d}_i + \beta)(\sum_{j=1}^{N}\mathcal{L}_{ij}\xi_j)^T P^{-1}BB^T P^{-1}(\sum_{j=1}^{N}\mathcal{L}_{ij}\xi_j).
\end{aligned}
\qquad (9.57)
$$

In light of Assumption 9.3, we can obtain that

$$\xi^T(\mathcal{L} \otimes P^{-1}B)F(x,t) \leq \sum_{j=1}^{N}(a_i + b_i\|x_i\|)\|B^T P^{-1}\sum_{j=1}^{N}\mathcal{L}_{ij}\xi_j\|. \tag{9.58}$$

In what follows, we consider three cases.
i) $(d_i + e_i\|x_i\|)\|K\sum_{j=1}^{N}\mathcal{L}_{ij}\xi_j\| > \kappa_i$, $i = 1, \cdots, N$.
In this case, we can get from (9.50) and (9.33) that

$$\xi^T(\mathcal{L} \otimes P^{-1}B)\bar{R}(\xi) = -\sum_{i=1}^{N}[d_i + e_i\|x_i\|]\|B^T P^{-1}\sum_{j=1}^{N}\mathcal{L}_{ij}\xi_j\|. \tag{9.59}$$

Substituting (9.57), (9.58), and (9.59) into (9.56) yields

$$\begin{aligned}
\dot{V}_6 \leq{} & \frac{1}{2}\xi^T[\mathcal{Y}\xi + 2(\mathcal{L} \otimes P^{-1})(G(x) - \mathbf{1} \otimes g(\bar{x}))] \\
& - \sum_{i=1}^{N}(\beta - a_i)\|B^T P^{-1}\sum_{j=1}^{N}\mathcal{L}_{ij}\xi_j\| \\
& - \frac{1}{2}\sum_{i=1}^{N}(\varphi_i\tilde{d}_i^2 + \psi_i\tilde{e}_i^2) + \frac{1}{2}\sum_{i=1}^{N}(\beta^2\varphi_i + b_i^2\psi_i) \\
\leq{} & \frac{1}{2}\xi^T[\mathcal{Y}\xi + 2(\mathcal{L} \otimes P^{-1})(G(x) - \mathbf{1} \otimes g(\bar{x}))] \\
& - \frac{1}{2}\sum_{i=1}^{N}(\varphi_i\tilde{d}_i^2 + \psi_i\tilde{e}_i^2) + \frac{1}{2}\sum_{i=1}^{N}(\beta^2\varphi_i + b_i^2\psi_i),
\end{aligned}$$

where $\mathcal{Y} \triangleq \mathcal{L} \otimes (P^{-1}A + A^T P^{-1}) - 2\beta\mathcal{L}^2 \otimes P^{-1}BB^T P^{-1}$ and we have used the facts that $\beta \geq \max_{i=1,\cdots,N} a_i$ and $-\tilde{d}_i^2 - \tilde{d}_i\beta \leq -\frac{1}{2}\tilde{d}_i^2 + \frac{1}{2}\beta^2$.
ii) $(d_i + e_i\|x_i\|)\|K\sum_{j=1}^{N}\mathcal{L}_{ij}\xi_j\| \leq \kappa_i$, $i = 1, \cdots, N$.
In this case, we can get from (9.50) and (9.33) that

$$\xi^T(\mathcal{L} \otimes P^{-1}B)\bar{R}(\xi) = -\sum_{i=1}^{N}\frac{(d_i + e_i\|x_i\|)^2}{\kappa_i}\|B^T P^{-1}\sum_{j=1}^{N}\mathcal{L}_{ij}\xi_j\|^2. \tag{9.60}$$

Then, it follows from (9.57), (9.58), (9.60), and (9.56) that

$$\begin{aligned}
\dot{V}_6 \leq{} & \frac{1}{2}\xi^T[\mathcal{Y}\xi + 2(\mathcal{L} \otimes P^{-1})(G(x) - \mathbf{1} \otimes g(\bar{x}))] \\
& - \frac{1}{2}\sum_{i=1}^{N}(\varphi_i\tilde{d}_i^2 + \psi_i\tilde{e}_i^2) + \frac{1}{2}\sum_{i=1}^{N}(\beta^2\varphi_i + b_i^2\psi_i + \frac{1}{2}\kappa_i),
\end{aligned} \tag{9.61}$$

where we have used the fact that $-\frac{(d_i+e_i\|x_i\|)^2}{\kappa_i}\|B^T P^{-1}\sum_{j=1}^{N}\mathcal{L}_{ij}\xi_j\|^2 + (d_i +$

$e_i\|x_i\|)\|B^T P^{-1} \sum_{j=1}^{N} \mathcal{L}_{ij}\xi_j\| \le \frac{1}{4}\kappa_i$, *for* $(d_i + e_i\|x_i\|)\|K \sum_{j=1}^{N} \mathcal{L}_{ij}\xi_j\| \le \kappa_i$, $i = 1, \cdots, N$.

iii) $(d_i + e_i\|x_i\|)\|K \sum_{j=1}^{N} \mathcal{L}_{ij}\xi_j\| > \kappa_i$, $i = 1, \cdots, l$, *and* $(d_i + e_i\|x_i\|)\|K \sum_{j=1}^{N} \mathcal{L}_{ij}\xi_j\| \le \kappa_i$, $i = l+1, \cdots, N$, *where* $2 \le l \le N-1$.

By following similar steps in the two cases above, it is not difficult to get that

$$\dot{V}_6 \le \frac{1}{2}\xi^T[\mathcal{Y}\xi + 2(\mathcal{L} \otimes P^{-1})(G(x) - \mathbf{1} \otimes g(\bar{x}))] + \frac{1}{4}\sum_{i=1}^{N-l}\kappa_i$$
$$- \frac{1}{2}\sum_{i=1}^{N}(\varphi_i \tilde{d}_i^2 + \psi_i \tilde{e}_i^2) + \frac{1}{2}\sum_{i=1}^{N}(\beta^2\varphi_i + b_i^2\psi_i).$$

Therefore, based on the above three cases, we can get that $\dot{V}_6$ satisfies (9.61) for all $\xi \in \mathbf{R}^{Nn}$. Note that (9.61) can be rewritten into

$$\dot{V}_6 \le -\delta V_6 + \delta V_6 + \frac{1}{2}\xi^T[\mathcal{Y}\xi + 2(\mathcal{L} \otimes P^{-1})(G(x) - \mathbf{1} \otimes g(\bar{x}))]$$
$$- \frac{1}{2}\sum_{i=1}^{N}(\varphi_i \tilde{d}_i^2 + \psi_i \tilde{e}_i^2) + \frac{1}{2}\sum_{i=1}^{N}(\beta^2\varphi_i + b_i^2\psi_i + \frac{1}{2}\kappa_i)$$
$$= -\delta V_6 + \frac{1}{2}\xi^T[\mathcal{Y}\xi + \delta(\mathcal{L} \otimes P^{-1})\xi + 2(\mathcal{L} \otimes P^{-1}) \tag{9.62}$$
$$\times (G(x) - \mathbf{1} \otimes g(\bar{x}))] - \frac{1}{2}\sum_{i=1}^{N}[(\varphi_i - \frac{\delta}{\nu_i})\tilde{d}_i^2$$
$$+ (\psi_i - \frac{\delta}{\epsilon_i})\tilde{e}_i^2)] + \frac{1}{2}\sum_{i=1}^{N}(\beta^2\varphi_i + b_i^2\psi_i + \frac{1}{2}\kappa_i).$$

Because $\beta\lambda_2 \ge 1$ and $0 < \delta \le \alpha$, by following similar steps in the proof of Theorem 57, we can show that

$$\xi^T[\mathcal{Y}\xi + \delta(\mathcal{L} \otimes P^{-1})\xi + 2(\mathcal{L} \otimes P^{-1})(G(x) - \mathbf{1} \otimes g(\bar{x}))] \le 0.$$

Further, by noting that $\delta \le \min_{i=1,\cdots,N}\{\varphi_i\nu_i, \psi_i\epsilon_i\}$, it follows from (9.62) that

$$\dot{V}_6 \le -\delta V_6 + \frac{1}{2}\sum_{i=1}^{N}(\beta^2\varphi_i + b_i^2\psi_i + \frac{1}{2}\kappa_i), \tag{9.63}$$

which implies that

$$V_6 \le [V_6(0) - \frac{1}{2\delta}\sum_{i=1}^{N}(\beta^2\varphi_i + b_i^2\psi_i + \frac{1}{2}\kappa_i)]e^{-\delta t}$$
$$+ \frac{1}{2\delta}\sum_{i=1}^{N}(\beta^2\varphi_i + b_i^2\psi_i + \frac{1}{2}\kappa_i). \tag{9.64}$$

Therefore, V_6 exponentially converges to the residual set $\mathcal{D}_2$ in (9.54) with a convergence rate faster than $e^{-\delta t}$, which, in light of $V_6 \geq \frac{\lambda_2}{2\lambda_{\max}(P)}\|\xi\|^2 + \sum_{i=1}^{N}\frac{\tilde{d}_i^2}{2\nu_i} + \sum_{i=1}^{N}\frac{\tilde{e}_i^2}{2\epsilon_i}$, implies that ξ, d_i, and e_i are uniformly ultimately bounded.

Next, if $\varrho \triangleq \min_{i=1,\cdots,N}\{\varphi_i\nu_i, \psi_i\epsilon_i\} < \alpha$, we can obtain a smaller residual set for ξ by rewriting (9.62) into

$$\dot{V}_6 \leq -\varrho V_6 + \frac{1}{2}\xi^T[\mathcal{Y}\xi + \alpha(\mathcal{L}\otimes P^{-1})\xi + 2(\mathcal{L}\otimes P^{-1})(G(x) - \mathbf{1}\otimes g(\bar{x}))]$$

$$-\frac{\alpha - \varrho}{2}\xi^T(\mathcal{L}\otimes P^{-1})\xi + \frac{1}{2}\sum_{i=1}^{N}(\beta^2\varphi_i + b_i^2\psi_i + \frac{1}{2}\kappa_i)$$

$$\leq -\varrho V_6 - \frac{\lambda_2(\alpha - \varrho)}{2\lambda_{\max}(P)}\|\xi\|^2 + \frac{1}{2}\sum_{i=1}^{N}(\beta^2\varphi_i + b_i^2\psi_i + \frac{1}{2}\kappa_i).$$

$$(9.65)$$

Obviously, it follows from (9.65) that $\dot{V}_6 \leq -\varrho V_6$ if $\|\xi\|^2 > \frac{\lambda_{\max}(P)}{\lambda_2(\alpha-\varrho)}\sum_{i=1}^{N}(\beta^2\varphi_i + b_i^2\psi_i + \frac{1}{2}\kappa_i)$. Then, we can get that if $\varrho \leq \alpha$ then ξ exponentially converges to the residual set $\mathcal{D}_3$ in (9.55) with a convergence rate faster than $e^{-\varrho t}$.

Remark 74 *In Theorem 58, the design of the adaptive protocol (9.49) relies on only the agent dynamics, requiring neither the minimal nonzero eigenvalue of $\mathcal{L}$ nor the upper bounds of of the uncertainties $f_i(x_i, t)$. Thus, the adaptive controller (9.49), contrary to the static protocol (9.30), can be computed implemented in a fully distributed fashion without requiring any global information. As explained in Section 8.1 of Chapter 8, we should choose reasonably small φ_i, ψ_i, and κ_i in practical implementations such that the consensus error ξ is satisfactorily small and at the same time the adaptive gains d_i and e_i are tolerable.*

Remark 75 *A special case of the uncertainties $f_i(x_i, t)$ satisfying Assumptions 9.2 and 9.3 is that there exist positive constants d_i such that $\|f_i(x_i, t)\| \leq d_i$, $i = 1, \cdots, N$. For this case, the proposed protocols (9.30) and (9.49) can be accordingly simplified. The adaptive protocol (9.49) in this case can be modified into*

$$u_i = \bar{d}_i K\sum_{j=1}^{N}a_{ij}(x_i - x_j) + \hat{r}_i(K\sum_{j=1}^{N}a_{ij}(x_i - x_j)),$$

$$\dot{\bar{d}}_i = \tau_i[-\varphi_i\bar{d}_i + (\sum_{j=1}^{N}a_{ij}(x_i - x_j)^T)\Gamma \qquad (9.66)$$

$$\times (\sum_{j=1}^{N}a_{ij}(x_i - x_j)) + \|K\sum_{j=1}^{N}a_{ij}(x_i - x_j)\|],$$

where the nonlinear functions $\hat{r}_i(\cdot)$ *are defined such that* $\hat{r}_i(w) =$
$$\begin{cases} \frac{w\bar{d}_i}{\|w\|} & \text{if } \bar{d}_i\|w\| > \kappa_i \\ \frac{w\bar{d}_i^2}{\kappa} & \text{if } \bar{d}_i\|w\| \leq \kappa_i \end{cases}$$ *and the rest of the variables are defined as in (9.49).*

Remark 76 *The case where there exists a leader whose control input is bounded and unknown to any follower can be also investigated. Distributed controllers based on the relative state information can be similarly designed by following the steps in Section 8.3 of Chapter 8 and in this section, when the communication graph contains a directed spanning tree with the leader as the root and the subgraph associated with the followers is undirected. The details are omitted here for conciseness.*

9.2.3   Adaptive Protocols for the Case without Uncertainties

Theorem 58 shows that the distributed adaptive protocol (9.49) can solve the consensus problem for the agents in (9.29) subject to matching uncertainties $f_i(x_i, t)$. For the special case where the agents do not contain any uncertainty, i.e., $f_i(x_i, t) = 0$, $i = 1, \cdots, N$, the adaptive protocol (9.49) can be simplified into the following form:

$$u_i = d_i K \sum_{j=1}^{N} a_{ij}(x_i - x_j),$$

$$\dot{d}_i = \nu_i [\sum_{j=1}^{N} a_{ij}(x_i - x_j)^T] \Gamma [\sum_{j=1}^{N} a_{ij}(x_i - x_j)], \ i = 1, \cdots, N. \tag{9.67}$$

Accordingly, Theorem 58 can be reduced as follows.

Theorem 59 *Suppose that* $\mathcal{G}$ *is connected. Then, for the agents in (9.29) with* $f_i(x_i, t) = 0$ *under the adaptive protocol (9.66) with* K *and* Γ *chosen as in Theorem 58, the consensus error* ξ *asymptotically converges to zero and the adaptive gains* d_i, $i = 1, \cdots, N$, *converges to some finite steady-state values.*

The proof of Theorem 59 can be completed by following the steps in the proof of Theorem 59. The protocol (9.66) is actually a node-based adaptive protocol, using the terms in Chapter 5. Similarly, distributed edge-based adaptive protocol can be constructed and the consensus results can be accordingly derived. The details are omitted here.

Consider the case where the agents in (9.29) with $f_i(x_i, t) = 0$ but subject to external disturbances. The dynamics of the agents are described by

$$\dot{x}_i = Ax_i + Bu_i + g(x_i) + \omega_i, \quad i = 1, \cdots, N, \tag{9.68}$$

where $\omega_i \in \mathbf{R}^n$ denotes bounded external disturbances. By the analysis in Section 4.3, we know that the distributed adaptive protocol (9.66) can cause parameter drift phenomenon (specifically, the adaptive gains d_i, $i = 1, \cdots, N$, grows unbounded), i.e., it is not robust in the presence of external disturbances. The σ-modification technique can be utilized to present a distributed robust adaptive protocol, described by

$$u_i = d_i K \sum_{j=1}^{N} a_{ij}(x_i - x_j),$$

$$\dot{d}_i = \nu_i[-\varphi_i d_i + (\sum_{j=1}^{N} a_{ij}(x_i - x_j)^T)\Gamma(\sum_{j=1}^{N} a_{ij}(x_i - x_j))], \ i = 1, \cdots, N.$$

$$(9.69)$$

Following the steps in the proof of Theorem 58, it is not difficult to show that the adaptive protocol (9.69) can guarantee the ultimate boundedness of the consensus error ξ and adaptive gains d_i, $i = 1, \cdots, N$, i.e., (9.69) is robust with respect to bounded disturbances.

9.2.4 Simulation Examples

In this subsection, a simulation example is provided for illustration.

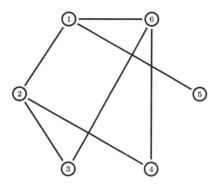

FIGURE 9.4: A leaderless communication graph.

Example 26 *Consider a network of uncertain agents described by (9.29),*

with $x_i = \begin{bmatrix} x_{i1} \\ x_{i2} \\ x_{i3} \\ x_{i4} \end{bmatrix}$, $A = \begin{bmatrix} 0 & 1 & 0 & 0 \\ -48.6 & -1.26 & 48.6 & 0 \\ 0 & 0 & 0 & 10 \\ 1.95 & 0 & -1.95 & 0 \end{bmatrix}$, $B = \begin{bmatrix} 0 & 21.6 & 0 & 0 \end{bmatrix}^T$, $g(x_i) =$

$\begin{bmatrix} 0 & 0 & 0 & -0.333\sin(x_{i1}) \end{bmatrix}^T$ *[129]. Clearly, $g(x_i)$ here satisfies (9.2) with a Lipschitz constant $\alpha = 0.333$. For illustration, the heterogeneous uncertainties are chosen as $f_1 = \sin(t)$, $f_2 = \sin(x_{25})/2$, $f_3 = \sin(t/2)$, $f_4 = \cos(t) + 1$, $f_5 = \sin(t+1)$ and $f_6 = \sin(x_{63} - 4)/2$.*

Here we use the adaptive protocol (9.66) to solve the global consensus problem. The communication graph is given as in FIGURE 9.4, which is connected. Solving the LMI (9.5) gives the feedback gain matrices of (9.66) as $K =$

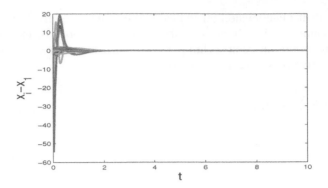

FIGURE 9.5: The consensus error $x_i - x_1$, $i = 2, \cdots, 6$, under (9.66).

$$-\begin{bmatrix} -2.8189 & -0.1766 & 1.9532 & -5.5117 \end{bmatrix} \; and \; \Gamma = \begin{bmatrix} 7.9464 & 0.4979 & -5.5061 & 15.5372 \\ 0.4979 & 0.0312 & -0.3450 & 0.9735 \\ -5.5061 & -0.3450 & 3.8151 & -10.7657 \\ 15.5372 & 0.9735 & -10.7657 & 30.3791 \end{bmatrix} \cdot S\text{-}$$

elect $\kappa_i = 0.5$, $\varphi_i = 0.05$, and $\tau_i = 1$, $i = 1, \cdots, 6$, in (9.66). The consensus error $x_i - x_1$, $i = 2, \cdots, 5$, under (9.66) designed as above are depicted in FIGURE 9.5, implying that consensus is achieved. The adaptive gains $\bar{d}_i$ in (9.66) are shown in FIGURE 9.6, which are bounded.

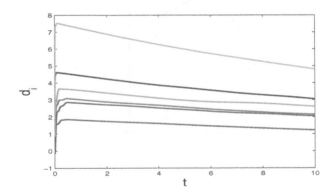

FIGURE 9.6: The adaptive gains $\bar{d}_i$ in (9.66).

9.3   Notes

The materials of this section are mainly based on [90, 96]. For more results on consensus of Lipschitz-type nonlinear multi-agent systems, please refer to [60,

84, 156, 181, 189]. In [181], the agents are characterized by first-order Lipschitz nonlinear systems. In [156, 189], the agents were assumed to be second-order systems with Lipschitz nonlinearity. [84] studied the global leader-follower consensus of coupled Lur'e systems with certain sector-bound nonlinearity, where the subgraph associated with the followers is required to be undirected. [60] was concerned with the second-order consensus problem of multi-agent systems with a virtual leader, where all agents and the virtual leader share the same intrinsic dynamics with a locally Lipschitz condition. Consensus tracking for multi-agent systems with Lipschitz-type nonlinear dynamics and switching directed topology was studied in [182].

There exists a quite large body of research papers on cooperative control of other types of nonlinear multi-agent systems. For instance, consensus of a network of Euler-Lagrange systems was studied in [1, 25, 79, 109, 176], a passivity-based design framework was proposed to deal with the consensus problem and other group coordination problems in [5, 8], and cooperative adaptive output regulation problems were addressed in [37, 163] for several classes of nonlinear multi-agent systems.

Bibliography

[1] A. Abdessameud and A. Tayebi. Attitude synchronization of a group of spacecraft without velocity measurements. *IEEE Transactions on Automatic Control*, 54(11):2642–2648, 2009.

[2] R. Agaev and P. Chebotarev. On the spectra of nonsymmetric Laplacian matrices. *Linear Algebra and its Applications*, 399(1):157–178, 2005.

[3] G. Alefeld and N. Schneider. On square roots of M-matrices. *Linear Algebra and its Applications*, 42:119–132, 1982.

[4] G. Antonelli. Interconnected dynamic systems: An overview on distributed control. *IEEE Control Systems Magazine*, 33(1):76–88, 2013.

[5] M. Arcak. Passivity as a design tool for group coordination. *IEEE Transactions on Automatic Control*, 52(8):1380–1390, 2007.

[6] M. Aung, A. Ahmed, M. Wette, D. Scharf, J. Tien, G. Purcell, M. Regehr, and B. Landin. An overview of formation flying technology development for the terrestrial planet finder mission. *Proceedings of the 2004 Aerospace Conference*, pp. 2667–2679, 2004.

[7] H. Bai, M. Arcak, and J.T. Wen. Rigid body attitude coordination without inertial frame information. *Automatica*, 44(12):3170–3175, 2008.

[8] H. Bai, M. Arcak, and J.T. Wen. *Cooperative Control Design: A Systematic, Passivity-based Approach*. Springer-Verlag, New York, NY, 2011.

[9] R. W Beard, J. Lawton, F.Y Hadaegh, A coordination architecture for spacecraft formation control. *IEEE Transactions on control systems technology*, 9(6):777–790, 2001.

[10] A. Berman and R.J. Plemmons. *Nonnegative Matrices in the Mathematical Sciences*. Academic Press, New York, NY, 1979.

[11] D.S. Bernstein. *Matrix Mathematics: Theory, Facts, and Formulas*. Princeton University Press, Princeton, NJ, 2009.

[12] A. Bidram, A. Davoudi, F.L. Lewis, and Z.H. Qu. Secondary control of microgrids based on distributed cooperative control of multi-agent systems. *IET Generation, Transmission and Distribution*, 7(8):822–831, 2013.

[13] V.S. Bokharaie, O. Mason, and M. Verwoerd. D-stability and delay-independent stability of homogeneous cooperative systems. *IEEE Transactions on Automatic Control*, 55(12):2882–2885, 2010.

[14] V. Borkar and P. Varaiya. Asymptotic agreement in distributed estimation. *IEEE Transactions on Automatic Control*, 27(3):650–655, 1982.

[15] S. Boyd, L. El Ghaoui, E. Feron, and V. Balakrishnan. *Linear Matrix Inequalities in System and Control Theory*. SIAM, Philadelphia, PA, 1994.

[16] Y.C. Cao and W. Ren. Containment control with multiple stationary or dynamic leaders under a directed interaction graph. *Proceedings of the 48th IEEE Conference on Decision and Control and the 28th Chinese Control Conference*, pp. 3014–3019, 2009.

[17] Y.C. Cao and W. Ren. Sampled-data discrete-time coordination algorithms for double-integrator dynamics under dynamic directed interaction. *International Journal of Control*, 83(3):506–515, 2010.

[18] Y.C. Cao, W. Ren, and M. Egerstedt. Distributed containment control with multiple stationary or dynamic leaders in fixed and switching directed networks. *Automatica*, 48(8):1586–1597, 2012.

[19] Y.C. Cao, D. Stuart, W. Ren, and Z.Y. Meng. Distributed containment control for multiple autonomous vehicles with double-integrator dynamics: Algorithms and experiments. *IEEE Transactions on Control Systems Technology*, 19(4):929–938, 2011.

[20] Y.C. Cao, W.W. Yu, W. Ren, and G.R. Chen. An overview of recent progress in the study of distributed multi-agent coordination. *IEEE Transactions on Industrial Informatics*, 9(1):427–438, 2013.

[21] Y.C. Cao and W. Ren. Distributed coordinated tracking with reduced interaction via a variable structure approach. *IEEE Transactions on Automatic Control*, 57(1):33–48, 2012.

[22] R. Carli, F. Bullo, and S. Zampieri. Quantized average consensus via dynamic coding/decoding schemes. *International Journal of Robust and Nonlinear Control*, 20(2):156–175, 2009.

[23] C.T. Chen. *Linear System Theory and Design*. Oxford University Press, New York, NY, 1999.

[24] G. Chen and F.L. Lewis. Distributed adaptive tracking control for synchronization of unknown networked lagrangian systems. *IEEE Transactions on Systems, Man, and Cybernetics, Part B: Cybernetics*, 41(3):805–816, 2011.

[25] S.J. Chung, U. Ahsun, and J.J.E. Slotine. Application of synchronization to formation flying spacecraft: Lagrangian approach. *Journal of Guidance, Control and Dynamics*, 32(2):512–526, 2009.

[26] M. Corless and G. Leitmann. Continuous state feedback guaranteeing uniform ultimate boundedness for uncertain dynamic systems. *IEEE Transactions on Automatic Control*, 26(5):1139–1144, 1981.

[27] J. Cortes, S. Martinez, T. Karatas, and F. Bullo. Coverage control for mobile sensing networks. *IEEE Transactions on Robotics and Automation*, 20(2):243–255, 2004.

[28] F. Cucker and J.G. Dong. Avoiding collisions in flocks. *IEEE Transactions on Automatic Control*, 55(5):1238–1243, 2010.

[29] F. Cucker and S. Smale. Emergent behavior in flocks. *IEEE Transactions on Automatic Control*, 52(5):852–862, 2007.

[30] A. Das and F.L. Lewis. Distributed adaptive control for synchronization of unknown nonlinear networked systems. *Automatica*, 46(12):2014–2021, 2010.

[31] A. Das and F.L. Lewis. Cooperative adaptive control for synchronization of second-order systems with unknown nonlinearities. *International Journal of Robust and Nonlinear Control*, 21(13): 1509-1524, 2011.

[32] P. DeLellis, M. diBernardo, and F. Garofalo. Novel decentralized adaptive strategies for the synchronization of complex networks. *Automatica*, 45(5):1312–1318, 2009.

[33] P. DeLellis, M.D. diBernardo, T.E. Gorochowski, and G. Russo. Synchronization and control of complex networks via contraction, adaptation and evolution. *IEEE Circuits and Systems Magazine*, 10(3):64–82, 2010.

[34] D.V. Dimarogonas, E. Frazzoli, and K.H. Johansson. Distributed event-triggered control for multi-agent systems. *IEEE Transactions on Automatic Control*, 57(5):1291–1297, 2012.

[35] D.V. Dimarogonas and K.J. Kyriakopoulos. Inverse agreement protocols with application to distributed multi-agent dispersion. *IEEE Transactions on Automatic Control*, 54(3):657–663, 2009.

[36] D.V. Dimarogonas, P. Tsiotras, and K.J. Kyriakopoulos. Leader–follower cooperative attitude control of multiple rigid bodies. *Systems and Control Letters*, 58(6):429–435, 2009.

[37] Z.T. Ding. Consensus output regulation of a class of heterogeneous nonlinear systems. *IEEE Transactions on Automatic Control*, 58(10):2648–2653, 2013.

[38] K.D. Do. Formation tracking control of unicycle-type mobile robots with limited sensing ranges. *IEEE Transactions on Control Systems Technology*, 16(3):527–538, 2008.

[39] W.J. Dong. Flocking of multiple mobile robots based on backstepping. *IEEE Transactions on Systems, Man, and Cybernetics, Part B: Cybernetics*, 41(2):414–424, 2011.

[40] W.J. Dong and J.A. Farrell. Cooperative control of multiple nonholonomic mobile agents. *IEEE Transactions on Automatic Control*, 53(6):1434–1448, 2008.

[41] Z.S. Duan, G.R. Chen, and L. Huang. Disconnected synchronized regions of complex dynamical networks. *IEEE Transactions on Automatic Control*, 54(4):845–849, 2009.

[42] Z.S. Duan, G.R. Chen, and L. Huang. Synchronization of weighted networks and complex synchronized regions. *Physics Letters A*, 372(21):3741–3751, 2008.

[43] C. Edwards and S.K. Spurgeon. *Sliding mode control: Theory and applications*. Taylor & Francis, London, 1998.

[44] J.A. Fax and R.M. Murray. Information flow and cooperative control of vehicle formations. *IEEE Transactions on Automatic Control*, 49(9):1465–1476, 2004.

[45] B. Fidan, J.M. Hendrickx, and B.D.O. Anderson. Closing ranks in rigid multi-agent formations using edge contraction. *International Journal of Robust and Nonlinear Control*, 20(18):2077–2092, 2010.

[46] L. Galbusera, G. Ferrari-Trecate, and R. Scattolini. A hybrid model predictive control scheme for containment and distributed sensing in multi-agent systems. *Systems and Control Letters*, 62(5):413–419, 2013.

[47] Y. Gao, L. Wang, G. Xie, and B. Wu. Consensus of multi-agent systems based on sampled-data control. *International Journal of Control*, 82(12):2193–2205, 2009.

[48] C. Godsil and G. Royle. *Algebraic Graph Theory*. Springer-Verlag, New York, NY, 2001.

[49] B. Grocholsky, J. Keller, V. Kumar, and G.J. Pappas. Cooperative air and ground surveillance. *IEEE Robotics and Automation Magazine*, 13(3):16–25, 2006.

[50] G.X. Gu, L. Marinovici, and F.L. Lewis. Consensusability of discrete-time dynamic multiagent systems. *IEEE Transactions on Automatic Control*, 57(8):2085–2089, 2012.

[51] V. Gupta, B. Hassibi, and R.M. Murray. A sub-optimal algorithm to synthesize control laws for a network of dynamic agents. *International Journal of Control*, 78(16):1302–1313, 2005.

[52] S.Y. Ha, T. Ha, and J.H. Kim. Emergent behavior of a cucker-smale type particle model with nonlinear velocity couplings. *IEEE Transactions on Automatic Control*, 55(7):1679–1683, 2010.

[53] V. Hahn and P.C. Parks. *Stability Theory*. Prentice-Hall, Englewood Cliffs, NJ, 1993.

[54] J.M. Hendrickx, B.D.O. Anderson, J. Delvenne, and V.D. Blondel. Directed graphs for the analysis of rigidity and persistence in autonomous agent systems. *International Journal of Robust and Nonlinear Control*, 17(10-11):960–981, 2007.

[55] K. Hengster-Movric, K.Y. You, F.L. Lewis, and L.H. Xie. Synchronization of discrete-time multi-agent systems on graphs using Riccati design. *Automatica*, 49(2):414–423, 2013.

[56] Y.G. Hong, J.P. Hu, and L. Gao. Tracking control for multi-agent consensus with an active leader and variable topology. *Automatica*, 42(7):1177–1182, 2006.

[57] Y.G. Hong, G.R. Chen, and L. Bushnell. Distributed observers design for leader-following control of multi-agent networks. *Automatica*, 44(3):846–850, 2008.

[58] R.A. Horn and C.R. Johnson. *Matrix Analysis*. Cambridge University Press, New York, NY, 1990.

[59] A. Howard, M.J. Matarić, and G.S. Sukhatme. Mobile sensor network deployment using potential fields: A distributed, scalable solution to the area coverage problem. *Proceedings of the 6th International Symposium on Distributed Autonomous Robotics Systems*, pp. 299–308, 2002.

[60] Y.B. Hu, H.S. Su, and J. Lam. Adaptive consensus with a virtual leader of multiple agents governed by locally Lipschitz nonlinearity. *International Journal of Robust and Nonlinear Control*, 23(9):978–990, 2013.

[61] I.I. Hussein and D.M. Stipanovic. Effective coverage control for mobile sensor networks with guaranteed collision avoidance. *IEEE Transactions on Control Systems Technology*, 15(4):642–657, 2007.

[62] P. A. Ioannou and J. Sun. *Robust Adaptive Control*. Prentice-Hall, New York, NY, 1996.

[63] P.A. Ioannou and P.V. Kokotovic. Instability analysis and improvement of robustness of adaptive control. *Automatica*, 20(5):583–594, 1984.

[64] T. Iwasaki and R.E. Skelton. All controllers for the general H_∞ control problem: LMI existence conditions and state space formulas. *Automatica*, 30(8):1307–1317, 1994.

[65] A. Jadbabaie, J. Lin, and A.S. Morse. Coordination of groups of mobile autonomous agents using nearest neighbor rules. *IEEE Transactions on Automatic Control*, 48(6):988–1001, 2003.

[66] I.S. Jeon, J.I. Lee, and M.J. Tahk. Homing guidance law for cooperative attack of multiple missiles. *Journal of Guidance, Control, and Dynamics*, 33(1):275–280, 2010.

[67] M. Ji, G. Ferrari-Trecate, M. Egerstedt, and A. Buffa. Containment control in mobile networks. *IEEE Transactions on Automatic Control*, 53(8):1972–1975, 2008.

[68] F.C. Jiang and L. Wang. Consensus seeking of high-order dynamic multi-agent systems with fixed and switching topologies. *International Journal of Control*, 85(2):404–420, 2010.

[69] V. Kapila, A.G. Sparks, J.M. Buffington, and Q. Yan. Spacecraft formation flying: Dynamics and control. *Journal of Guidance, Control, and Dynamics*, 23(3):561–564, 2000.

[70] O. Katsuhiko. *Modern Control Engineering*. Prentice Hall, Upper Saddle River, NJ, 1996.

[71] H.K. Khalil. *Nonlinear Systems*. Prentice Hall, Englewood Cliffs, NJ, 2002.

[72] M. Krstić, I. Kanellakopoulos, and P.V. Kokotovic. *Nonlinear and Adaptive Control Design*. John Wiley & Sons, New York, 1995.

[73] J. Larson, C. Kammer, K.Y. Liang, and K.H. Johansson. Coordinated route optimization for heavy-duty vehicle platoons. *Proceedings of the 16th International IEEE Annual Conference on Intelligent Transportation Systems*, pp. 1196–1202, 2013.

[74] F.L. Lewis, H.W. Zhang, K. Hengster-Movric, and A. Das. *Cooperative Control of Multi-Agent Systems: Optimal and Adaptive Design Approaches*. Springer-Verlag, London, 2014.

[75] T. Li, M.Y. Fu, L.H. Xie, and J.F. Zhang. Distributed consensus with limited communication data rate. *IEEE Transactions on Automatic Control*, 56(2):279–292, 2011.

[76] W. Li and C.G. Cassandras. Distributed cooperative coverage control of sensor networks. *Proceedings of the 44th IEEE Conference on Decision and Control and 2005 European Control Conference*, pp. 2542–2547, 2005.

[77] Z.K. Li, Z.S. Duan, and F.L. Lewis. Distributed robust consensus control of multi-agent systems with heterogeneous matching uncertainties. *Automatica*, 50(3):883–889, 2014.

[78] Z.K. Li, G.H. Wen, Z.S. Duan, and W. Ren. Designing fully distributed consensus protocols for linear multi-agent systems with directed communication graphs. *IEEE Transactions on Automatic Control*, in press, 2014.

[79] Z.K. Li and Z.S. Duan. Distributed adaptive attitude synchronization of multiple spacecraft. *Science China Technological Sciences*, 54(8):1992–1998, 2011.

[80] Z.K. Li and Z.S. Duan. Distributed adaptive consensus protocols for linear multi-agent systems with directed graphs in the presence of external disturbances. *Proceedings of the 33th Chinese Control Conference*, pp. 1632–1637, 2014.

[81] Z.K. Li, Z.S. Duan, and G.R. Chen. H_∞ consensus regions of multi-agent systems. *Proceedings of the 29th Chinese Control Conference*, pp. 4601–4606, 2010.

[82] Z.K. Li, Z.S. Duan, and G.R. Chen. Consensus of discrete-time linear multi-agent systems with observer-type protocols. *Discrete and Continuous Dynamical Systems-Series B*, 16(2):489–505, 2011.

[83] Z.K. Li, Z.S. Duan, and G.R. Chen. Dynamic consensus of linear multi-agent systems. *IET Control Theory and Applications*, 5(1):19–28, 2011.

[84] Z.K. Li, Z.S. Duan, and G.R. Chen. Global synchronised regions of linearly coupled Lur'e systems. *International Journal of Control*, 84(2):216–227, 2011.

[85] Z.K. Li, Z.S. Duan, and G.R. Chen. On H_∞ and H_2 performance regions of multi-agent systems. *Automatica*, 47(4):797–803, 2011.

[86] Z.K. Li, Z.S. Duan, G.R. Chen, and L. Huang. Consensus of multiagent systems and synchronization of complex networks: A unified viewpoint. *IEEE Transactions on Circuits and Systems I: Regular Papers*, 57(1):213–224, 2010.

[87] Z.K. Li, Z.S. Duan, and L. Huang. H_∞ control of networked multi-agent systems. *Journal of Systems Science and Complexity*, 22(1):35–48, 2009.

[88] Z.K. Li, Z.S. Duan, W. Ren, and G. Feng. Containment control of linear multi-agent systems with multiple leaders of bounded inputs using distributed continuous controllers. *International Journal of Robust and Nonlinear Control*, in press, 2014.

[89] Z.K. Li, Z.S. Duan, L.H. Xie, and X.D. Liu. Distributed robust control of linear multi-agent systems with parameter uncertainties. *International Journal of Control*, 85(8):1039–1050, 2012.

[90] Z.K. Li, X.D. Liu, M.Y. Fu, and L.H. Xie. Global H_∞ consensus of multi-agent systems with Lipschitz non-linear dynamics. *IET Control Theory and Applications*, 6(13):2041–2048, 2012.

[91] Z.K. Li, X.D. Liu, P. Lin, and W. Ren. Consensus of linear multi-agent systems with reduced-order observer-based protocols. *Systems and Control Letters*, 60(7):510–516, 2011.

[92] Z.K. Li, X.D. Liu, W. Ren, and L.H. Xie. Distributed tracking control for linear multi-agent systems with a leader of bounded unknown input. *IEEE Transactions on Automatic Control*, 58(2):518–523, 2013.

[93] Z.K. Li, W. Ren, X.D. Liu, and M.Y. Fu. Consensus of multi-agent systems with general linear and Lipschitz nonlinear dynamics using distributed adaptive protocols. *IEEE Transactions on Automatic Control*, 58(7):1786–1791, 2013.

[94] Z.K. Li, W. Ren, X.D. Liu, and M.Y. Fu. Distributed containment control of multi-agent systems with general linear dynamics in the presence of multiple leaders. *International Journal of Robust and Nonlinear Control*, 23(5):534–547, 2013.

[95] Z.K. Li, W. Ren, X.D. Liu, and L.H. Xie. Distributed consensus of linear multi-agent systems with adaptive dynamic protocols. *Automatica*, 49(7):1986–1995, 2013.

[96] Z.K. Li, Y. Zhao, and Z.S. Duan. Distributed robust global consensus of a class of Lipschitz nonlinear multi-agent systems with matching uncertainties. *Asian Journal of Control*, in press, 2014.

[97] P. Lin and Y.M. Jia. Consensus of second-order discrete-time multi-agent systems with nonuniform time-delays and dynamically changing topologies. *Automatica*, 45(9):2154–2158, 2009.

[98] P. Lin and Y.M. Jia. Robust H_∞ consensus analysis of a class of second-order multi-agent systems with uncertainty. *IET Control Theory and Applications*, 4(3):487–498, 2010.

[99] P. Lin and Y.M. Jia. Distributed robust H_∞ consensus control in directed networks of agents with time-delay. *Systems and Control Letters*, 57(8):643–653, 2008.

[100] P. Lin, K.Y. Qin, Z.K. Li, and W. Ren. Collective rotating motions of second-order multi-agent systems in three-dimensional space. *Systems and Control Letters*, 60(6):365–372, 2011.

[101] C. Liu, Z.S. Duan, G.R. Chen, and L. Huang. Analyzing and controlling the network synchronization regions. *Physica A*, 386(1):531–542, 2007.

[102] Y. Liu and Y.M. Jia. H_∞ consensus control of multi-agent systems with switching topology: A dynamic output feedback protocol. *International Journal of Control*, 83(3):527–537, 2010.

[103] J. Löfberg. YALMIP: A toolbox for modeling and optimization in MAT-LAB. *Proceedings of 2004 IEEE International Symposium on Computer Aided Control Systems Design*, pp. 284–289, 2004.

[104] Y. Lou and Y.G. Hong. Target containment control of multi-agent systems with random switching interconnection topologies. *Automatica*, 48(5):879–885, 2012.

[105] J.H. Lv, X.H. Yu, G.R. Chen, and D.Z. Cheng. Characterizing the synchronizability of small-world dynamical networks. *IEEE Transactions on Circuits and Systems I: Regular Papers*, 51(4):787–796, 2004.

[106] C.Q. Ma and J.F. Zhang. Necessary and sufficient conditions for consensusability of linear multi-sgent systems. *IEEE Transactions on Automatic Control*, 55(5):1263–1268, 2010.

[107] R.N. Madan. *Chua's Circuit: A Paradigm for Chaos*. World Scientific, Singapore, 1993.

[108] P. Massioni and M. Verhaegen. Distributed control for identical dynamically coupled systems: A decomposition approach. *IEEE Transactions on Automatic Control*, 54(1):124–135, 2009.

[109] J. Mei, W. Ren, and G. Ma. Distributed coordinated tracking with a dynamic leader for multiple Euler-Lagrange systems. *IEEE Transactions on Automatic Control*, 56(6):1415–1421, 2011.

[110] J. Mei, W. Ren, and G. Ma. Distributed containment control for Lagrangian networks with parametric uncertainties under a directed graph. *Automatica*, 48(4):653–659, 2012.

[111] Z.Y. Meng, W. Ren, and Z. You. Distributed finite-time attitude containment control for multiple rigid bodies. *Automatica*, 46(12):2092–2099, 2010.

[112] M. Mcsbahi and M. Egerstedt. *Graph Theoretic Methods in Multiagent Networks*. Princeton University Press, Princeton, NJ, 2010.

[113] N. Michael, J. Fink, and V. Kumar. Cooperative manipulation and transportation with aerial robots. *Autonomous Robots*, 30(1):73–86, 2011.

[114] U. Munz, A. Papachristodoulou, and F. Allgower. Robust consensus controller design for nonlinear relative degree two multi-agent systems with communication constraints. *IEEE Transactions on Automatic Control*, 56(1):145–151, 2011.

[115] R.M. Murray. Recent research in cooperative control of multivehicle systems. *Journal of Dynamic Systems, Measurement, and Control*, 129:571–583, 2007.

[116] W. Ni and D.Z. Cheng. Leader-following consensus of multi-agent systems under fixed and switching topologies. *Systems and Control Letters*, 59(3-4):209–217, 2010.

[117] K. Ogata. *Modern Control Engineering*. Prentice Hall, Englewood Cliffs, NJ, 1996.

[118] A. Okubo. Dynamical aspects of animal grouping: Swarms, schools, flocks, and herds. *Advances in Biophysics*, 22:1–94, 1986.

[119] R. Olfati-Saber. Flocking for multi-agent dynamic systems: Algorithms and theory. *IEEE Transactions on Automatic Control*, 51(3):401–420, 2006.

[120] R. Olfati-Saber, J.A. Fax, and R.M. Murray. Consensus and cooperation in networked multi-agent systems. *Proceedings of the IEEE*, 95(1):215–233, 2007.

[121] R. Olfati-Saber and R.M. Murray. Consensus problems in networks of agents with switching topology and time-delays. *IEEE Transactions on Automatic Control*, 49(9):1520–1533, 2004.

[122] A. Papachristodoulou, A. Jadbabaie, and U. Munz. Effects of delay in multi-agent consensus and oscillator synchronization. *IEEE Transactions on Automatic Control*, 55(6):1471–1477, 2010.

[123] M. Pavone and E. Frazzoli. Decentralized policies for geometric pattern formation and path coverage. *Journal of Dynamic Systems, Measurement, and Control*, 129(5):633–643, 2007.

[124] L.M. Pecora and T.L. Carroll. Master stability functions for synchronized coupled systems. *Physical Review Letters*, 80(10):2109–2112, 1998.

[125] Z.H. Peng, D. Wang, and H.W. Zhang. Cooperative tracking and estimation of linear multi-agent systems with a dynamic leader via iterative learning. *International Journal of Control*, 87(6):1163–1171, 2014.

[126] Z.H. Peng, D. Wang, H.W. Zhang, G. Sun, and H. Wang. Distributed model reference adaptive control for cooperative tracking of uncertain dynamical multi-agent systems. *IET Control Theory and Applications*, 7(8):1079–1087, 2013.

[127] Z.H. Qu. *Cooperative Control of Dynamical Systems: Applications to Autonomous Vehicles.* Springer-Verlag, London, UK, 2009.

[128] R. Rajamani. Observers for Lipschitz nonlinear systems. *IEEE Transactions on Automatic Control,* 43(3):397–401, 1998.

[129] R. Rajamani and Y.M. Cho. Existence and design of observers for nonlinear systems: Relation to distance to unobservability. *International Journal of Control,* 69(5):717–731, 1998.

[130] W. Ren. On consensus algorithms for double-integrator dynamics. *IEEE Transactions on Automatic Control,* 53(6):1503–1509, 2008.

[131] W. Ren. Consensus tracking under directed interaction topologies: Algorithms and experiments. *IEEE Transactions on Control Systems Technology,* 18(1):230–237, 2010.

[132] W. Ren and E.M. Atkins. Distributed multi-vehicle coordinated control via local information exchange. *International Journal of Robust and Nonlinear Control,* 17(10-11):1002–1033, 2007.

[133] W. Ren and R.W. Beard. Consensus seeking in multiagent systems under dynamically changing interaction topologies. *IEEE Transactions on Automatic Control,* 50(5):655–661, 2005.

[134] W. Ren, R.W. Beard, and E.M. Atkins. Information consensus in multi-vehicle cooperative control. *IEEE Control Systems Magazine,* 27(2):71–82, 2007.

[135] W. Ren, K.L. Moore, and Y. Chen. High-order and model reference consensus algorithms in cooperative control of multivehicle systems. *Journal of Dynamic Systems, Measurement, and Control,* 129(5):678–688, 2007.

[136] W. Ren. Consensus strategies for cooperative control of vehicle formations. *IET Control Theory and Applications,* 1(2):505–512, 2007.

[137] W. Ren. Collective motion from consensus with Cartesian coordinate coupling. *IEEE Transactions on Automatic Control,* 54(6):1330–1335, 2009.

[138] W. Ren and R.W. Beard. *Distributed Consensus in Multi-Vehicle Cooperative Control.* Springer-Verlag, London, 2008.

[139] W. Ren and Y.C. Cao. *Distributed Coordination of Multi-Agent Networks: Emergent Problems, Models, and Issues.* Springer-Verlag, London, 2010.

[140] C.W. Reynolds. Flocks, herds, and schools: A distributed behavioral model. *Computer Graphics,* 21(4):25–34, 1987.

[141] E.J. Rodríguez-Seda, J.J. Troy, C.A. Erignac, P. Murray, D.M. Stipanovic, and M.W. Spong. Bilateral teleoperation of multiple mobile agents: coordinated motion and collision avoidance. *IEEE Transactions on Control Systems Technology*, 18(4):984–992, 2010.

[142] A. Sarlette, R. Sepulchre, and N.E. Leonard. Autonomous rigid body attitude synchronization. *Automatica*, 45(2):572–577, 2009.

[143] L. Scardovi and R. Sepulchre. Synchronization in networks of identical linear systems. *Automatica*, 45(11):2557–2562, 2009.

[144] L. Schenato, B. Sinopoli, M. Franceschetti, K. Poolla, and S.S. Sastry. Foundations of control and estimation over lossy networks. *Proceedings of the IEEE*, 95(1):163–187, 2007.

[145] L. Schenato and F. Fiorentin. Average timesynch: A consensus-based protocol for clock synchronization in wireless sensor networks. *Automatica*, 47(9):1878–1886, 2011.

[146] J.H. Seo, H. Shim, and J. Back. Consensus of high-order linear systems using dynamic output feedback compensator: Low gain approach. *Automatica*, 45(11):2659–2664, 2009.

[147] R. Sepulchre, D.A. Paley, and N.E. Leonard. Stabilization of planar collective motion: All-to-all communication. *IEEE Transactions on Automatic Control*, 52(5):811–824, 2007.

[148] R. Sepulchre, D.A. Paley, and N.E. Leonard. Stabilization of planar collective motion with limited communication. *IEEE Transactions on Automatic Control*, 53(3):706–719, 2008.

[149] G.S. Seyboth, D.V. Dimarogonas, and K.H. Johansson. Event-based broadcasting for multi-agent average consensus. *Automatica*, 49(1):245–252, 2013.

[150] D. Shevitz and B. Paden. Lyapunov stability theory of nonsmooth systems. *IEEE Transactions on Automatic Control*, 39(9):1910–1914, 1994.

[151] D.D. Šiljak. *Decentralized Control of Complex Systems*. Academic Press, New York, NY, 1991.

[152] B. Sinopoli, L. Schenato, M. Franceschetti, K. Poolla, M.I. Jordan, and S.S. Sastry. Kalman filtering with intermittent observations. *IEEE Transactions on Automatic Control*, 49(9):1453–1464, 2004.

[153] F. Sivrikaya and B. Yener. Time synchronization in sensor networks: A survey. *IEEE Network*, 18(4):45–50, 2004.

[154] J.J.E. Slotine and W. Li. *Applied Nonlinear Control*. Prentice Hall, Englewood Cliffs, NJ, 1991.

[155] R.S. Smith and F.Y. Hadaegh. Control of deep-space formation-flying spacecraft; Relative sensing and switched information. *Journal of Guidance, Control and Dynamics*, 28(1):106–114, 2005.

[156] Q. Song, J.D. Cao, and W.W. Yu. Second-order leader-following consensus of nonlinear multi-agent systems via pinning control. *Systems and Control Letters*, 59(9):553-562, 2010.

[157] E. Sontag and A. Teel. Changing supply functions in input/state stable systems. *IEEE Transactions on Automatic Control*, 40(8):1476–1478, 1995.

[158] J.F. Sturm. Using SeDuMi 1.02, a MATLAB toolbox for optimization over symmetric cones. *Optimization Methods and Software*, 11(1):625–653, 1999.

[159] H.S. Su, G.R. Chen, X.F. Wang, and Z.L. Lin. Adaptive second-order consensus of networked mobile agents with nonlinear dynamics. *Automatica*, 47(2):368–375, 2011.

[160] H.S. Su, X.F. Wang, and G.R. Chen. A connectivity-preserving flocking algorithm for multi-agent systems based only on position measurements. *International Journal of Control*, 82(7):1334–1343, 2009.

[161] H.S. Su, X.F. Wang, and W. Yang. Flocking in multi-agent systems with multiple virtual leaders. *Asian Journal of control*, 10(2):238–245, 2008.

[162] H.S. Su, X.F. Wang, and Z.L. Lin. Flocking of multi-agents with a virtual leader. *IEEE Transactions on Automatic Control*, 54(2):293–307, 2009.

[163] Y.F. Su and J. Huang. Cooperative adaptive output regulation for a class of nonlinear uncertain multi-agent systems with unknown leader. *Systems and Control Letters*, 62(6):461–467, 2013.

[164] C. Tan and G. P. Liu. Consensus of networked multi-agent systems via the networked predictive control and relative outputs. *Journal of the Franklin Institute*, 349(7):2343–2356, 2012.

[165] H.G. Tanner, A. Jadbabaie, and G.J. Pappas. Flocking in fixed and switching networks. *IEEE Transactions on Automatic Control*, 52(5):863–868, 2007.

[166] Y.P. Tian and C.L. Liu. Robust consensus of multi-agent systems with diverse input delays and asymmetric interconnection perturbations. *Automatica*, 45(5):1347–1353, 2009.

[167] M. Tillerson, G. Inalhan, and J.P. How. Co-ordination and control of distributed spacecraft systems using convex optimization techniques. *International Journal of Robust and Nonlinear Control*, 12(2-3):207–242, 2002.

[168] L.N. Trefethen and D. Bau. *Numerical Linear Algebra*. SIAM, Philadelphia, PA, 1997.

[169] H.L. Trentelman, K. Takaba, and N. Monshizadeh. Robust synchronization of uncertain linear multi-agent systems. *IEEE Transactions on Automatic Control*, 58(6):1511–1523, 2013.

[170] J.N. Tsitsiklis, D.P. Bertsekas, M. Athans, Distributed asynchronous deterministic and stochastic gradient optimization algorithms. *IEEE Transactions on Automatic Control*, 31(9):803–812, 1986.

[171] S. E. Tuna. LQR-based coupling gain for synchronization of linear systems. *arXiv preprint arXiv:0801.3390*, 2008.

[172] S.E. Tuna. Synchronizing linear systems via partial-state coupling. *Automatica*, 44(8):2179–2184, 2008.

[173] S.E. Tuna. Conditions for synchronizability in arrays of coupled linear systems. *IEEE Transactions on Automatic Control*, 54(10):2416–2420, 2009.

[174] T. Vicsek, A. Czirók, E. Ben-Jacob, I. Cohen, and O. Shochet. Novel type of phase transition in a system of self-driven particles. *Physical Review Letters*, 75(6):1226–1229, 1995.

[175] C. Wang, X. Wang, and H. Ji. Leader-following consensus for an integrator-type nonlinear multi-agent systems using distributed adaptive protocol. *Proceedings of the 10th IEEE International Conference on Control and Automation*, pp. 1166–1171, 2011.

[176] H.L. Wang. Passivity based synchronization for networked robotic systems with uncertain kinematics and dynamics. *Automatica*, 49(3):755–761, 2013.

[177] J.Y. Wang, Z.S. Duan, Z.K. Li, and G.H. Wen. Distributed H_∞ and H_2 consensus control in directed networks. *IET Control Theory and Applications*, 8(3):193–201, 2014.

[178] X. Wang, A. Saberi, A.A. Stoorvogel, H.F. Grip, and T. Yang. Consensus in the network with uniform constant communication delay. *Automatica*, 49(8):2461–2467, 2013.

[179] G.H. Wen, Z.S. Duan, Z.K. Li, and G.R. Chen. Flocking of multi-agent dynamical systems with intermittent nonlinear velocity measurements. *International Journal of Robust and Nonlinear Control*, 22(16):1790–1805, 2012.

[180] G.H. Wen, G.Q. Hu, W.W. Yu, and G.R. Chen. Distributed H_∞ consensus of higher order multiagent systems with switching topologies. *IEEE Transactions on Circuits and Systems II: Express Briefs*, 61(5):359-363, 2014.

[181] G.H. Wen, Z.S. Duan, Z.K. Li, G.R. Chen. Consensus and its L_2-gain performance of multi-agent systems with intermittent information transmissions. *International Journal of Control*, 85(4):384-396, 2012.

[182] G.H. Wen, Z.S. Duan, G.R. Chen, W.W. Yu. Consensus tracking of multi-agent systems with Lipschitz-type node dynamics and switching topologies. *IEEE Transactions on Circuits and Systems I: Regular Papers*, 61(2):499-511, 2014.

[183] G. Wheeler, C.Y. Su, and Y. Stepanenko. A sliding mode controller with improved adaptation laws for the upper bounds on the norm of uncertainties. *Automatica*, 34(12):1657–1661, 1998.

[184] E. Yong. Autonomous drones flock like birds. *Nature News*, doi:10.1038/nature.2014.14776, 2014.

[185] K.Y. You and L.H. Xie. Network topology and communication data rate for consensusability of discrete-time multi-agent systems. *IEEE transactions on automatic control*, 56(10):2262–2275, 2011.

[186] K.Y. You, Z.K. Li, and L.H. Xie. Consensus condition for linear multi-agent systems over randomly switching topologies. *Automatica*, 49(10):3125–3132, 2013.

[187] K.D. Young, V.I. Utkin, and U. Ozguner. A control engineer's guide to sliding mode control. *IEEE Transactions on Control Systems Technology*, 7(3):328–342, 1999.

[188] C.B. Yu, B.D.O. Anderson, S. Dasgupta, and B. Fidan. Control of minimally persistent formations in the plane. *SIAM Journal on Control and Optimization*, 48(1):206–233, 2009.

[189] W.W. Yu, G.R. Chen, M. Cao, and J. Kurths. Second-order consensus for multiagent systems with directed topologies and nonlinear dynamics. *IEEE Transactions on Systems, Man, and Cybernetics, Part B: Cybernetics*, 40(3):881–891, 2010.

[190] W.W. Yu, W. Ren, W.X. Zheng, G.R. Chen, and J.H. Lv. Distributed control gains design for consensus in multi-agent systems with second-order nonlinear dynamics. *Automatica*, 49(7):2107–2115, 2013.

[191] T. Yucelen and W.M. Haddad. Low-frequency learning and fast adaptation in model reference adaptive control. *IEEE Transactions on Automatic Control*, 58(4):1080–1085, 2013.

[192] T. Yucelen and M. Egerstedt. Control of multiagent systems under persistent disturbances. *Proceedings of the 2012 American Control Conference*, pp. 5264–5269, 2012.

[193] T. Yucelen and E.N. Johnson. Control of multivehicle systems in the presence of uncertain dynamics. *International Journal of Control*, 86(9):1540–1553, 2013.

[194] M.M. Zavlanos, H.G. Tanner, A. Jadbabaie, and G.J. Pappas. Hybrid control for connectivity preserving flocking. *IEEE Transactions on Automatic Control*, 54(12):2869–2875, 2009.

[195] H.W. Zhang, F.L. Lewis, and A. Das. Optimal design for synchronization of cooperative systems: State feedback, observer, and output feedback. *IEEE Transactions on Automatic Control*, 56(8):1948–1952, 2011.

[196] H.W. Zhang and F.L. Lewis. Adaptive cooperative tracking control of higher-order nonlinear systems with unknown dynamics. *Automatica*, 48(7):1432–1439, 2012.

[197] Y. Zhao, Z.S. Duan, G.H. Wen, and G.R. Chen. Distributed H_∞ consensus of multi-agent systems: A performance region-based approach. *International Journal of Control*, 85(3):332–341, 2012.

[198] Y. Zhao, G.H. Wen, Z.S. Duan, X. Xu, and G.R. Chen. A new observer-type consensus protocol for linear multi-agent dynamical systems. *Asian Journal of Control*, 15(2):571–582, 2013.

[199] B. Zhou, C.C. Xu, and G.R. Duan. Distributed and truncated reduced-order observer based output feedback consensus of multi-agent systems. *IEEE Transactions on Automatic Control*, 59(8): 2264–2270, 2014

[200] K.M. Zhou and J.C. Doyle. *Essentials of Robust Control*. Prentice Hall, Upper Saddle River, NJ, 1998.

[201] F. Zhu and Z. Han. A note on observers for Lipschitz nonlinear systems. *IEEE Transactions on automatic control*, 47(10):1751–1754, 2002.

[202] W. Zhu and D.Z. Cheng. Leader-following consensus of second-order agents with multiple time-varying delays. *Automatica*, 46(12):1994–1999, 2010.

Index

σ-modification technique, 115, 130, 148

adaptive consensus protocol, 125
  edge-based, 95, 104
  node-based, 96, 104
  relative-output, 104
  robust, 130
  robust edge-based, 115
  robust node-based, 115
algebraic connectivity, 14
  generalized, 14, 88
algebraic Riccati equation, 33, 34, 99, 130

Barbalat's Lemma, 17
boundary layer, 145, 148

Chua's circuit, 199
Comparison Lemma, 17
consensus, 4
  H_2, 83
  H_∞, 75, 87
  global, 209
  global H_∞, 212
  global leader-follower, 215
  leader-follower, 5, 111, 124
  leader-follower H_∞, 86
  robust leader-follower, 197
  robust leaderless, 184
consensus region, 24
  H_2, 83
  H_∞, 78
  bounded, 26
  bounded H_∞, 78
  disconnected, 27
  disconnected H_∞, 79
  discrete-time, 57

discrete-time bounded, 61, 62
discrete-time disconnected, 57
unbounded, 31, 32, 38
unbounded H_2, 84
unbounded H_∞, 80
containment control, 5, 161
  robust, 205
containment controller
  continuous adaptive, 176
  continuous static, 171
  dynamic, 161, 165
  static, 164, 168
convergence rate, 33, 34
convex hull, 11, 161
cycle, 11

decay rate, 33
distributed tracking, 5, 139, 197
distributed tracking controller
  continuous adaptive, 148
  continuous static, 145
  discontinuous adaptive, 142
  discontinuous static, 139
  output adaptive, 152

edge, 11

Fiedler eigenvalue, 14
Finsler's Lemma, 18
flocking, 2, 7
formation control, 6, 46, 69

Gershgorin's disc theorem, 15
graph
  balanced, 11, 12
  connected, 11
  directed, 11
  strongly connected, 11, 12, 14, 87

undirected, 11, 12
  weighted, 11
graph theory, 11

Hill's equations, 48
Hurwitz, 10

in-degree, 11

Kronecker product, 10

LaSalle's Invariance Principle, 16
LaSalle–Yoshizawa Theorem, 17
linear matrix inequality, 81, 97, 105,
  116, 140, 162, 164, 186, 202,
  209, 212
Lipschitz nonlinearity, 208, 219

matching uncertainty, 185, 219
matrix
  M-, 11, 125
  adjacency, 12
  degree, 13
  doubly stochastic, 15
  Hermite, 10
  Laplacian, 13
  normalized Laplacian, 15
  orthogonal, 10
  row-stochastic, 15
  stochastic, 15, 55
  unitary, 10
modified algebraic Riccati equation,
  62, 166, 168

neighbor, 11
node, 11

observer-type consensus protocol, 35,
  41, 65, 67
orthogonal projection, 10

path
  directed, 11
  undirected, 11
performance limit
  H_2, 83, 85
  H_∞, 75, 81

reduced-order observer-based consensus protocol, 42, 68
Reynolds rules, 7

satellite formation flying, 47
Schur Complement Lemma, 18
Schur stability, 55
spanning tree, 12, 14
stable
  neutrally, 10
  Schur, 10
subgraph, 11, 139, 171
switching communication graph, 44,
  102

uniformly ultimately bounded, 17,
  116, 120, 131, 145, 148, 153,
  173, 186, 226

Young's Inequality, 18

Printed and bound by CPI Group (UK) Ltd, Croydon, CR0 4YY

18/10/2024

01776264-0004